NONLINEAR SYSTEM THEORY

JOHNS HOPKINS SERIES

IN INFORMATION SCIENCES AND SYSTEMS

S. Rao Kosaraju and
Wilson J. Rugh, Editors

NONLINEAR SYSTEM THEORY

The Volterra/Wiener Approach

Wilson J. Rugh

THE JOHNS HOPKINS UNIVERSITY PRESS

Baltimore and London

Wilson J. Rugh is Professor and Chairman of the Department of Electrical Engineering at The Johns Hopkins University.

Printed in the United States of America
The Johns Hopkins University Press, Baltimore, Maryland 21218
The Johns Hopkins Press Ltd., London

Library of Congress Cataloging in Publication Data

Rugh, Wilson J.
Nonlinear system theory.
(Johns Hopkins series in information sciences and systems)
Includes index.
1. System analysis. 2. Nonlinear theories. I. Title.
II. Title: The Volterra/Wiener approach. III. Series.

QA402.R86 003.19 80-8874
ISBN O-8018-2549-0 AACR2

To My Parents

Contents

Preface

When confronted with a nonlinear systems engineering problem, the first approach usually is to linearize; in other words, to try to avoid the nonlinear aspects of the problem. It is indeed a happy circumstance when a solution can be obtained in this way. When it cannot, the tendency is to try to avoid the situation altogether, presumably in the hope that the problem will go away. Those engineers who forge ahead are often viewed as foolish, or worse. Nonlinear systems engineering is regarded not just as a difficult and confusing endeavor; it is widely viewed as dangerous to those who think about it for too long.

This skepticism is to an extent justifiable. When compared with the variety of techniques available in linear system theory, the tools for analysis and design of nonlinear systems are limited to some very special categories. First, there are the relatively simple techniques, such as phase-plane analysis, which are graphical in nature and thus of limitcd gcncrality. Thcn, thcrc arc thc rathcr gcncral (and subtlc) tcchniques based on the theory of differential equations, functional analysis, and operator theory. These provide a language, a framework, and existence/uniqueness proofs, but often little problem-specific information beyond these basics. Finally, there is simulation, sometimes ad nauseam, on the digital computer.

I do not mean to say that these techniques or approaches are useless. Certainly phase-plane analysis describes nonlinear phenomena such as limit cycles and multiple equilibria of second-order systems in an efficient manner. The theory of differential equations has led to a highly developed stability theory for some classes of nonlinear systems. (Though, of course, an engineer cannot live by stability alone.) Functional analysis and operator theoretic viewpoints are philosophically appealing, and undoubtedly will become more applicable in the future. Finally, everyone is aware of the occasional success story emanating from the local computer center.

What I do mean to say is that a theory is needed that occupies the middle ground in generality and applicability. Such a theory can be of great importance for it can serve as a starting point, both for more esoteric mathematical studies and for the development of engineering techniques. Indeed, it can serve as a bridge or communication link between these two activities.

In the early 1970s it became clear that the time was ripe for a middle-of-the-road formulation for nonlinear system theory. It seemed that such a formulation should use some aspects of differential- (or difference-) equation descriptions, and transform representations, as well as some aspects of operator-theoretic descriptions. The question was whether, by making structural assumptions and ruling out pathologies, a reasonably simple, reasonably general, nonlinear system theory could be developed. Hand in hand with this viewpoint was the feeling that many of the approaches useful for linear systems ought to be extensible to the nonlinear theory. This is a key point if the theory is to be used by practitioners as well as by researchers.

These considerations led me into what has come to be called the Volterra/Wiener representation for nonlinear systems. Articles on this topic had been appearing sporadically in the engineering literature since about 1950, but it seemed to be time for an investigation that incorporated viewpoints that in recent years proved so successful in linear system theory. The first problem was to specialize the topic, both to avoid the vagueness that characterized some of the literature, and to facilitate the extension of linear system techniques. My approach was to consider those systems that are composed of feedback-free interconnections of linear dynamic systems and simple static nonlinear elements.

Of course, a number of people recognized the needs outlined above. About the same time that I began working with Volterra/Wiener representations, others achieved a notable success in specializing the structure of nonlinear differential equations in a profitable way. It was shown that bilinear state equations were amenable to analysis using many of the tools associated with linear state equations. In addition, the Volterra/Wiener representation corresponding to bilinear state equations turned out to be remarkably simple.

These topics, interconnection-structured systems, bilinear state equations, Volterra/Wiener representations, and their various interleavings form recurring themes in this book. I believe that from these themes will be forged many useful engineering tools for dealing with nonlinear systems in the future. But a note of caution is appropriate. Nonlinear systems do not yield easily to analysis, especially in the sense that for a given analytical method it is not hard to find an inscrutable system. Worse, it is not always easy to ascertain beforehand when methods based on the Volterra/Wiener representation are appropriate. The folk wisdom is that if the nonlinearities are mild, then the Volterra/Wiener methods should be tried. Unfortunately, more detailed characterization tends to destroy this notion before capturing it, at least in a practical sense.

So, in these matters I ask some charity from the reader. My only recommendation is the merely obvious one to keep all sorts of methods in mind. Stability questions often will call for application of methods based on the theory of differential equations. Do not forget the phase plane or the computer center, for they are sure to be useful in their share of situations. At the same time I urge the reader to question and reflect upon the possibilities for application of the Volterra/Wiener methods discussed herein.

The theory is incomplete, and likely to remain so for some time. But I hope to convince that, though the sailing won't be always smooth, the wind is up and the tide fair for this particular passage into nonlinear system theory - and that the engineering tools to be found will make the trip worthwhile.

This text represents my first attempt to write down in an organized fashion the nonlinear system theory alluded to above. As such, the effort has been somewhat frustrating since the temptation always is to view gaps in the development as gaps, and not as research opportunities. In particular the numerous research opportunities have forced certain decisions concerning style and content. Included are topics that appear to be a good bet to have direct and wide applicability to engineering problems. Others for which the odds seem longer are mentioned and referenced only. As to style I eschew the trappings of rigor and adopt a more mellifluous tone. The material is presented informally, but in such a way that the reader probably can formalize the treatment relatively easily once the main features are grasped. As an aid to this process each chapter contains a Remarks and References section that points the way to the research literature. (Historical comments that unavoidably have crept into these sections are general indications, not the result of serious historical scholarship.)

The search for simple physical examples has proven more enobling than productive. As a result, the majority of examples in the text illustrate calculations or technical features rather than applications. The same can be said about the problems included in each chapter. The problems are intended to illuminate and breed familiarity with the subject matter. Although the concepts involved in the Volterra/Wiener approach are not difficult, the formulas become quite lengthy and tend to have hidden features. Therefore, I recommend that consideration of the problems be an integral part of reading the book. For the most part the problems do not involve extending the presented material in significant ways. Nor are they designed to be overly difficult or open-ended. My view is that the diligent reader will be able to pose these kinds of problems with alacrity.

The background required for the material in this book is relatively light if some discretion is exercised. For the stationary system case, the presumed knowledge of linear system theory is not much beyond the typical third- or fourth-year undergraduate course that covers both state-equation and transfer-function concepts. However, a dose of the oft-prescribed mathematical maturity will help, particularly in the more abstract material concerning realization theory. As background for some of the material concerning nonstationary systems, I recommend that the more-or-less typical material in a first-year graduate course in linear system theory be studied, at least concurrently. Finally, some familiarity with the elements of stochastic processes is needed to appreciate fully the material on random process inputs.

I would be remiss indeed if several people who have worked with me in the nonlinear systems area were not mentioned. Winthrop W. Smith, Stephen L. Baumgartner, Thurman R. Harper, Edward M. Wysocki, Glenn E. Mitzel, and Steven J. Clancy all worked on various aspects of the material as graduate students at The Johns Hopkins University. Elmer G. Gilbert of the University of Michigan contributed much to my understanding of the theory during his sabbatical visit to The Hopkins, and in numerous subsequent discussions. Arthur E. Frazho of Purdue University has been most helpful in clarifying my presentation of his realization theory. William H. Huggins at Johns Hopkins introduced me to the computer text processor, and guided me through a sometimes stormy author-computer relationship. It is a pleasure to express my gratitude to these colleagues for their contributions.

NONLINEAR SYSTEM THEORY

CHAPTER 1

Input/Output Representations in the Time Domain

The Volterra/Wiener representation for nonlinear systems is based on the Volterra series functional representation from mathematics. Though it is a mathematical tool, the application to system input/output representation can be discussed without first going through the mathematical development. I will take this ad hoc approach, with motivation from familiar linear system representations, and from simple examples of nonlinear systems. In what will become a familiar pattern, linear systems will be reviewed first. Then homogeneous nonlinear systems (one-term Volterra series), polynomial systems (finite Volterra series), and finally Volterra systems (infinite series) will be discussed in order.

This chapter is devoted largely to terminology, introduction of notation, and basic manipulations concerning nonlinear system representations. A number of different ways of writing the Volterra/Wiener representation will be reviewed, and interrelationships between them will be established. In particular, there are three special forms for the representation that will be treated in detail: the symmetric, triangular, and regular forms. Each of these has advantages and disadvantages, but all will be used in later portions of the book. Near the end of the chapter I will discuss the origin and justification of the Volterra series as applied to system representation. Both the intuitive and the more mathematical aspects will be reviewed.

1.1 Linear Systems

Consider the input/output behavior of a system that can be described as single-input, single-output, linear, stationary, and causal. I presume that the reader is familiar with the convolution representation

$$y(t) = \int_{-\infty}^{\infty} h(\sigma)u(t-\sigma)\, d\sigma \tag{1}$$

where $u(t)$ is the input signal, and $y(t)$ is the output signal. The impulse response $h(t)$, herein called the *kernel*, is assumed to satisfy $h(t) = 0$ for $t < 0$.

There are several technical assumptions that should go along with (1). Usually it is assumed that $h(t)$ is a real-valued function defined for $t \in (-\infty, \infty)$, and piecewise continuous except possibly at $t = 0$ where an impulse (generalized) function can occur. Also the input signal is a real-valued function defined for $t \in (-\infty, \infty)$; usually assumed to be piecewise continuous, although it also can contain impulses. Finally, the matter of impulses aside, these conditions imply that the output signal is a continuous, real-valued function defined for $t \in (-\infty, \infty)$.

More general settings can be adopted, but they are unnecessary for the purposes here. In fact, it would be boring beyond the call of duty to repeat these technical assumptions throughout the sequel. Therefore, I will be casual and leave these issues understood, except when a particularly cautious note should be sounded.

It probably is worthwhile for the reader to verify that the system descriptors used above are valid for the representation (1). Of course linearity is obvious from the properties of the integral. It is only slightly less easy to see that the one-sided assumption on $h(t)$ corresponds to causality; the property that the system output at a given time cannot depend on future values of the input. Finally, simple inspection shows that the response to a delayed version of the input $u(t)$ is the delayed version of the response to $u(t)$, and thus that the system represented by (1) is stationary. Stated more precisely, if the response to $u(t)$ is $y(t)$, then the response to $u(t-T)$ is $y(t-T)$, for any $T \geqslant 0$, and hence the system is stationary.

The one-sided assumption on $h(t)$ implies that the infinite lower limit in (1) can be replaced by 0. Considering only input signals that are zero prior to $t = 0$, and often this will be the case, allows the upper limit in (1) to be replaced by t. The advantage in keeping infinite limits is that in the many changes of integration variables that will be performed on such expressions, there seldom is a need to change the limits. One of the disadvantages is that some manipulations are made to appear more subtle than they are. For example, when the order of multiple integrations is interchanged, I need only remind that the limits actually are finite to proceed with impunity.

A change of the integration variable shows that (1) can be rewritten as

$$y(t) = \int_{-\infty}^{\infty} h(t-\sigma)u(\sigma)\, d\sigma \tag{2}$$

In this form the one-sided assumption on $h(t)$ implies that the upper limit can be lowered to t, while a one-sided assumption on $u(t)$ would allow the lower limit to be raised to 0. The representation (1) will be favored for stationary systems - largely because the kernel is displayed with unmolested argument, contrary to the form in (2).

To diagram a linear system from the input/output point of view, the labeling shown in Figure 1.1 will be used. In this block diagram the system is denoted by its kernel. If the kernel is unknown, then Figure 1.1 is equivalent to the famous linear black box.

If the assumption that the system is stationary is removed, then the following input/output representation is appropriate. Corresponding to the real-valued function

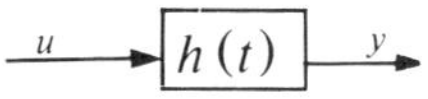

Figure 1.1. A stationary linear system.

$h(t,\sigma)$ defined for $t \in (-\infty,\infty)$, $\sigma \in (-\infty,\infty)$, with $h(t,\sigma) = 0$ if $\sigma > t$, write

$$y(t) = \int_{-\infty}^{\infty} h(t,\sigma)u(\sigma)\, d\sigma \tag{3}$$

As before, it is easy to check that this represents a linear system, and that the special assumption on $h(t,\sigma)$ corresponds to causality. Only the delay-invariance property that corresponds to stationarity has been dropped. Typically $h(t,\sigma)$ is allowed to contain impulses for $\sigma = t$, but otherwise is piecewise continuous for $t \geqslant \sigma \geqslant 0$. Of course, the range of integration in (3) can be narrowed as discussed before.

Comparison of (2) and (3) makes clear the fact that a stationary linear system can be regarded as a special case of a nonstationary linear system. Therefore, it is convenient to call the kernel $h(t,\sigma)$ in (3) *stationary* if there exists a kernel g(t) such that

$$g(t-\sigma) = h(t,\sigma) \tag{4}$$

An easy way to check for stationarity of $h(t,\sigma)$ is to check the condition $h(0,\sigma-t) = h(t,\sigma)$. If this is satisfied, then setting $g(t) = h(0,-t)$ verifies (4) since $g(t-\sigma) = h(0,\sigma-t) = h(t,\sigma)$.

A (possibly) nonstationary linear system will be diagramed using the representation (3) as shown in Figure 1.2.

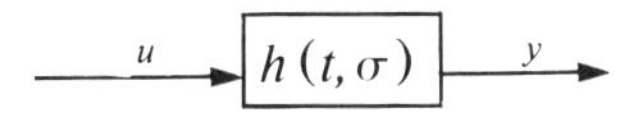

Figure 1.2. A nonstationary linear system.

1.2 Homogeneous Nonlinear Systems

The approach to be taken to the input/output representation of nonlinear systems involves a simple generalization of the representations discussed in Section 1.1. The more difficult, somewhat unsettled, and in a sense philosophical questions about the generality and usefulness of the representation will be postponed. For the moment I will write down the representation, discuss some of its properties, and give enough examples to permit the claim that it is interesting.

Corresponding to the real-valued function of n variables $h_n(t_1,\ldots,t_n)$ defined for $t_i \in (-\infty,\infty)$, $i = 1,\ldots,n$, and such that $h_n(t_1,\ldots,t_n) = 0$ if any $t_i < 0$, consider the input/output relation

$$y(t) = \int_{-\infty}^{\infty} \cdots \int_{-\infty}^{\infty} h_n(\sigma_1,\ldots,\sigma_n)u(t-\sigma_1) \cdots u(t-\sigma_n)\, d\sigma_1 \cdots d\sigma_n \tag{5}$$

The resemblance to the linear system representations of the previous section is clear. Furthermore the same kinds of technical assumptions that are appropriate for the convolution representation for linear systems in (1) are appropriate here. Indeed, (5) often is called generalized convolution, although I won't be using that term.

Probably the first question to be asked is concerned with the descriptors that can be associated to a system represented by (5). It is obvious that the assumption that $h_n(t_1,\ldots,t_n)$ is one-sided in each variable corresponds to causality. The system is not linear, but it is a stationary system as a check of the delay invariance property readily shows.

A system represented by (5) will be called a *degree-n homogeneous system.* The terminology arises because application of the input $\alpha u(t)$, where α is a scalar, yields the output $\alpha^n y(t)$, where $y(t)$ is the response to $u(t)$. Note that this terminology includes the case of a linear system as a degree-1 homogeneous system. Just as in the linear case, $h_n(t_1,\ldots,t_n)$ will be called the *kernel* associated with the system.

For simplicity of notation I will collapse the multiple integration and, when no confusion is likely to arise, drop the subscript on the kernel to write (5) as

$$y(t) = \int_{-\infty}^{\infty} h(\sigma_1,\ldots,\sigma_n)u(t-\sigma_1) \cdots u(t-\sigma_n)\, d\sigma_1 \cdots d\sigma_n \tag{6}$$

Just as in the linear case, the lower limit(s) can be replaced by 0 because of the one-sided assumption on the kernel. If it is assumed also that the input signal is one-sided, then all the upper limit(s) can be replaced by t. Finally, a change of each variable of integration shows that (6) can be rewritten in the form

$$y(t) = \int_{-\infty}^{\infty} h(t-\sigma_1,\ldots,t-\sigma_n)u(\sigma_1) \cdots u(\sigma_n)\, d\sigma_1 \cdots d\sigma_n \tag{7}$$

At this point it should be no surprise that a stationary degree-n homogeneous system will be diagramed as shown in Figure 1.3. Again the system box is labeled with the kernel.

$$\xrightarrow{\;u\;} \boxed{h(t_1,\ldots,t_n)} \xrightarrow{\;y\;}$$

Figure 1.3. A stationary degree-n homogeneous system.

There are at least two generic ways in which homogeneous systems can arise in engineering applications. The first involves physical systems that naturally are structured in terms of interconnections of linear subsystems and simple nonlinearities. In particular I will consider situations that involve stationary linear subsystems, and nonlinearities that can be represented in terms of multipliers. For so-called *interconnection structured systems* such as this, it is often easy to derive the overall system kernel from the subsystem kernels simply by tracing the input signal through the system diagram. (In this case subscripts will be used to denote different subsystems since all kernels are single variable.)

Example 1.1 Consider the multiplicative connection of three linear subsystems, shown in Figure 1.4. The linear subsystems can be described by

$$y_i(t) = \int_{-\infty}^{\infty} h_i(\sigma)u(t-\sigma)\, d\sigma, \quad i = 1,2,3$$

and thus the overall system is described by

$$\begin{aligned} y(t) &= y_1(t)y_2(t)y_3(t) \\ &= \int_{-\infty}^{\infty} h_1(\sigma)u(t-\sigma)\, d\sigma \int_{-\infty}^{\infty} h_2(\sigma)u(t-\sigma)\, d\sigma \int_{-\infty}^{\infty} h_3(\sigma)u(t-\sigma)\, d\sigma \\ &= \int_{-\infty}^{\infty} h_1(\sigma_1)h_2(\sigma_2)h_3(\sigma_3)u(t-\sigma_1)u(t-\sigma_2)u(t-\sigma_3)\, d\sigma_1 d\sigma_2 d\sigma_3 \end{aligned}$$

Clearly, a kernel for this degree-3 homogeneous system is

$$h(t_1,t_2,t_3) = h_1(t_1)h_2(t_2)h_3(t_3)$$

A second way in which homogeneous systems can arise begins with a state equation description of a nonlinear system. To illustrate, consider a compartmental model wherein each variable $x_i(t)$ represents a population, chemical concentration, or other quantity of interest. If the rate of change of $x_i(t)$ depends linearly on other $x_j(t)$'s,

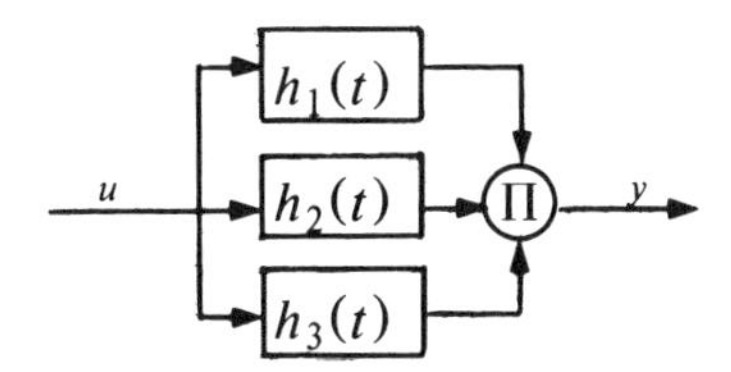

Figure 1.4. An interconnection structured system.

but with a scalar parametric control signal, then $\dot{x}_i(t)$ will contain terms of the form $du(t)x_j(t)$. Nonlinear compartmental models of this type lead to the study of so-called *bilinear state equations*

$$\dot{x}(t) = Ax(t) + Dx(t)u(t) + bu(t)$$
$$y(t) = cx(t), \quad t \geqslant 0, \; x(0) = x_0$$

where $x(t)$ is the $n \times 1$ state vector, and $u(t)$ and $y(t)$ are the scalar input and output signals. Such state equations will be discussed in detail later on, so for now a very simple case will be used to indicate the connection to homogeneous systems.

Example 1.2 Consider a nonlinear system described by the differential equation

$$\dot{x}(t) = Dx(t)u(t) + bu(t)$$
$$y(t) = cx(t), \quad t \geqslant 0, \; x(0) = 0$$

where $x(t)$ is a 2×1 vector, $u(t)$ and $y(t)$ are scalars, and

$$D = \begin{bmatrix} 0 & 0 \\ 1 & 0 \end{bmatrix}, \quad b = \begin{bmatrix} 1 \\ 0 \end{bmatrix}, \quad c = [0 \quad 1]$$

It can be shown that a differential equation of this general form has a unique solution for all $t \geqslant 0$ for a piecewise continuous input signal. I leave it as an exercise to verify that this solution can be written in the form

$$x(t) = \int_0^t e^{D\int_{\sigma_2}^t u(\sigma_1)\,d\sigma_1} bu(\sigma_2)\,d\sigma_2$$

where, of course, the matrix exponential is given by

$$e^{D\int_{\sigma_2}^t u(\sigma_1)\,d\sigma_1} = I + D\int_{\sigma_2}^t u(\sigma_1)\,d\sigma_1 + \frac{1}{2!}D^2[\int_{\sigma_2}^t u(\sigma_1)\,d\sigma_1]^2 + \cdots$$

For the particular case at hand, $D^2 = 0$ so that

$$e^{D\int_{\sigma_2}^t u(\sigma_1)\,d\sigma_1} = \begin{bmatrix} 1 & 0 \\ \int_{\sigma_2}^t u(\sigma_1)\,d\sigma_1 & 1 \end{bmatrix}$$

Thus the input/output relation can be written in the form

$$y(t) = \int_0^t ce^{D\int_{\sigma_2}^t u(\sigma_1)\,d\sigma_1} bu(\sigma_2)\, d\sigma_2$$

$$= \int_0^t \int_{\sigma_2}^t u(\sigma_1)u(\sigma_2)\, d\sigma_1 d\sigma_2$$

From this expression it is clear that the system is homogeneous and of degree 2. To put the input/output representation into a more familiar form, the *unit step function*

$$\delta_{-1}(t) = \begin{cases} 0, & t < 0 \\ 1, & t \geqslant 0 \end{cases}$$

can be introduced to write

$$y(t) = \int_0^t \int_0^t \delta_{-1}(\sigma_1 - \sigma_2)u(\sigma_1)u(\sigma_2)\, d\sigma_1 d\sigma_2$$

Thus, a kernel for the system is

$$h(t_1,t_2) = \delta_{-1}(t_1 - t_2)$$

There will be occasion in later chapters to consider homogeneous systems that may not be stationary. Such a system is represented by the input/output expression

$$y(t) = \int_{-\infty}^{\infty} h(t,\sigma_1,\ldots,\sigma_n)u(\sigma_1) \cdots u(\sigma_n)\, d\sigma_1 \cdots d\sigma_n \tag{8}$$

It is assumed that the kernel satisfies $h(t,\sigma_1,\ldots,\sigma_n) = 0$ when any $\sigma_i > t$ so that the system is causal. Of course, this permits all the upper limits to be replaced by t. If one-sided inputs are considered, then the lower limits can be raised to 0.

As a simple example of a nonstationary homogeneous system, the reader can rework Example 1.1 under the assumption that the linear subsystems are nonstationary. But I will consider here a case where the nonstationary representation quite naturally arises from a stationary interconnection structured system.

Example 1.3 The interconnection shown in Figure 1.5 is somewhat more complicated than that treated in Example 1.1. As suggested earlier, a good way to find a kernel is to begin with the input signal and find expressions for each labeled signal, working toward the output. The signal $v(t)$ can be written as

$$v(t) = \int_{-\infty}^{t} h_3(t-\sigma_3)u(\sigma_3)\, d\sigma_3\; u(t)$$

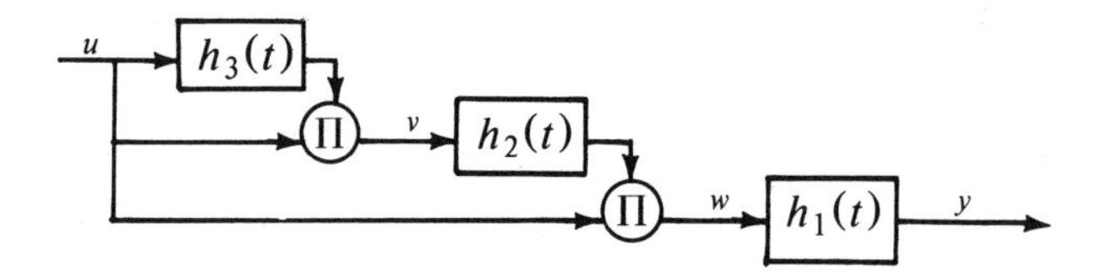

Figure 1.5. The system considered in Example 1.4.

Similarly

$$w(t) = \int_{-\infty}^{t} h_2(t-\sigma_2)v(\sigma_2)\, d\sigma_2\, u(t)$$

$$= \int_{-\infty}^{t} h_2(t-\sigma_2) \int_{-\infty}^{\sigma_2} h_3(\sigma_2-\sigma_3)u(\sigma_3)\, d\sigma_3\, u(\sigma_2)\, d\sigma_2\, u(t)$$

$$= \int_{-\infty}^{t} \int_{-\infty}^{\sigma_2} h_2(t-\sigma_2)h_3(\sigma_2-\sigma_3)u(\sigma_3)u(\sigma_2)\, d\sigma_3 d\sigma_2\, u(t)$$

The output signal is given by

$$y(t) = \int_{-\infty}^{t} h_1(t-\sigma_1)w(\sigma_1)\, d\sigma_1$$

$$= \int_{-\infty}^{t} \int_{-\infty}^{\sigma_1} \int_{-\infty}^{\sigma_2} h_1(t-\sigma_1)h_2(\sigma_1-\sigma_2)h_3(\sigma_2-\sigma_3)u(\sigma_1)u(\sigma_2)u(\sigma_3)\, d\sigma_3 d\sigma_2 d\sigma_1$$

Thus a kernel for this degree-3 system can be written in the form

$$h(t,\sigma_1,\sigma_2,\sigma_3) = h_1(t-\sigma_1)h_2(\sigma_1-\sigma_2)h_3(\sigma_2-\sigma_3)\delta_{-1}(\sigma_2-\sigma_3)\delta_{-1}(\sigma_1-\sigma_2)$$

Because of the usual one-sided assumptions on the linear subsystem kernels, the step functions might be regarded as superfluous. More importantly, a comparison of Examples 1.1 and 1.3 indicates that different forms of the kernel are more natural for different system structures.

Comparing the representation (8) for nonstationary systems to the representation (7) for stationary systems leads to the definition that a kernel $h(t,\sigma_1,\ldots,\sigma_n)$ is *stationary* if there exists a kernel $g(t_1,\ldots,t_n)$ such that the relationship

$$g(t-\sigma_1,\ldots,t-\sigma_n) = h(t,\sigma_1,\ldots,\sigma_n) \tag{9}$$

holds for all $t, \sigma_1, \ldots, \sigma_n$. Usually it is convenient to check for stationarity by checking the functional relationship

$$h(0, \sigma_1 - t, \ldots, \sigma_n - t) = h(t, \sigma_1, \ldots, \sigma_n) \tag{10}$$

for if this is satisfied, then (9) is obtained by setting

$$g(t_1, \ldots, t_n) = h(0, -t_1, \ldots, -t_n) \tag{11}$$

Therefore, when (10) is satisfied I can write, in place of (8),

$$y(t) = \int_{-\infty}^{\infty} g(t-\sigma_1, \ldots, t-\sigma_n) u(\sigma_1) \cdots u(\sigma_n)\, d\sigma_1 \cdots d\sigma_n \tag{12}$$

Performing this calculation for Example 1.3 gives a stationary kernel for the system in Figure 1.5:

$$g(t_1, t_2, t_3) = h_1(t_1) h_2(t_2 - t_1) h_3(t_3 - t_2) \delta_{-1}(t_3 - t_2) \delta_{-1}(t_2 - t_1)$$

As mentioned in Section 1.1, in the theory of linear systems it is common to allow impulse (generalized) functions in the kernel. For example, in (1) suppose $h(t) = g(t) + g_0 \delta_0(t)$, where $g(t)$ is a piecewise continuous function and $\delta_0(t)$ is a unit impulse at $t = 0$. Then the response to an input $u(t)$ is

$$\begin{aligned} y(t) &= \int_{-\infty}^{\infty} h(t-\sigma) u(\sigma)\, d\sigma \\ &= \int_{-\infty}^{\infty} g(t-\sigma) u(\sigma)\, d\sigma + \int_{-\infty}^{\infty} g_0 \delta_0(t-\sigma) u(\sigma)\, d\sigma \\ &= \int_{\infty}^{\infty} g(t-\sigma) u(\sigma)\, d\sigma + g_0 u(t) \end{aligned} \tag{13}$$

That is, the impulse in the kernel corresponds to what might be called a *direct transmission term* in the input/output relation. Even taking the input $u(t) = \delta_0(t)$ causes no problems in this set-up. The resulting impulse response is

$$\begin{aligned} y(t) &= \int_{-\infty}^{\infty} g(t-\sigma) \delta_0(\sigma)\, d\sigma + \int_{-\infty}^{\infty} g_0 \delta_0(t-\sigma) \delta_0(\sigma)\, d\sigma \\ &= g(t) + g_0 \delta_0(t) \end{aligned} \tag{14}$$

Unfortunately these issues are much more devious for homogeneous systems of degree $n > 1$. For such systems, impulse inputs cause tremendous problems when a direct transmission term is present. To see why, notice that such a term must be of degree n, and so it leads to undefined objects of the form $\delta_0^n(t)$ in the response. Since impulsive inputs must be ruled out when direct transmission terms are present, it seems prudent to display such terms explicitly. However, there are a number of different kinds of terms that share similar difficulties in the higher degree cases, and the equations I am presenting are sufficiently long already. For example, consider a degree-2 system with input/output relation

$$y(t) = \int_{-\infty}^{\infty} \int_{-\infty}^{\infty} g(t-\sigma_1,t-\sigma_2)u(\sigma_1)u(\sigma_2)\, d\sigma_1 d\sigma_2$$

$$+ \int_{-\infty}^{\infty} g_1(t-\sigma_1)u^2(\sigma_1)\, d\sigma_1 + g_0 u^2(t) \tag{15}$$

Adopting a loose terminology, I will call both of the latter two terms direct transmission terms. Allowing impulses in the kernel means that the representation

$$y(t) = \int_{-\infty}^{\infty} h(t-\sigma_1,t-\sigma_2)u(\sigma_1)u(\sigma_2)\, d\sigma_1 d\sigma_2$$

suffices with

$$h(t_1,t_2) = g(t_1,t_2) + g_1(t_1)\delta_0(t_1-t_2) + g_0\delta_0(t_1)\delta_0(t_2) \tag{16}$$

The dangers not withstanding, impulses will be allowed in the kernel to account for the various direct transmission terms. But as a matter of convention, a kernel is assumed to be impulse free unless stated otherwise. I should point out that, as indicated by the degree-2 case, the impulses needed for this purpose occur only for values of the kernel's arguments satisfying certain patterns of equalities.

Example 1.4 A simple system for computing the integral-square value of a signal is shown in Figure 1.6. This system is described by

$$y(t) = \int_{-\infty}^{\infty} \delta_{-1}(t-\sigma)u^2(\sigma)\, d\sigma$$

so that a standard-form degree-2 homogeneous representation is

$$y(t) = \int_{-\infty}^{\infty} \int_{-\infty}^{\infty} \delta_{-1}(t-\sigma_1)\delta_0(\sigma_1-\sigma_2)u(\sigma_1)u(\sigma_2)\, d\sigma_2 d\sigma_1$$

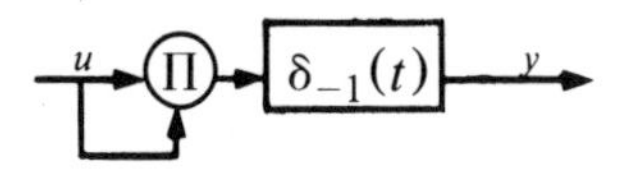

Figure 1.6. An integral-square computer.

If the input signal is one-sided, then the representation can be simplified to

$$y(t) = \int_0^t \int_0^t \delta_0(\sigma_1 - \sigma_2) u(\sigma_1) u(\sigma_2)\, d\sigma_2 d\sigma_1$$

On the other hand, a simple system for computing the square-integral of a signal is shown in Figure 1.7. This system is described by

$$y(t) = [\int_{-\infty}^{\infty} \delta_{-1}(t-\sigma) u(\sigma)\, d\sigma]^2$$

$$= \int_{-\infty}^{\infty} \int_{-\infty}^{\infty} \delta_{-1}(t-\sigma_1)\delta_{-1}(t-\sigma_2) u(\sigma_1) u(\sigma_2)\, d\sigma_1 d\sigma_2$$

If the input signal is one-sided, then the representation simplifies to

$$y(t) = \int_0^t \int_0^t u(\sigma_1) u(\sigma_2)\, d\sigma_1 d\sigma_2$$

Comparison of these two systems indicates that direct transmission terms (impulsive kernels) arise from unintegrated input signals in the nonlinear part of the system.

A kernel describing a degree-n homogeneous system will be called *separable* if it can be expressed in the form

$$h(t_1, \ldots, t_n) = \sum_{i=1}^{m} v_{1i}(t_1) v_{2i}(t_2) \cdots v_{ni}(t_n) \tag{17}$$

or

$$h(t, \sigma_1, \ldots, \sigma_n) = \sum_{i=1}^{m} v_{0i}(t) v_{1i}(\sigma_1) \cdots v_{ni}(\sigma_n) \tag{18}$$

where each $v_{ji}(.)$ is a continuous function. It will be called *differentiably separable* if each $v_{ji}(.)$ is differentiable. Almost all of the kernels of interest herein will be differentiably separable. Although explicit use of this terminology will not occur until much later, it will become clear from examples and problems that separability is a routinely occurring property of kernels.

The reader probably has noticed from the examples that more than one kernel can be used to describe a given system. For instance, the kernel derived in Example 1.1

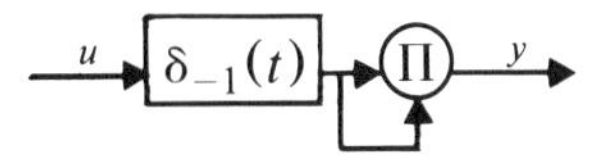

Figure 1.7. A square-integral computer.

can be rewritten in several ways simply by reordering the variables of integration. This feature not only is disconcerting at first glance, it also leads to serious difficulties when system properties are described in terms of properties of the kernel. Therefore, it becomes important in many situations to impose uniqueness by working with special, restricted forms for the kernel. Three such special forms will be used in the sequel: the symmetric kernel, the triangular kernel, and the regular kernel. I now turn to the introduction of these forms.

A *symmetric kernel* in the stationary case satisfies

$$h_{sym}(t_1,\ldots,t_n) = h_{sym}(t_{\pi(1)},\ldots,t_{\pi(n)}) \tag{19}$$

or, in the nonstationary case,

$$h_{sym}(t,\sigma_1,\ldots,\sigma_n) = h_{sym}(t,\sigma_{\pi(1)},\ldots,\sigma_{\pi(n)}) \tag{20}$$

where $\pi(.)$ denotes any permutation of the integers $1,\ldots,n$. It is easy to show that without loss of generality the kernel of a homogeneous system can be assumed to be symmetric. In fact any given kernel, say $h(t_1,\ldots,t_n)$ in (6), can be replaced by a symmetric kernel simply by setting

$$h_{sym}(t_1,\ldots,t_n) = \frac{1}{n!}\sum_{\pi(.)} h(t_{\pi(1)},\ldots,t_{\pi(n)}) \tag{21}$$

where the indicated summation is over all $n!$ permutations of the integers 1 through n. To see that this replacement does not affect the input/output relation, consider the expression

$$\int_{-\infty}^{\infty} h_{sym}(\sigma_1,\ldots,\sigma_n)u(t-\sigma_1)\cdots u(t-\sigma_n)\,d\sigma_1\cdots d\sigma_n =$$

$$\frac{1}{n!}\sum_{\pi(.)}\int_{-\infty}^{\infty} h(\sigma_{\pi(1)},\ldots,\sigma_{\pi(n)})u(t-\sigma_{\pi(1)})\cdots u(t-\sigma_{\pi(n)})\,d\sigma_{\pi(1)}\cdots d\sigma_{\pi(n)} \tag{22}$$

Introducing the change of variables (actually, just a relabeling) $\tau_i = \sigma_{\pi(i)}$, $i = 1,\ldots,n$, in every term of the summation in (22) shows that all terms are identical. Thus summing the $n!$ identical terms on the right side shows that the two kernels yield the same input/output behavior.

Often a kernel of interest is partially symmetric in the sense that not all terms of the summation in (21) are distinct. In this situation the symmetric version of the kernel can be obtained by summing over those permutations that give distinct summands, and replacing the $n!$ by the number of such permutations. A significant reduction in the number of terms is often the result.

Example 1.5 Consider a degree-3 kernel that has the form

$$h(t_1,t_2,t_3) = g(t_1)g(t_2)g(t_3)f(t_1+t_2)$$

Incidently, note that this is not a separable kernel unless $f(t_1+t_2)$ can be written as a sum of terms of the form $f_1(t_1)f_2(t_2)$. To symmetrize this kernel, (21) indicates that six terms must be added. However, the first three factors in this particular case are symmetric, and there are only three permutations that will yield distinct forms of the last factor; namely $f(t_1+t_2)$, $f(t_1+t_3)$, and $f(t_2+t_3)$. Thus, the symmetric form of the given kernel is

$$h_{sym}(t_1,t_2,t_3) = \frac{1}{3}g(t_1)g(t_2)g(t_3)[f(t_1+t_2)+f(t_1+t_3)+f(t_2+t_3)]$$

Again I emphasize that although the symmetric version of a kernel usually contains more terms than an asymmetric version, it does offer a standard form for the kernel. In many cases system properties can be related more simply to properties of the symmetric kernel than to properties of an asymmetric kernel.

The second special form of interest is the *triangular kernel.* The kernel in (8), $h(t,\sigma_1,\ldots,\sigma_n)$, is triangular if it satisfies the additional property that $h(t,\sigma_1,\ldots,\sigma_n) = 0$ when $\sigma_{i+j} > \sigma_j$ for i,j positive integers. A triangular kernel will be indicated by the subscript "tri" when convenient. For such a kernel the representation (8) can be written in the form

$$y(t) = \int_{-\infty}^{t}\int_{-\infty}^{\sigma_1}\cdots\int_{-\infty}^{\sigma_{n-1}} h_{tri}(t,\sigma_1,\ldots,\sigma_n)u(\sigma_1)\cdots u(\sigma_n)\,d\sigma_n\cdots d\sigma_1 \tag{23}$$

Sometimes this special form of the input/output relation will be maintained for triangular kernels, but often I will raise all the upper limits to ∞ or t and leave triangularity understood. On some occasions the triangularity of the kernel will be emphasized by appending unit step functions. In this manner (23) becomes

$$y(t) = \int_{-\infty}^{\infty} h_{tri}(t,\sigma_1,\ldots,\sigma_n)\delta_{-1}(\sigma_1-\sigma_2)\delta_{-1}(\sigma_2-\sigma_3)$$

$$\cdots\delta_{-1}(\sigma_{n-1}-\sigma_n)u(\sigma_1)\cdots u(\sigma_n)\,d\sigma_n\cdots d\sigma_1 \tag{24}$$

Notice that there is no need to use precisely this definition of triangularity. For example, if $h_{tri}(t,\sigma_1,\ldots,\sigma_n) = 0$ when $\sigma_j > \sigma_{i+j}$, then the suitable triangular representation is

$$y(t) = \int_{-\infty}^{t}\int_{-\infty}^{\sigma_{n-1}}\cdots\int_{-\infty}^{\sigma_2} h_{tri}(t,\sigma_1,\ldots\sigma_n)u(\sigma_1)\cdots u(\sigma_n)\,d\sigma_1\cdots\sigma_n \tag{25}$$

Stated another way, a triangular kernel

$$h_{tri}(t,\sigma_1,\ldots,\sigma_n) = h_{tri}(t,\sigma_1,\ldots,\sigma_n)\delta_{-1}(\sigma_1-\sigma_2)\cdots\delta_{-1}(\sigma_{n-1}-\sigma_n) \tag{26}$$

remains triangular for any permutation of the arguments $\sigma_1,\ldots,\sigma_n$. A permutation of arguments simply requires that the integration be performed over the appropriate triangular domain, and this domain can be made clear by the appended step functions. However, I will stick to the ordering of variables indicated in (23) and (26) most of the time.

Now assume that the triangular kernel in (26) in fact is stationary. Then let

$$g_{tri}(\sigma_1, \ldots, \sigma_n) = h_{tri}(0, -\sigma_1, \ldots, -\sigma_n)\delta_{-1}(\sigma_2 - \sigma_1) \cdots \delta_{-1}(\sigma_n - \sigma_{n-1}) \quad (27)$$

so that

$$g_{tri}(t-\sigma_1, \ldots, t-\sigma_n) = h_{tri}(t, \sigma_1, \ldots, \sigma_n)\delta_{-1}(\sigma_1 - \sigma_2) \cdots \delta_{-1}(\sigma_{n-1} - \sigma_n) \quad (28)$$

and the input/output relation in (23) becomes

$$y(t) = \int_{-\infty}^{t} \int_{-\infty}^{\sigma_1} \cdots \int_{-\infty}^{\sigma_{n-1}} g_{tri}(t-\sigma_1, \ldots, t-\sigma_n) u(\sigma_1) \cdots u(\sigma_n) \, d\sigma_n \cdots d\sigma_1 \quad (29)$$

Or, performing the usual variable change,

$$y(t) = \int_{-\infty}^{t} \int_{-\infty}^{\sigma_n} \cdots \int_{-\infty}^{\sigma_2} g_{tri}(\sigma_1, \ldots, \sigma_n) u(t-\sigma_1) \cdots u(t-\sigma_n) \, d\sigma_1 \cdots d\sigma_n \quad (30)$$

an expression that emphasizes that in (27) triangularity implies $g_{tri}(t_1, \ldots, t_n) = 0$ if $t_i > t_{i+j}$. But, again, this is not the only choice of triangular domain. In fact, for a degree-n kernel there are $n!$ choices for the triangular domain, corresponding to the $n!$ permutations of variables in the inequality $t_{\pi(1)} \geqslant t_{\pi(2)} \geqslant \cdots \geqslant t_{\pi(n)} \geqslant 0$. So there is flexibility here: pick the domain you like, or like the domain you pick.

To present examples of triangular kernels, I need only review some of the earlier examples. Notice that the nonstationary kernel obtained in Example 1.3 actually is in the triangular form (24). Also the input/output representation obtained in Example 1.2 can be written in the form

$$y(t) = \int_0^t \int_0^{\sigma_1} u(\sigma_1) u(\sigma_2) \, d\sigma_2 d\sigma_1$$

This corresponds to the triangular kernel $h_{tri}(t, \sigma_1, \sigma_2) = \delta_{-1}(\sigma_1 - \sigma_2)$ in (24), or to the triangular kernel $g_{tri}(t_1, t_2) = \delta_{-1}(t_2 - t_1)$ in (29).

The relationship between symmetric and triangular kernels should clarify the features of both. Assume for the moment that only impulse-free inputs are allowed. To symmetrize a triangular kernel it is clear that the procedure of summing over all permutations of the indices applies. However, in this case the summation is merely a patching process since no two of the terms in the sum will be nonzero at the same point, except along lines of equal arguments such as $\sigma_i = \sigma_j$, $\sigma_i = \sigma_j = \sigma_k$, and so on. And since the integrations are not affected by changes in integrand values along a line, this aspect can be ignored. On the other hand, for the symmetric kernel $h_{sym}(t, \sigma_1, \ldots, \sigma_n)$ I can write the input/output relation as a sum of $n!$ n-fold integrations over the $n!$ triangular domains in the first orthant. Since each of these integrations is identical, the triangular form is given by

$$h_{tri}(t, \sigma_1, \ldots, \sigma_n) = n! h_{sym}(t, \sigma_1, \ldots, \sigma_n)\delta_{-1}(\sigma_1 - \sigma_2)\delta_{-1}(\sigma_2 - \sigma_3) \cdots \delta_{-1}(\sigma_{n-1} - \sigma_n) \quad (31)$$

In the stationary case the symmetric kernel $h_{sym}(t_1,\ldots,t_n)$ yields the triangular kernel corresponding to (30) as

$$g_{tri}(t_1,\ldots,t_n) = n!h_{sym}(t_1,\ldots,t_n)\delta_{-1}(t_2-t_1)\cdots\delta_{-1}(t_n-t_{n-1}) \tag{32}$$

Of course, these formulas imply that either of these special forms is (essentially) uniquely specified by the other.

Example 1.6 For the stationary, symmetric degree-2 kernel

$$h_{sym}(t_1,t_2) = e^{t_1+t_2}\,e^{min[t_1,t_2]}$$

a corresponding triangular kernel is

$$h_{tri}(t_1,t_2) = 2e^{2t_1+t_2}\delta_{-1}(t_2-t_1)$$

It is instructive to recompute the symmetric form. Following (21),

$$\begin{aligned} h_{sym}(t_1,t_2) &= \frac{1}{2}[2e^{2t_1+t_2}\delta_{-1}(t_2-t_1) + 2e^{2t_2+t_1}\delta_{-1}(t_1-t_2)] \\ &= e^{t_1+t_2}[e^{t_1}\delta_{-1}(t_2-t_1) + e^{t_2}\delta_{-1}(t_1-t_2)] \end{aligned}$$

Now this is almost the symmetric kernel I started with. Almost, because for $t_1 = t_2$ the original symmetric kernel is e^{3t_1}, while the symmetrized triangular kernel is $2e^{3t_1}$. This is precisely the point of my earlier remark. To wit, values of the kernel along equal argument lines can be changed without changing the input/output representation. In fact they must be changed to make circular calculations yield consistent answers.

Now consider what happens when impulse inputs are allowed, say $u(t) = \delta_0(t)$. In terms of the (nonstationary) symmetric kernel, the response is $y(t) = h_{sym}(t,0,\ldots,0)$, and in terms of the triangular kernel, $y(t) = h_{tri}(t,0,\ldots,0)$. Thus, it is clear that in this situation (31) is not consistent. Of course, the difficulty is that when impulse inputs are allowed, the value of a kernel along lines of equal arguments can affect the input/output behavior. For a specific example, reconsider the stationary kernels in Example 1.6 with an impulse input.

Again, the problem here is that the value of the triangular kernel along equal argument lines is defined to be equal to the value of the symmetric kernel. This can be fixed by more careful definition of the triangular kernel. Specifically, what must be done is to adjust the definition so that the triangular kernel gets precisely its fair share of the value of the symmetric kernel along equal-argument lines. A rather fancy "step function" can be defined to do this, but at considerable expense in simplicity. My vote is cast for simplicity, so impulse inputs henceforth are disallowed in the presence of these issues, and kernel values along lines will be freely adjusted when necessary. (This luxury is not available in the discrete-time case discussed in Chapter 6, and a careful definition of the triangular kernel which involves a fancy step function is used there. The reader inclined to explicitness is invited to transcribe those definitions to the continuous-time case at hand.)

The third special form for the kernel actually involves a special form for the entire input/output representation. This new form is most easily based on the triangular kernel. Intuitively speaking, it shifts the discontinuity of the triangular kernel out of the picture and yields a smooth kernel over all of the first orthant. This so-called *regular kernel* will be used only in the stationary system case, and only for one-sided input signals.

Suppose $h_{tri}(t_1,\ldots,t_n)$ is a triangular kernel that is zero outside of the domain $t_1 \geqslant t_2 \geqslant \cdots \geqslant t_n \geqslant 0$. Then the corresponding input/output representation can be written in the form

$$y(t) = \int_{-\infty}^{\infty} h_{tri}(\sigma_1,\ldots,\sigma_n)u(t-\sigma_1)\cdots u(t-\sigma_n)\,d\sigma_1\cdots d\sigma_n$$

where the unit step functions are dropped and the infinite limits are retained just to make the bookkeeping simpler. Now make the variable change from σ_1 to $\tau_1 = \sigma_1 - \sigma_2$. Then the input/output representation is

$$y(t) = \int_{-\infty}^{\infty} h_{tri}(\tau_1+\sigma_2,\sigma_2,\ldots,\sigma_n)u(t-\tau_1-\sigma_2)u(t-\sigma_2)$$

$$\cdots u(t-\sigma_n)\,d\tau_1 d\sigma_2\cdots d\sigma_n$$

Now replace σ_2 by $\tau_2 = \sigma_2 - \sigma_3$ to obtain

$$y(t) = \int_{-\infty}^{\infty} h_{tri}(\tau_1+\tau_2+\sigma_3,\tau_2+\sigma_3,\sigma_3,\ldots,\sigma_n)$$

$$u(t-\tau_1-\tau_2-\sigma_3)u(t-\tau_2-\sigma_3)u(t-\sigma_4)\cdots u(t-\sigma_n)\,d\tau_1 d\tau_2 d\sigma_3\cdots d\sigma_n$$

Continuing this process gives

$$y(t) = \int_{-\infty}^{\infty} h_{tri}(\tau_1+\cdots+\tau_n,\tau_2+\cdots+\tau_n,\ldots,\tau_{n-1}+\tau_n,\tau_n)$$

$$u(t-\tau_1-\cdots-\tau_n)u(t-\tau_2-\cdots-\tau_n)\cdots u(t-\tau_n)\,d\tau_1\cdots d\tau_n$$

(In continuing the process, each variable change can be viewed as a change of variable in one of the iterated integrals. Thus the Jacobian of the overall change of variables is unity, as is easily verified. This is a general feature of variable changes in the sequel.) Letting

$$h_{reg}(t_1,\ldots,t_n) = h_{tri}(t_1+\cdots+t_n,t_2+\cdots+t_n,\ldots,t_n) \tag{33}$$

be the regular kernel, I can write

$$y(t) = \int_{-\infty}^{\infty} h_{reg}(\tau_1,\ldots,\tau_n)u(t-\tau_1-\cdots-\tau_n)u(t-\tau_2-\cdots-\tau_n)$$

$$\cdots u(t-\tau_n)\, d\tau_1 \cdots d\tau_n \tag{34}$$

where $h_{reg}(t_1,\ldots,t_n)$ is zero outside of the first orthant, $t_1,\ldots,t_n \geqslant 0$. As mentioned above, the usual discontinuities encountered along the lines $t_{j-1} = t_j$, and so on, in the triangular kernel occur along the edges $t_j = 0$ of the domain of the regular kernel.

It should be clear from (33) that the triangular kernel corresponding to a given regular kernel is

$$h_{tri}(t_1,\ldots,t_n) = h_{reg}(t_1-t_2,t_2-t_3,\ldots,t_{n-1}-t_n,t_n)$$
$$\delta_{-1}(t_1-t_2)\delta_{-1}(t_2-t_3)\cdots\delta_{-1}(t_{n-1}-t_n),\ t_1,\ldots,t_n \geqslant 0 \tag{35}$$

Thus (33) and (35), in conjunction with the earlier discussion of the relationship between the triangular and symmetric kernels, show how to obtain the symmetric kernel from the regular kernel, and vice versa.

I noted earlier that particular forms for the kernel often are natural for particular system structures. Since the regular kernel is closely tied to the triangular kernel, it is not surprising that when one is convenient, the other probably is also (restricting attention, of course, to the case of stationary systems with one-sided inputs). This can be illustrated by reworking Example 1.3 in a slightly different way.

Example 1.7 Using an alternative form of the linear system convolution representation, the calculations in Example 1.3 proceed as follows. Clearly the system is stationary, and one-sided input signals are assumed implicitly. First the signal $v(t)$ can be written in the form

$$v(t) = \int_{-\infty}^{\infty} h_3(\sigma_3)u(t-\sigma_3)\, d\sigma_3\, u(t)$$

Then

$$w(t) = \int_{-\infty}^{\infty} h_2(\sigma_2)v(t-\sigma_2)\, d\sigma_2\, u(t)$$

$$= \int_{-\infty}^{\infty}\int_{-\infty}^{\infty} h_2(\sigma_2)h_3(\sigma_3)u(t-\sigma_2-\sigma_3)u(t-\sigma_2)\, d\sigma_2 d\sigma_3\, u(t)$$

and

$$y(t) = \int_{-\infty}^{\infty} h_1(\sigma_1)w(t-\sigma_1)\, d\sigma_1$$

$$= \int_{-\infty}^{\infty}\int_{-\infty}^{\infty}\int_{-\infty}^{\infty} h_1(\sigma_1)h_2(\sigma_2)h_3(\sigma_3)u(t-\sigma_1-\sigma_2-\sigma_3)$$
$$u(t-\sigma_1-\sigma_2)u(t-\sigma_1)\, d\sigma_1 d\sigma_2 d\sigma_3$$

Now a simple interchange of the integration variables σ_1 and σ_3 gives

$$y(t) = \int_{-\infty}^{\infty} h_3(\sigma_1)h_2(\sigma_2)h_1(\sigma_3)u(t-\sigma_1-\sigma_2-\sigma_3)u(t-\sigma_2-\sigma_3)u(t-\sigma_3)\, d\sigma_1 d\sigma_2 d\sigma_3$$

which is in the regular form (34).

It is worthwhile to run through the triangular and regular forms for a very specific case. This will show some of the bookkeeping that so far has been hidden by the often implicit causality and one-sided input assumptions, and the infinite limits. Also, it will emphasize the special starting point for the derivation of the regular kernel representation.

Example 1.8 A triangular kernel representation for the input/output behavior of the bilinear state equation in Example 1.2 has been found to be

$$y(t) = \int_0^t \int_0^{\sigma_1} u(\sigma_1)u(\sigma_2)\, d\sigma_2 d\sigma_1$$

Explicitly incorporating the one-sidedness of the input signal, causality, and triangularity into the kernel permits rewriting this in the form

$$y(t) = \int_{-\infty}^{\infty}\int_{-\infty}^{\infty} \delta_{-1}(t-\sigma_1)\delta_{-1}(t-\sigma_2)\delta_{-1}(\sigma_1-\sigma_2)\delta_{-1}(\sigma_1)\delta_{-1}(\sigma_2)u(\sigma_1)u(\sigma_2)\, d\sigma_2 d\sigma_1$$

This expression can be simplified by removing the redundant step functions. Then replace the variables of integration σ_1 and σ_2 by $t-\sigma_1$ and $t-\sigma_2$, respectively, to obtain

$$y(t) = \int_{-\infty}^{\infty}\int_{-\infty}^{\infty} \delta_{-1}(\sigma_1)\delta_{-1}(\sigma_2-\sigma_1)\delta_{-1}(t-\sigma_1)\delta_{-1}(t-\sigma_2)u(t-\sigma_1)u(t-\sigma_2)\, d\sigma_2 d\sigma_1$$

Now the kernel clearly is triangular, and nonzero on the domain $\sigma_2 \geqslant \sigma_1 \geqslant 0$. Interchanging the two integration variables gives an input/output expression in terms of a triangular kernel with domain $\sigma_1 \geqslant \sigma_2 \geqslant 0$:

$$y(t) = \int_{-\infty}^{\infty}\int_{-\infty}^{\infty} \delta_{-1}(\sigma_2)\delta_{-1}(\sigma_1-\sigma_2)\delta_{-1}(t-\sigma_1)\delta_{-1}(t-\sigma_2)u(t-\sigma_1)u(t-\sigma_2)\, d\sigma_2 d\sigma_1$$

This is the starting point for computing the regular kernel representation. Replace σ_1 with $\tau_1 = \sigma_1-\sigma_2$, and then σ_2 with τ_2 to obtain

$$y(t) = \int_{-\infty}^{\infty}\int_{-\infty}^{\infty} \delta_{-1}(\tau_1)\delta_{-1}(\tau_2)\delta_{-1}(t-\tau_1-\tau_2)\delta_{-1}(t-\tau_2)u(t-\tau_1-\tau_2)u(t-\tau_2)\, d\tau_2 d\tau_1$$

This input/output representation is in regular form, and if the one-sidedness of the input is left understood, the regular kernel is

$$h_{reg}(t_1,t_2) = \delta_{-1}(t_1)\delta_{-1}(t_2)$$

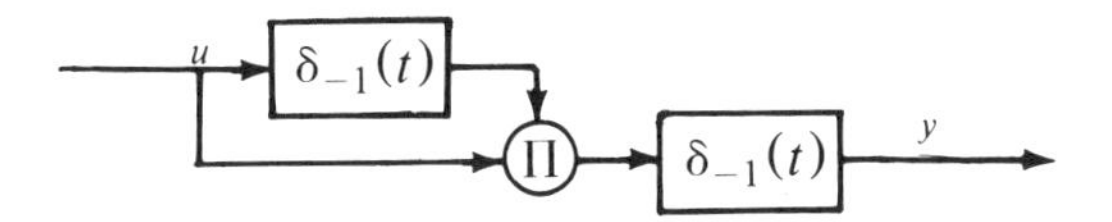

Figure 1.8. Interconnection representation for Example 1.2.

Furthermore, putting together this result and a slight variation of Example 1.7 shows that the bilinear state equation in Example 1.2 can be represented by the interconnection of integrators and multipliers shown in Figure 1.8.

Incidently, it is obvious that the triangular, symmetric, and regular forms all collapse to the same thing for homogeneous systems of degree 1. Therefore, when compared to linear system problems, it should be expected that a little more foresight and artistic judgement are needed to pose nonlinear systems problems in a convenient way. This is less an inherited ability than a matter of experience, and by the time you reach the back cover such judgements will be second-nature.

1.3 Polynomial and Volterra Systems

A system described by a finite sum of homogeneous terms of the form

$$y(t) = \sum_{n=1}^{N} \int_{-\infty}^{\infty} h_n(\sigma_1,\ldots,\sigma_n) u(t-\sigma_1) \cdots u(t-\sigma_n)\, d\sigma_1 \cdots d\sigma_n \tag{36}$$

will be called a *polynomial system of degree N,* assuming $h_N(t_1,\ldots,t_N) \neq 0$. If a system is described by an infinite sum of homogeneous terms, then it will be called a *Volterra system.* Of course, the same terminology is used if the homogeneous terms are nonstationary. By adding a degree-0 term, say $y_0(t)$, systems that have nonzero responses to identically zero inputs can be represented.

Note that, as special cases, static nonlinear systems described by a polynomial or power series in the input:

$$y(t) = a_1 u(t) + \cdots + a_N u^N(t)$$

$$y(t) = \sum_{n=1}^{\infty} a_n u^n(t) \tag{37}$$

are included. Simply take $h_n(t_1,\ldots,t_n) = a_n \delta_0(t_1) \cdots \delta_0(t_n)$ in (36). Further examples of polynomial systems are easy to generate from interconnection structured systems. The simplest case is a cascade connection of a linear system followed by a polynomial nonlinearity. If the nonlinearity is described by an infinite power series, a Volterra system is the result.

Since the Volterra system representation is an infinite series, there must be associated convergence conditions to guarantee that the representation is meaningful. Usually these conditions involve a bound on the time interval and a bound for $u(t)$ on this interval. These bounds typically depend upon each other in a roughly inverse way. That is, as the time interval is made larger, the input bound must be made smaller, and vice versa. The calculations required to find suitable bounds often are difficult.

Example 1.9 The following is possibly the simplest type of convergence argument for a Volterra system of the form (36) with $N = \infty$. Suppose that for all t

$$|u(t)| \leqslant K$$

and

$$\int_{-\infty}^{\infty} |h_n(\sigma_1, \ldots, \sigma_n)| \, d\sigma_1 \cdots d\sigma_n \leqslant a_n$$

Then since the absolute value of a sum is bounded by the sum of the absolute values,

$$\begin{aligned}
|y(t)| &\leqslant \sum_{n=1}^{\infty} \Big| \int_{-\infty}^{\infty} h_n(\sigma_1, \ldots, \sigma_n) u(t-\sigma_1) \cdots u(t-\sigma_n) \, d\sigma_1 \cdots d\sigma_n \Big| \\
&\leqslant \sum_{n=1}^{\infty} \int_{-\infty}^{\infty} |h_n(\sigma_1, \ldots, \sigma_n)| |u(t-\sigma_1)| \cdots |u(t-\sigma_n)| \, d\sigma_1 \cdots d\sigma_n \\
&\leqslant \sum_{n=1}^{\infty} a_n K^n
\end{aligned}$$

Convergence of the power series on the right side implies convergence of the series defining the Volterra system. In this case the time interval is infinite, but of course the convergence condition is quite restrictive.

In the sequel I will be concerned for the most part with polynomial- or homogeneous-system representations, thereby leaping over convergence in a single bound. Of course, convergence is a background issue in that a polynomial system that is a truncation of a Volterra system may be a good approximation only if the Volterra system representation converges. When Volterra systems are considered, the infinite series will be treated informally in that the convergence question will be ignored. All this is not to slight the importance of the issue. Indeed, convergence is crucial when the Volterra series representation is to be used for computation. The view adopted here is more a consequence of the fact that convergence properties often must be established using particular features of the problem at hand. A simple example will illustrate the point.

Example 1.10 Consider the Volterra system

$$y(t) = \int_0^t cbu(\sigma)\, d\sigma + \int_0^t \int_0^{\sigma_1} cDbu(\sigma_1)u(\sigma_2)\, d\sigma_2 d\sigma_1$$

$$+ \int_0^t \int_0^{\sigma_1} \int_0^{\sigma_2} cD^2bu(\sigma_1)u(\sigma_2)u(\sigma_3)\, d\sigma_3 d\sigma_2 d\sigma_1 + \cdots$$

where c, b, and D are $1 \times n$, $n \times 1$, and $n \times n$ matrices, respectively. Factoring out the c and b, and using a simple identity to rewrite the triangular integrations gives

$$y(t) = c\{ \int_0^t u(\sigma)d\sigma + \frac{1}{2!}D[\int_0^t u(\sigma)d\sigma]^2 + \frac{1}{3!}D^2[\int_0^t u(\sigma)d\sigma]^3 + \cdots \}b$$

Now arguments similar to those used to investigate convergence of the matrix exponential can be applied. The result is that this Volterra system converges uniformly on any finite time interval as long as the input is piecewise continuous—a much, in fact infinitely, better result than would be obtained using the approach in Example 1.9. Incidentaly, this Volterra system representation corresponds to the bilinear state equation

$$\dot{x}(t) = Dx(t)u(t) + bu(t)$$

$$y(t) = cx(t)$$

a particular case of which was discussed in Example 1.2. I suggest that the reader discover this by differentiating the vector Volterra system representation for $x(t)$:

$$x(t) = \int_0^t bu(\sigma)\, d\sigma + \int_0^t \int_0^{\sigma_1} Dbu(\sigma_1)u(\sigma_2)\, d\sigma_2 d\sigma_1 + \cdots$$

1.4 Interconnections of Nonlinear Systems

Three basic interconnections of nonlinear systems will be considered: additive and multiplicative parallel connections, and the cascade connection. Of course, additive parallel and cascade connections are familiar from linear system theory, since linearity is preserved. The multiplicative parallel connection probably is unfamiliar, but it should seem to be a natural thing to do in a nonlinear context. The results will be described in terms of stationary systems. I leave to the reader the light task of showing that little is changed when the nonstationary case is considered.

Interconnections of homogeneous systems will be discussed first. No special form assumptions are made for the kernels because the triangular or symmetric forms are not preserved under all the interconnections. Furthermore, the regular kernel representation will be ignored for the moment. To describe interconnections of polynomial or Volterra systems, a general operator notation will be introduced later in the

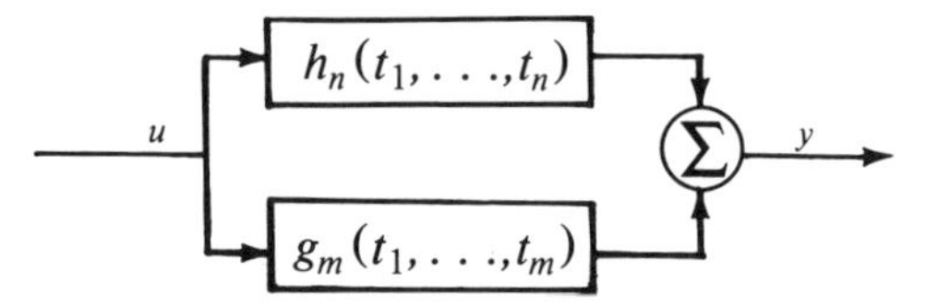

Figure 1.9. An additive parallel connection.

section. This operator notation always can be converted back to the usual kernel expressions, but often much ink is saved by postponing this conversion as long as possible.

The basic additive connection of two homogeneous systems is shown in Figure 1.9. The overall system is described by

$$y(t) = \int_{-\infty}^{\infty} h_n(\sigma_1,\ldots,\sigma_n)u(t-\sigma_1)\cdots u(t-\sigma_n)\, d\sigma_1 \cdots d\sigma_n$$

$$+ \int_{-\infty}^{\infty} g_m(\sigma_1,\ldots,\sigma_m)u(t-\sigma_1)\cdots u(t-\sigma_m)\, d\sigma_1 \cdots d\sigma_m \tag{38}$$

When $m = n$ it is clear that this is a degree-n homogeneous system with kernel

$$f_n(t_1,\ldots,t_n) = h_n(t_1,\ldots,t_n) + g_n(t_1,\ldots,t_n) \tag{39}$$

And if both kernels $h_n(t_1,\ldots,t_n)$ and $g_n(t_1,\ldots,t_n)$ are symmetric (triangular), then the kernel $f_n(t_1,\ldots,t_n)$ will be symmetric (triangular). When $m \neq n$ the overall system is a polynomial system of degree $N = max[n,m]$.

The second connection of interest is the parallel multiplicative connection shown in Figure 1.10. The mathematical description of the overall system is

$$y(t) = [\int_{-\infty}^{\infty} h_n(\sigma_1,\ldots,\sigma_n)u(t-\sigma_1)\cdots u(t-\sigma_n)\, d\sigma_1 \cdots d\sigma_n]$$

$$[\int_{-\infty}^{\infty} g_m(\sigma_1,\ldots,\sigma_m)u(t-\sigma_1)\cdots u(t-\sigma_m)\, d\sigma_1 \cdots d\sigma_m]$$

$$= \int_{-\infty}^{\infty} [\, h_n(\sigma_1,\ldots,\sigma_n)g_m(\sigma_{n+1},\ldots,\sigma_{n+m})]u(t-\sigma_1)$$

$$\cdots u(t-\sigma_{n+m})d\sigma_1 \cdots d\sigma_{n+m} \tag{40}$$

Thus the multiplicative connection yields a homogeneous system of degree $n + m$ with kernel

$$f_{n+m}(t_1,\ldots,t_{n+m}) = h_n(t_1,\ldots,t_n)\, g_m(t_{n+1},\ldots,t_{n+m}) \tag{41}$$

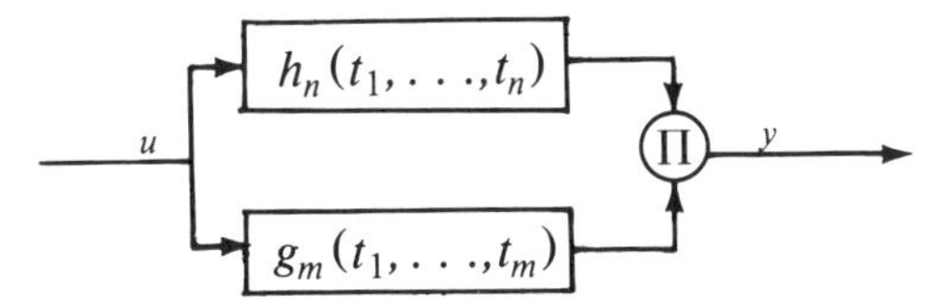

Figure 1.10. A multiplicative parallel connection.

In general, neither symmetry nor triangularity is preserved in this case. Note the relabeling of variables in this development for it is quite common. Distinct labels for the variables must be maintained to avoid confusion and preserve sanity.

The cascade connection of two systems is shown in Figure 1.11. The customary, though usually unstated, assumption is made that the two systems do not interact with each other. That is, there is no loading effect. To obtain a description for this connection, write

$$y(t) = \int_{-\infty}^{\infty} g_m(\sigma_1, \ldots, \sigma_m) v(t-\sigma_1) \cdots v(t-\sigma_m)\, d\sigma_1 \cdots d\sigma_m \tag{42}$$

where, for $j = 1, \ldots, m$,

$$v(t-\sigma_j) = \int_{-\infty}^{\infty} h_n(\sigma_{m+(j-1)n+1}, \ldots, \sigma_{m+jn}) u(t-\sigma_j-\sigma_{m+(j-1)n+1}) \cdots u(t-\sigma_j-\sigma_{m+jn})\, d\sigma_{m+(j-1)n+1} \cdots d\sigma_{m+jn} \tag{43}$$

Of course, I have chosen the labeling of variables in (43) to make the end result look nice. Substituting (43) into (42) gives

$$\begin{aligned} y(t) = \int_{-\infty}^{\infty} g_m(\sigma_1, \ldots, \sigma_m) &\Big[\int_{-\infty}^{\infty} h_n(\sigma_{m+1}, \ldots, \sigma_{m+n}) u(t-\sigma_1-\sigma_{m+1}) \\ &\cdots u(t-\sigma_1-\sigma_{m+n}) d\sigma_{m+1} \cdots d\sigma_{m+n}\Big] \\ &\cdots \Big[\int_{-\infty}^{\infty} h_n(\sigma_{m+(m-1)n+1}, \ldots, \sigma_{m+mn}) u(t-\sigma_m-\sigma_{m+(m-1)n+1}) \\ &\cdots u(t-\sigma_m-\sigma_{m+mn})\, d\sigma_{m+(m-1)n+1} \cdots d\sigma_{m+mn}\Big]\, d\sigma_1 \cdots d\sigma_m \end{aligned} \tag{44}$$

u → $h_n(t_1, \ldots, t_n)$ → v → $g_m(t_1, \ldots, t_m)$ → y

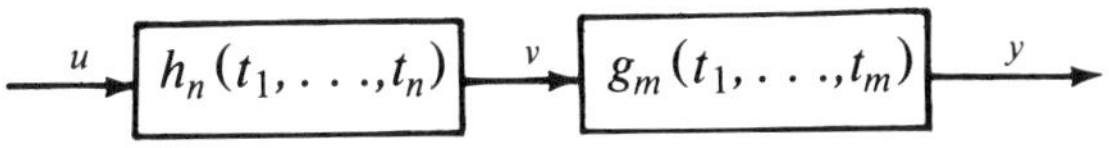

Figure 1.11. Cascade connection of two systems.

Now, in the bracketed terms replace each variable of integration $\sigma_{m+(j-1)n+i}$ by the variable $\tau_{(j-1)n+i} = \sigma_{m+(j-1)n+i} + \sigma_j$, $i = 1,\ldots,n$, $j = 1,\ldots,m$. Then moving the outer m-fold integration to the inside gives

$$y(t) = \int_{-\infty}^{\infty} [\int_{-\infty}^{\infty} g_m(\sigma_1,\ldots,\sigma_m) h_n(\tau_1-\sigma_1,\ldots,\tau_n-\sigma_1)$$

$$\cdots h_n(\tau_{(m-1)n+1}-\sigma_m,\ldots,\tau_{mn}-\sigma_m)\, d\sigma_1 \cdots d\sigma_m] u(t-\tau_1)$$

$$\cdots u(t-\tau_{mn})\, d\tau_1 \cdots d\tau_{mn} \tag{45}$$

Thus, the cascade connection yields a homogeneous system of degree mn with kernel

$$f_{mn}(t_1,\ldots,t_{mn}) = \int_{-\infty}^{\infty} g_m(\sigma_1,\ldots,\sigma_m) h_n(t_1-\sigma_1,\ldots,t_n-\sigma_1)$$

$$\cdots h_n(t_{(m-1)n+1}-\sigma_m,\ldots,t_{mn}-\sigma_m)\, d\sigma_1 \cdots d\sigma_m \tag{46}$$

It almost is needless to say that symmetry or triangularity usually is lost in this connection. This means that $f_{mn}(t_1,\ldots,t_n)$ must be symmetrized or triangularized as a separate operation.

I should pause at this point to comment that double-subscripted integration variables can be used in derivations such as the above. However, it is usually more convenient in the long run to work with single-subscripted variables, and the results look better.

When applying the cascade-connection formula, and other convolution-like formulas to specific systems, some caution must be exercised in order to account properly for causality. The use of infinite limits and implicit causality assumptions is an invitation to disaster for the careless. I invite the reader to work the following example in a cavalier manner just to see what can happen.

Example 1.11 Consider the cascade connection shown in Figure 1.11 with

$$h_1(t_1) = e^{-t_1},\quad g_2(t_1,t_2) = \delta_0(t_1-t_2)$$

The kernels can be rewritten in the form

$$h_1(t_1) = e^{-t_1}\delta_{-1}(t_1),\quad g_2(t_1,t_2) = \delta_0(t_1-t_2)\delta_{-1}(t_1)\delta_{-1}(t_2)$$

to incorporate explicitly the causality conditions. Then for $t_1,t_2 \geq 0$, (46) gives the kernel of the overall system as

$$f(t_1,t_2) = \int_{-\infty}^{\infty}\int_{-\infty}^{\infty} \delta_0(\sigma_1-\sigma_2)\delta_{-1}(\sigma_1)\delta_{-1}(\sigma_2)e^{-(t_1-\sigma_1)}$$

$$\delta_{-1}(t_1-\sigma_1)e^{-(t_2-\sigma_2)}\delta_{-1}(t_2-\sigma_2)\; d\sigma_1\sigma_2$$

$$= \int_{-\infty}^{\infty} \delta_{-1}(\sigma_1)\delta_{-1}(\sigma_1)e^{-(t_1-\sigma_1)}\delta_{-1}(t_1-\sigma_1)e^{-(t_2-\sigma_1)}\delta_{-1}(t_2-\sigma_1)\; d\sigma_1$$

This expression can be simplified by using the integration limits to account for the constraints imposed by the unit step functions. Then, for $t_1,\ t_2 \geqslant 0$,

$$f(t_1,t_2) = \int_0^{min[t_1,t_2]} e^{-(t_1-\sigma_1)}e^{-(t_2-\sigma_1)}\; d\sigma_1$$

$$= e^{-t_1-t_2}\int_0^{min[t_1,t_2]} e^{2\sigma_1}\; d\sigma_1$$

$$= \frac{1}{2}e^{-t_1-t_2}[e^{2min[t_1,t_2]} - 1]$$

Another way to write this result is

$$f(t_1,t_2) = \frac{1}{2}e^{-|t_1-t_2|} - \frac{1}{2}e^{-t_1-t_2},\quad t_1,t_2 \geqslant 0$$

Of course, these same subtleties can arise in linear system theory. They may have escaped notice in part because only single integrals are involved, and there is little need for the notational simplicity of infinite limits - and in part because the use of the Laplace transform takes care of convolution in a neat way. The title of Chapter 2 should be reassuring in this regard.

When the regular kernel representation is used, the interconnection rules are more difficult to derive. Of course, the analysis of the additive parallel connection is the exception. If the two regular representations are of the same degree, then the regular kernel for the additive connection is simply the sum of the subsystem regular kernels. If they are not of the same degree, a polynomial system is the result, with the two homogeneous subsystems given in the regular representation. For cascade and multiplicative-parallel interconnections, I suggest the following simple but tedious procedure. For each subsystem compute the triangular kernel from the regular kernel. Then use the rules just derived to find a kernel for the overall system. Finally, symmetrize this kernel and use the result of Problem 1.15 to compute the corresponding regular kernel.

For interconnections of polynomial or Volterra systems, a general operator notation will be used to avoid carrying a plethora of kernels, integration variables, and so forth, through the calculations. At the end of the calculation the operator notation can be replaced by the underlying description in terms of subsystem kernels. However, for some types of problems this last step need not be performed. For example, to determine if two block diagrams represent the same input/output behavior it simply must be checked that the two diagrams are described by the same overall operator. I should note that convergence issues in the Volterra-system case are discussed in Appendix 1.1, and will be ignored completely in the following development.

The notation

$$y(t) = H[u(t)] \tag{47}$$

denotes a system H with input $u(t)$ and output $y(t)$. Often the time argument will be dropped, and (47) will be written simply as

$$y = H[u] \tag{48}$$

(Though nonstationary systems are being ignored here, notice that for such a system the time argument probably should be displayed, for example, $y(t) = H[t,u(t)]$.) It is convenient to have a special notation for a degree-n homogeneous system, and so a subscript will be used for this purpose:

$$y = H_n[u] \tag{49}$$

Then a polynomial system H can be written in the form

$$y = H[u] = \sum_{n=1}^{N} H_n[u] \tag{50}$$

with a similar notation for Volterra systems. The convenience of this slightly more explicit notation is that conversion to the kernel notation is most easily accomplished in a homogeneous-term-by-homogeneous-term fashion.

Considering system interconnections at this level of notation is a simple matter, at least in the beginning. The additive parallel connection of two systems, H and G, gives the system $H + G$, which is described by

$$y = H[u] + G[u] = (H + G)[u] \tag{51}$$

As usual, the addition of mathematical operators (systems) is defined via the addition in the range space (addition of output signals). In a similar manner, the multiplicative parallel connection of the systems H and G gives the system HG described by

$$y = H[u]G[u] = (HG)[u] \tag{52}$$

Notice that both of these operations are commutative and associative. That is

$$GH = HG, \quad (FG)H = F(GH)$$

$$G + H = H + G, \quad (F + G) + H = F + (G + H) \tag{53}$$

Furthermore, the multiplication of systems is distributive with respect to addition:

$$F(G + H) = FG + FH \tag{54}$$

In terms of the notation in (50), it is perfectly clear that

$$H + G = (H_1 + G_1) + (H_2 + G_2) + \cdots \tag{55}$$

Using (54) the multiplication of two systems is described by

$$\begin{aligned} HG &= (H_1 + H_2 + \cdots)(G_1 + G_2 + \cdots) \\ &= (H_1 + H_2 + \cdots)G_1 + (H_1 + H_2 + \cdots)G_2 + \cdots \\ &= H_1G_1 + (H_2G_1 + H_1G_2) + (H_3G_1 + H_2G_2 + H_1G_3) + \cdots \end{aligned} \tag{56}$$

The terms in (55) and (56) have been grouped according to degree since

$$\begin{aligned} \textit{degree}\ (H_m + G_m) &= m \\ \textit{degree}\ (H_m G_n) &= m + n \end{aligned} \tag{57}$$

Now it is a simple matter to replace the expressions in (55) and (56) by the corresponding kernel representations.

So far it has been good clean fun, but the cascade connection is a less easy topic. A system H followed in cascade by a system G yields the overall system $G*H$, where the "$*$" notation is defined by

$$y = G[H[u]] = (G*H)[u] \tag{58}$$

But a little more technical caution should be exercised at this point. In particular I have not mentioned the domain and range spaces of the operator representations. For the multiplicative and additive parallel connections, these can be chosen both for convenience and for fidelity to the actual system setting. However, for the composition of operators in (58) it must be guaranteed that the range space of H is contained in the domain of G. Having been duly mentioned, this condition and others like it will be assumed.

The cascade operation is not commutative except in special cases—one being the case where only degree-1 systems are involved:

$$G_1*H_1 = H_1*G_1 \tag{59}$$

The cascade operation is distributive with respect to addition and multiplication only in the particular orderings

$$\begin{aligned} (G + H)*F &= G*F + H*F \\ (GH)*F &= (G*F)(H*F) \end{aligned} \tag{60}$$

and in the special case of the alternative ordering:

$$F_1*(G + H) = F_1*G + F_1*H \tag{61}$$

These results can be established easily by resorting to the corresponding kernel representations, and the rules derived earlier. Also it is obvious from the earlier results that

$$\textit{degree}\ (G_n*H_m) = \textit{degree}\ (H_m*G_n) = mn \tag{62}$$

To consider cascade connections of polynomial or Volterra systems in terms of the notation in (50) requires further development. Using the notations

$$w = \sum_{n=1}^{\infty} w_n = \sum_{n=1}^{\infty} H_n[u]$$

$$y = \sum_{n=1}^{\infty} y_n = \sum_{n=1}^{\infty} G_n[w] \tag{63}$$

the objective is to find for the cascade system $G*H$ an operator expression of the form

$$y = \sum_{n=1}^{\infty} F_n[u] \tag{64}$$

where each homogeneous operator F_n is specified in terms of the H_n's and G_n's. It is convenient in this regard to consider the input signal $\alpha u(t)$, where α is an arbitrary real number. Then $H_n[\alpha u] = \alpha^n w_n$, and

$$w = \sum_{n=1}^{\infty} \alpha^n w_n \tag{65}$$

so that

$$y = \sum_{m=1}^{\infty} G_m[\sum_{n=1}^{\infty} \alpha^n w_n] \tag{66}$$

The general term of interest is

$$G_m[\sum_{n=1}^{\infty} \alpha^n w_n] \tag{67}$$

and to analyze this further it is necessary to bring in the kernel representation for G_m. Letting $g_{sym}(t_1,\ldots,t_m)$ be the symmetric kernel corresponding to G_m, a simple computation gives

$$G_m[\sum_{n=1}^{\infty} \alpha^n w_n] = \sum_{n_1=1}^{\infty} \cdots \sum_{n_m=1}^{\infty} \alpha^{n_1+\cdots+n_m} \hat{G}_m[(w_{n_1},\ldots,w_{n_m})] \tag{68}$$

where the new operator is defined by

$$\hat{G}_m[(w_{n_1},\ldots,w_{n_m})] = \int_{-\infty}^{\infty} g_{sym}(\sigma_1,\ldots,\sigma_m) w_{n_1}(t-\sigma_1)$$

$$\cdots w_{n_m}(t-\sigma_m)\, d\sigma_1 \cdots d\sigma_m \tag{69}$$

Note that $\hat{G}_m[(w_n,\ldots,w_n)]$ is the usual degree-m operator $G_m[w_n]$, and that $\hat{G}_m[(w_{n_1},\ldots,w_{n_m})]$ is symmetric in its arguments by the symmetry of the kernel. These properties will be used shortly.

Substituting (68) into (66) gives

$$y = \sum_{m=1}^{\infty} \sum_{n_1=1}^{\infty} \cdots \sum_{n_m=1}^{\infty} \alpha^{n_1+\cdots+n_m} \hat{G}_m[(w_{n_1},\ldots,w_{n_m})] \tag{70}$$

Thus, to determine the operators F_n in (64), coefficients of like powers of α in (70) and in the expression

$$y = \sum_{n=1}^{\infty} F_n[\alpha u] = \sum_{n=1}^{\infty} \alpha^n F_n[u] \tag{71}$$

must be equated. Then the various terms involving the notation $\hat{G}_n[(w_{n_1},\ldots,w_{n_m})]$ must be broken into their component parts involving the operators G_m and H_n. This is a complicated process in general, so I will do just the first few terms. Equating coefficients of α in (70) and (71) gives

$$F_1[u] = G_1[w_1] = G_1[H_1[u]]$$

Thus the degree-1 portion of the overall cascade connection is described by the operator

$$F_1 = G_1 * H_1 \tag{72}$$

Equating coefficients of α^2 gives

$$\begin{aligned} F_2[u] &= G_1[w_2] + \hat{G}_2[(w_1,w_1)] \\ &= G_1[w_2] + G_2[w_1] \\ &= G_1[H_2[u]] + G_2[H_1[u]] \end{aligned}$$

Thus,

$$F_2 = G_1 * H_2 + G_2 * H_1 \tag{73}$$

Kernels for F_1 and F_2 are easily calculated from kernels for H_1, H_2, G_1, and G_2 by using (72) and (73) in conjunction with (46).

More interesting things begin to happen when F_3 is sought. Equating coefficients of α^3 gives

$$\begin{aligned} F_3[u] &= G_1[w_3] + \hat{G}_2[(w_1,w_2)] + \hat{G}_2[(w_2,w_1)] + \hat{G}_3[(w_1,w_1,w_1)] \\ &= G_1[w_3] + 2\hat{G}_2[(w_1,w_2)] + G_3[w_1] \end{aligned}$$

But now a simple calculation involving the kernel representation shows that

$$G_2[w_1+w_2] = G_2[w_1] + G_2[w_2] + 2\hat{G}_2[(w_1,w_2)] \tag{74}$$

Thus

$$F_3[u] = G_1[w_3] + G_3[w_1] - G_2[w_1] - G_2[w_2] + G_2[w_1+w_2] \tag{75}$$

and the degree-3 operator for the overall system is

$$F_3 = G_1{*}H_3 + G_3{*}H_1 - G_2{*}H_1 - G_2{*}H_2 + G_2{*}(H_1 + H_2) \tag{76}$$

On the face of it, it might not be clear that (76) yields a degree-3 operator. Obviously, degree-2 and degree-4 terms are present, but it happens that these add out in the end. See Problem 1.16.

Though the way to proceed probably is clear by now, I will do one more just for the experience. Equating coefficients of α^4 in (70) and (71) gives

$$F_4[u] = G_1[w_4] + G_2[w_2] + 2\hat{G}_2[(w_1,w_3)] + 3\hat{G}_3[(w_1,w_1,w_2)] + G_4[w_1] \tag{77}$$

Using an expression of the form (74), the term $2\hat{G}_2[(w_1,w_3)]$ can be replaced. Also it is a simple calculation using the kernel representation to show that

$$3!\hat{G}_3[(w_1,w_1,w_2)] = G_3[2w_1+w_2] - 2G_3[w_1+w_2] - 6G_3[w_1] + G_3[w_2]$$

Thus, the degree-4 operator for the overall system is

$$F_4 = G_1{*}H_4 + G_2{*}H_2 + G_4{*}H_1 + G_2{*}(H_1+H_3) - G_2{*}H_1 - G_2{*}H_3$$
$$+ \frac{1}{2}G_3{*}(2H_1+H_2) - G_3{*}(H_1+H_2) - 3G_3{*}H_1 + \frac{1}{2}G_3{*}H_2 \tag{78}$$

Just as in (76), the use of (78) to compute a kernel for F_4 is straightforward for the $G_i{*}H_j$ terms, but a bit more complicated for the $G_i{*}(H_j + H_k)$ terms.

So far in this section the feedback connection has been studiously avoided. The time-domain analysis of nonlinear feedback systems in terms of kernel representations is quite unenlightening when compared to transform-domain techniques to be discussed later on. (This is similar to the linear case. Who ever analyzes linear feedback systems in terms of the impulse response?) However, the situation is far from simple regardless of the representation used. Even a cursory look at a feedback system from the operator viewpoint will point up some of the difficulties; in fact, it will raise some rather deep issues that the reader may wish to pursue.

The feedback interconnection of nonlinear systems is diagramed in operator notation in Figure 1.12.

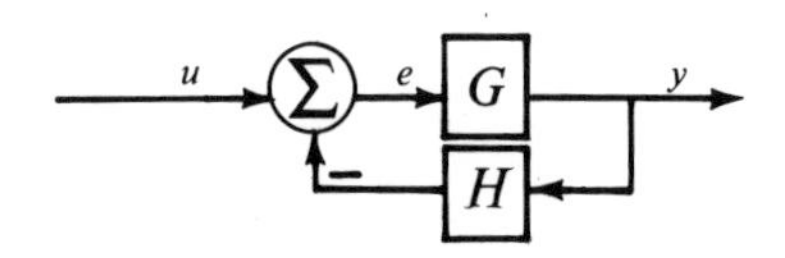

Figure 1.12. A nonlinear feedback system.

The equations describing this system are

$$y = G[e] \tag{79}$$

$$e = u - H[y] \tag{80}$$

It is of interest to determine first if these equations specify an "error system" operator $e = E[u]$. From (79) and (80) it is clear that such an operator must satisfy the equation

$$E[u] = u - H[G[E[u]]] \tag{81}$$

or, in operator form,

$$E = I - H*G*E \tag{82}$$

where I is the identity operator, $I[u] = u$. Equation (82) can be rewritten in the form

$$E + H*G*E = (I + H*G)*E = I \tag{83}$$

Thus, a sufficient condition for the existence of a solution E is that $(I + H*G)^{-1}$ exist, in which case

$$E = (I + H*G)^{-1} \tag{84}$$

(If the inverse does not exist, then it can be shown that (79) and (80) do not have a solution for e, or that there are multiple solutions for e. Thus, the sufficient condition is necessary as well.)

Of course, the reason the operator E is of interest is that (79) then gives an overall (closed-loop) operator representation of the form $y = F[u]$ for the system according to

$$y = G[e] = G[E[u]]$$

Thus

$$F = G*E = G*(I + H*G)^{-1} \tag{85}$$

an expression that should have a familiar appearance. It should be noted that for these developments to be of system-theoretic interest, the indicated inverse must not only exist, it must also represent a causal system.

To complete the discussion, it remains to give methods for computing the homogeneous terms and corresponding kernels for F. The general approach is to combine the first equality in (85) with (82) to write

$$F = G*(I - H*F) \tag{86}$$

But from this point on, a simple example might be more enlightening than a general calculation.

Example 1.12 Consider the feedback system in Figure 1.12 where

$$G = G_1 + G_3, \quad H = H_1$$

In this case (86) becomes

$$F = (G_1 + G_3)*(I - H_1*F)$$

Writing F as a sum of homogeneous operators and using distributive properties gives the equation

$$F_1 + F_2 + F_3 + \cdots = G_1*(I - H_1*F_1 - H_1*F_2 - \cdots$$
$$+ G_3*(I - H_1*F_1 - H_1*F_2 - H_1*F_3 - \cdots)$$

To find an equation for F_1, the degree-1 terms on both sides are equated:

$$F_1 = G_1 - G_1*H_1*F_1$$

Thus

$$(I + G_1*H_1)*F_1 = G_1$$

and, granting the existence of the operator inverse, F_1 is given by

$$F_1 = (I + G_1*H_1)^{-1}*G_1$$

The terms in this development can be rearranged using the commutativity properties of degree-1 operators to give the equivalent expression

$$F_1 = G_1*(I + G_1*H_1)^{-1}$$

But either way, the result is of interest only if $(I + G_1*H_1)^{-1}$ represents a causal system.

Now, equating degree-2 terms on both sides of the original equation gives

$$F_2 = -G_1*H_1*F_2$$

Since the invertibility of $I + G_1*H_1$ has been assumed, this equation implies

$$F_2 = 0$$

Equating the degree-3 terms in the original equation gives

$$F_3 = -G_1*H_1*F_3 + G_3*(I - H_1*F_1)$$

The reader might well wonder how to conclude that the degree-3 terms in $G_3*(I - H_1*F_1 - H_1*F_2 - \cdots)$ are those indicated, since the tempting distributive law does not hold. In fact, the justification involves retreating to the time-domain representations using symmetric kernels, and showing that the omitted terms are of degree greater than 3. Leaving this verification to the reader, the degree-3 terms can be rearranged to give

$$(I + G_1*H_1)*F_3 = G_3*(I - H_1*F_1)$$

Solving yields, again with the operator inverse treated casually,

$$F_3 = (I + G_1*H_1)^{-1}*G_3*(I - H_1*F_1)$$

This can be rewritten in different ways using commutativity properties, and of course substitution can be made for F_1 if desired. The higher-degree terms can be calculated in a similar fashion.

Inspection of the homogeneous terms computed in this example indicates an interesting feature of the operator inverse in (85). Namely, only a linear operator inverse is required to compute the homogeneous terms in the closed-loop operator representation. This is a general feature, as the following brief development will show.

Suppose H is an operator representation for a nonlinear system. Then $G^{(p)}$ is called a pth*-degree postinverse* of H if

$$F = G^{(p)}*H = I + F_{p+1} + F_{p+2} + \cdots \tag{87}$$

In other words, $G^{(p)}$ can be viewed as a polynomial truncation of H^{-1}, assuming of course that H^{-1} exists. The expression (87) can be used to determine $G^{(p)}$ in a homogeneous-term-by-homogeneous-term fashion by using the cascade formulas developed earlier. That is, (87) can be written, through degree 3, in the form

$$\begin{aligned} G^{(p)}*H &= (G_1^{(p)} + G_2^{(p)} + G_3^{(p)} + \cdots)*(H_1 + H_2 + H_3 + \cdots) \\ &= (G_1^{(p)}*H_1) + (G_2^{(p)}*H_1 + G_1^{(p)}*H_2) \\ &+ [G_3^{(p)}*H_1 + G_1^{(p)}*H_3 - G_2^{(p)}*H_1 - G_2^{(p)}*H_2 + G_2^{(p)}*(H_1 + H_2)] + \cdots \end{aligned}$$

where the terms have been grouped according to degree. Assuming $p \geqslant 3$, the first condition to be satisfied is

$$G_1^{(p)}*H_1 = I$$

To solve this operator equation, H_1 must be invertible, and furthermore H_1^{-1} must correspond to a causal system for the result to be of interest in system theory. Often these restrictions are not satisfied, but it is hard to be explicit about conditions in terms of operator representations. (If H_1 can be described in terms of a proper rational transfer function representation $H_1(s)$, the reader is no doubt aware that for causal invertibility it is necessary and sufficient that $H_1(\infty) \neq 0$.) At any rate, I will assume the restrictions are satisfied, and write

$$G_1^{(p)} = H_1^{-1} \tag{88}$$

with the remark that for feedback applications inverses of the form $(I + H_1)^{-1}$ are required, and invertibility here is less of a problem.

The second condition to be satisfied by the pth-degree postinverse of H is

$$G_2^{(p)}*H_1 + G_1^{(p)}*H_2 = 0$$

Solving for $G_2^{(p)}$ is a simple matter, giving

$$G_2^{(p)} = -H_1^{-1}*H_2*H_1^{-1} \tag{89}$$

Notice that no further assumptions were required to solve for $G_2^{(p)}$. The final condition that will be treated explicitly is

$$G_3^{(p)}*H_1 + G_1^{(p)}*H_3 - G_2^{(p)}*H_1 - G_2^{(p)}*H_2 + G_2^{(p)}*(H_1+H_2) = 0$$

Regarding $G_1^{(p)}$ and $G_2^{(p)}$ as known gives

$$G_3^{(p)} = [-G_1^{(p)}*H_3 + G_2^{(p)}*H_1 + G_2^{(p)}*H_2 - G_2^{(p)}*(H_1 + H_2)]*H_1^{-1}$$
$$= -G_1^{(p)}*H_3*H_1^{-1} + G_2^{(p)} + G_2^{(p)}*H_2*H_1^{-1} - G_2^{(p)}*(I + H_2*H_1^{-1}) \tag{90}$$

and, again, the only inverse required is H_1^{-1}. The higher-degree homogeneous terms in a pth-degree postinverse can be calculated similarly. Furthermore, a *p^{th}-degree preinverse* can be defined, and it can be shown to be identical to the pth-degree postinverse. This is left to Problem 1.19. If the inverse of an operator exists, and the Volterra series representation is convergent, then the pth-degree inverses are polynomial approximations to the inverse that will be accurate for inputs sufficiently small.

To conclude this discussion, three comments are pertinent to the topic of feedback connections. The first is that even a simple feedback connection yields a Volterra system, with the complexity of the higher-degree terms increasing at a rapid rate. The second is that the operator inverses will sink of their own weight unless buoyed by an appropriate amount of rigor. Finally, efficient methods for computing the kernels corresponding to operator inverses have not been discussed, and in that sense the development here needs to be completed. I will return to this problem in Chapter 2, and in the meantime the references given in Remark 1.4 can be consulted for further discussion.

1.5 Heuristic and Mathematical Aspects

One justification or, loosely speaking, derivation of the Volterra series representation is based on a very intuitive approach to nonlinear system description. It is natural to view the output $y(t)$ of a nonlinear system at a particular time t as depending (in a nonlinear way) on all values of the input at times prior to t. That is, $y(t)$ depends on $u(t-\sigma)$ for all $\sigma \geqslant 0$. It is convenient, though not necessary, to regard t as the present instant, and then restate this as: the present output depends on all past input values. At any rate, this viewpoint leads to the following idea. If $u(t-\sigma)$ for all $\sigma \geqslant 0$ can be characterized by a set of quantities $u_1(t), u_2(t), \ldots$, then the output $y(t)$ can be represented as a nonlinear function of these quantities,

$$y(t) = f(u_1(t), u_2(t), \ldots) \tag{91}$$

The first step in pursuing this line of thought is to find a characterization for the past of an input signal. So suppose that t is fixed and the input $u(t-\sigma)$, $0 \leqslant \sigma < \infty$, is an element of the Hilbert space of square-integrable functions $L^2(0,\infty)$. That is,

$$\int_0^\infty u^2(t-\sigma)\, d\sigma < \infty$$

Furthermore, suppose that $\phi_1(\sigma), \phi_2(\sigma), \ldots$ is an orthonormal basis for this space:

$$\int_0^\infty \phi_i(\sigma)\phi_j(\sigma)\, d\sigma = \begin{cases} 1, & i = j \\ 0, & i \neq j \end{cases}$$

Then the value of the input signal at any time in the past can be written in the form

$$u(t-\sigma) = \sum_{i=1}^{\infty} u_i(t)\phi_i(\sigma) \tag{92}$$

where

$$u_i(t) = \int_0^{\infty} u(t-\sigma)\phi_i(\sigma)\, d\sigma \tag{93}$$

Although t is considered to be fixed, this development yields a characterization of the past of $u(t)$ in terms of $u_1(t), u_2(t), \ldots$ regardless of t.

With this characterization in hand, expand the function $f(u_1(t), u_2(t), \ldots)$ into a power series so that the output at any time t is

$$y(t) = a + \sum_{i=1}^{\infty} a_i u_i(t) + \sum_{i_1=1}^{\infty}\sum_{i_2=1}^{\infty} a_{i_1 i_2} u_{i_1}(t) u_{i_2}(t) + \cdots \tag{94}$$

(Of course, all the infinite sums are truncated in practice to obtain an approximate representation.) To see what all this has to do with the Volterra/Wiener representation, simply substitute (93) into (94) to obtain

$$y(t) = a + \int_0^{\infty} \sum_{i=1}^{\infty} a_i \phi(\sigma_1) u(t-\sigma_1)\, d\sigma_1$$

$$+ \int_0^{\infty}\int_0^{\infty} \sum_{i_1=1}^{\infty}\sum_{i_2=1}^{\infty} a_{i_1 i_2}\phi_{i_1}(\sigma_1)\phi_{i_2}(\sigma_2) u(t-\sigma_1) u(t-\sigma_2)\, d\sigma_1 d\sigma_2 + \cdots \tag{95}$$

With the obvious definition of the kernels in terms of the orthonormal functions $\phi_i(\sigma)$, this is precisely the kind of representation that has been discussed.

Though this demonstration is amusing, and somewhat enlightening, it seems appropriate at least to outline a more rigorous mathematical justification of the Volterra series representation. The development to be reviewed follows the style of the Weierstrass Theorem and deals with approximation of stationary nonlinear systems by stationary polynomial systems. It is assumed at the outset that the input signal space U and the output signal space Y are contained in normed linear function spaces so that a norm $\|.\|$ is available. Then the input/output behavior of a system is viewed as an operator $F:U \to Y$, and the object is to find a polynomial approximation to F. (The reader uninterested in mathematics, or unfamiliar with some of the terms just used, can skip directly to Section 1.6.)

The Weierstrass Theorem states that if $f(t)$ is a continuous, real-valued function on the closed interval $[t_1,t_2]$, then given any $\epsilon > 0$ there exists a real polynomial $p(t)$ such that $|f(t) - p(t)| < \epsilon$ for all $t \in [t_1,t_2]$. A generalization of this result known as the Stone-Weierstrass Theorem can be stated as follows. Suppose X is a compact space and Φ is an algebra of continuous, real-valued functions on X that separates points of X and that contains the constant functions. Then for any continuous, real-valued function f on X and any $\epsilon > 0$ there exists a function $p \in \Phi$ such that

$|f(x) - p(x)| < \epsilon$ for all $x \in X$. (The algebra Φ separates points if for any two distinct elements $x_1, x_2 \in X$ there exists a $p \in \Phi$ such that $p(x_1) - p(x_2) \neq 0$.) It is this generalization that leads rather easily to the representation of interest in the stationary-system case.

The set-up is as follows. One choice for the input space is the set $U \subset L_2(0,T)$ of functions satisfying: (a) there exists a constant $K > 0$ such that for all $u(t) \in U$,

$$\int_0^T |u(t)|^2 \, dt \leqslant K$$

(b) for every $\epsilon > 0$ there exists a $\delta > 0$ such that for all $u(t) \in U$ and all $|\tau| < \delta$,

$$\int_0^T |u(t+\tau) - u(t)|^2 \, dt < \epsilon$$

It can be shown that this is a compact space (see Remark 1.5). A property that follows easily using these conditions is that if $u(t) \in U$, then for any $t_1 \in [0,T]$, $u_1(t) \in U$, where

$$u_1(t) = \begin{cases} 0, & 0 \leqslant t \leqslant t_1 \\ u(t-t_1), & t_1 \leqslant t \leqslant T \end{cases}$$

This property is important, for the Stone-Weierstrass Theorem then gives results when the output space is $C[0,T]$; the space of continuous, real-valued functions on $[0,T]$ with the maximum absolute value norm. To see how this works, suppose that F is a continuous operator that is stationary and causal, and $F: U \to C[0,T]$. Suppose also that $P: U \to C[0,T]$ is a continuous, stationary, causal operator such that for all $u(t) \in U$,

$$|F[u] - P[u]| \, |_{t=T} < \epsilon$$

Then for any $u(t)$ there is a $t_1 \in [0,T]$ such that

$$\begin{aligned} \|F[u] - P[u]\| &= \max_{t \in [0,T]} |F[u] - P[u]| \\ &= |F[u] - P[u]| \, |_{t=T-t_1} = |F[u_1] - P[u_1]| \, |_{t=T} < \epsilon \end{aligned}$$

Therefore, F and P can be viewed as real-valued functions by looking at $F[u]$ and $P[u]$ only at $t = T$. But the bounds to be obtained will apply for all $t \in [0,T]$, that is, will apply for $F[u]$ and $P[u]$ as elements of $C[0,T]$.

Now the last step is to choose appropriately the algebra Φ of stationary, causal, and continuous operators from U into $C[0,T]$. For this take the algebra generated by $P_1[u] = 1$, and all operators of the form

$$P_2[u] = \int_0^t h(\sigma) u(t-\sigma) \, d\sigma$$

via repeated addition, scalar multiplication, and multiplication. Stationarity and causality of the operators in Φ is obvious. It is assumed that each $h(t)$ is such that

$$\int_0^T |h(t)|^2\, dt < \infty$$

for continuity of the operators. That Φ separates points in U is a technical calculation that will be omitted. The algebra Φ therefore consists of operators of the form

$$y(t) = P[u(t)] = h_0 + \sum_{i=1}^{N_1} \int_0^t h_{1,i}(\sigma)u(t-\sigma)\, d\sigma$$

$$+ \sum_{i=1}^{N_2} \sum_{j=1}^{N_3} \int_0^t \int_0^t h_{2,i}(\sigma_1)h_{3,j}(\sigma_2)u(t-\sigma_1)u(t-\sigma_2)\, d\sigma_1 d\sigma_2 + \cdots$$

which, with the obvious kernel definitions, is the set of stationary, causal, polynomial systems.

Everything is now in a form suitable for the Stone-Weierstrass Theorem. Thus, if a system input/output behavior can be represented by a continuous, stationary, causal operator, $F:U \to C[0,T]$, then given any $\epsilon > 0$ there is a continuous, stationary and causal operator P such that for all $u(t) \in U$

$$|F[u] - P[u]||_{t=T} < \epsilon$$

or,

$$\|F[u] - P[u]\| < \epsilon$$

That is, there is a polynomial system that approximates F to within ϵ. Furthermore, it is clear from the construction of Φ that the kernels of the polynomial system will be separable.

While this is a powerful result, I should point out that the main drawback is in the restrictive input space U. The compactness requirement rules out many of the more natural choices of U. For example, the unit ball in $L_2(0,T)$, the set of all $u(t)$ such that $\|u(t)\| \leqslant 1$, is not compact. At any rate, further discussion of the mathematical aspects of the Volterra representation can be found in Appendix 1.2, and the references cited in Section 1.6.

1.6 Remarks and References

Remark 1.1 The system representations discussed herein were introduced in mathematics by V. Volterra around the turn of the century in the very beginning of functional analysis. Volterra used the graphic terminology "function of lines," and the notation

$$F\|[u\overset{b}{\underset{a}{(t)}})]\|$$

(sometimes with the interval $[a,b]$ left understood) to describe what later came to be called a functional. He defined the notion of derivatives of a function of lines, and developed multiple integral representations. Then Volterra extended Taylor's Theorem to obtain expressions of the form

$$F[[f(t) + u(t)]] = F[[f(t)]] + \int_a^b F'[[f(t),\sigma_1]]u(\sigma_1)\, d\sigma_1$$

$$+ \int_a^b \int_a^b F''[[f(t),\sigma_1,\sigma_2]]u(\sigma_1)u(\sigma_2)\, d\sigma_1 d\sigma_2 + \cdots$$

the terms of which were called homogeneous of degree n. An overview of this work can be found in

Theory of Functionals and of Integral and Integro-differential Equations, Dover, New York, 1958.

This is an English translation of a book first published in Spanish in 1927. More detailed accounts can be found in various volumes by Volterra, Frechet, and others in the *Collection of Monographs on the Theory of Functions,* E. Borel editor, published by Gauthier-Villars, Paris. Volterra's earliest discussion of these ideas apparently resides in several notes in *R. C. Accademia dei Lincei,* in 1887.

Remark 1.2 The first use of Volterra's representation in nonlinear system theory occurs in the work of N. Wiener in the early 1940s. This work, which deals with the response of a nonlinear system to white noise inputs, will be discussed in Chapter 5. However, the heuristic justification of the Volterra series representation given at the beginning of Section 1.5 follows Wiener's viewpoint. Several technical reports from the Research Laboratory of Electronics (RLE) at the Massachusetts Institute of Technology (MIT) contain subsequent work on the Volterra functional representation applied to nonlinear systems. For the material introduced in this chapter, the most appropriate of these reports are the two listed below (with National Technical Information Service order numbers shown parenthetically).

M. Brilliant, "Theory of the Analysis of Nonlinear Systems," MIT RLE Technical Report No. 345, 1958 (AD216-209).

D. George, "Continuous Nonlinear Systems," MIT RLE Technical Report No. 355, 1959 (AD246-281).

Another early report of interest is from Cambridge University, and is reprinted in

J. Barrett, "The Use of Functionals in the Analysis of Nonlinear Physical Systems," *Journal of Electronics and Control,* Vol. 15, pp. 567-615, 1963.

Scattered through the literature in the 1960s are a number of articles that introduce the Volterra series representation for nonlinear systems. It is safe to say that many of these articles redevelop material essentially contained in the reports listed above, undoubtedly because these early reports were not published in the widely available literature. In recent years, two books have appeared dealing with the Volterra/Wiener representation for nonlinear systems. These are

P. Marmarelis, V. Marmarelis, *Analysis of Physiological Systems,* Plenum, New York, 1978.

M. Schetzen, *The Volterra and Wiener Theories of Nonlinear Systems,* John Wiley, New York, 1980.

The latter book is based on the early MIT work, while the former concentrates on applications of the Wiener theory in biomedical engineering. Both books contain introductory material on the Volterra series.

Remark 1.3 The symmetric form for kernels has been used since the beginning of work in this area. It is very natural from a mathematical point of view, and symmetric representations were used by Volterra. However, the use of triangular kernels is much more recent, beginning with the paper

A. Isidori, A. Ruberti, "Realization Theory of Bilinear Systems," in *Geometric Methods in System Theory,* D. Mayne and R. Brockett, eds., D. Reidel, Dordrecht, Holland, pp. 83-130, 1973.

These authors also have used the regular kernel (in an implicit manner) to solve the realization problem for bilinear state equations, a topic that will be discussed in Chapter 4. The first explicit discussion of the regular kernel representation appears in

G. Mitzel, S. Clancy, W. Rugh, "On Transfer Function Representations for Homogeneous Nonlinear Systems," *IEEE Transactions on Automatic Control,* Vol. AC-24, pp. 242-249, 1979.

While the various forms in which kernels and input/output representations can be written might seem needlessly confusing, each of them has important properties that will be encountered in due course. I feel that it is important to become familiar with these forms early in the program, for the right choice of representation can make a particular topic much easier.

Remark 1.4 Many, if not most, of the early reports and papers on Volterra series methods discuss the interconnection of systems. See for example the reports by George and Brilliant mentioned in Remark 1.1. Discussions of operator representations, interconnection rules, and the feedback connection in particular can be found in

G. Zames, "Functional Analysis Applied to Nonlinear Feedback Systems," *IEEE Transactions on Circuit Theory,* Vol. CT-10, pp. 392-404, 1963.

G. Zames, "Realizability Conditions for Nonlinear Feedback Systems," *IEEE Transactions on Circuit Theory,* Vol. CT-11, pp. 186-194, 1964.

and in

J. Willems, *The Analysis of Feedback Systems,* MIT Press, Cambridge, Massachusetts, 1971.

C. Desoer, M. Vidyasagar, *Feedback Systems: Input/Output Properties,* Academic Press, New York, 1975.

A recent review of the difficult and subtle issues surrounding operator inverses is given in

W. Porter, "An Overview of Polynomic System Theory," *Proceedings of the IEEE,* Vol. 64, pp. 18-23, 1976.

The closely related question of inverses for polynomial or Volterra systems is discussed in

M. Schetzen, "Theory of pth-Order Inverses of Nonlinear Systems," *IEEE Transactions on Circuits and Systems,* Vol. CAS-23, pp. 285-291, 1976.

A. Halme, J. Orava, "Generalized Polynomial Operators for Nonlinear Systems Analysis," *IEEE Transactions on Automatic Control,* Vol. AC-17, pp. 226-228, 1972.

All these more complicated issues aside, a warning is appropriate for even the simple problems involving parallel and cascade connections. The experience that everybody probably has in linear block diagram manipulation tends to encourage an overexuberant approach to problems involving interconnections of nonlinear systems. This easily can lead to wrong answers that are seductive in their simplicity. Sober and serious attention to the specific commutative and distributive properties of nonlinear interconnections is the only way to avoid being caught in a compromising position.

Remark 1.5 The Stone-Weierstrass approach to the question of approximate polynomial system representations dates back at least to the report by Brilliant mentioned in Remark 1.1. The presentation in Section 1.5 follows

P. Gallman, "Representation of Nonlinear Systems via the Stone-Weierstrass Theorem," *Automatica,* Vol. 12, pp. 619-622, 1976.

A proof that the set of inputs U used in that development is a compact space can be found in

L. Liusternik, V. Sobolev, *Elements of Functional Analysis,* Unger, New York, 1961.

A closely related line of work on Weierstrass approaches begins with the paper

P. Prenter, "A Weierstrass Theorem for Real Separable Hilbert Spaces," *Journal of Approximation Theory,* Vol. 3, pp.341-351, 1970.

and acquires the causality results so important for system theory in

W. Porter, T. Clark, "Causality Structure and the Weierstrass Theorem," *Journal of Mathematical Analysis and Applications,* Vol. 52, pp.351-363, 1975.

A constructive approach to the Weierstrass-type system approximation theorems is developed in

W. Porter, "Approximation by Bernstein Systems," *Mathematical Systems Theory,* Vol. 11, pp. 259-274, 1978.

The approximating systems are linear systems followed by polynomial nonlinearities, and therefore this paper can be viewed as a confirmation of the heuristic discussions in Section 1.5 and Problem 1.18. Another viewpoint toward the approximation properties of polynomial systems is developed in

W. Root, "On the Modeling of Systems for Identification, Part I: ϵ-Representations of Classes of Systems," *SIAM Journal on Control,* Vol. 13, pp. 927-944, 1975.

Finally, references on approximation by systems describable by bilinear state equations are given in Remark 4.6 of Section 4.6.

Remark 1.6 A somewhat different integral representation for nonlinear systems can be written in the form

$$y(t) = \int_{-\infty}^{\infty} h[t-\sigma, u(\sigma)] \, d\sigma$$

for stationary systems. Various properties are discussed in

L. Zadeh, "A Contribution to the Theory of Nonlinear Systems," *Journal of the Franklin Institute,* Vol. 255, pp. 387-408, 1953.

A. Gersho, "Nonlinear Systems with a Restricted Additivity Property," *IEEE Transactions on Circuit Theory,* Vol CT-16, pp.150-154, 1969.

The basic approximation properties for such a representation and the relationship to the Volterra type representation are presented in the paper by Gallman cited in Remark 1.5. Suffice it to say here that the core of the matter is another application of the Stone-Weierstrass Theorem.

1.7 Problems

1.1. A system that has the property that the response to $\alpha u(t)$ is $\alpha y(t)$ is a degree-1 homogeneous system. Does it follow that the system is linear? In other words, is the response to $\alpha_1 u_1(t) + \alpha_2 u_2(t)$ given by $\alpha_1 y_1(t) + \alpha_2 y_2(t)$?

1.2. Find two degree-2, homogeneous, interconnection structured systems that have the same response to the input $u(t) = \delta_0(t)$, but different responses to the input $u(t) = \delta_{-1}(t)$.

1.3. Find a kernel for the system shown below.

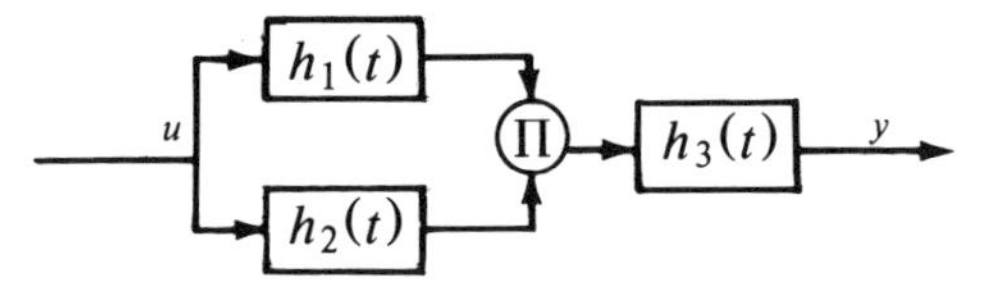

1.4. Symmetrize the kernel

$$h(t_1,t_2,t_3) = h_1(t_1+t_2+t_3)h_2(t_1+t_2)h_3(t_1)$$

1.5. Write the kernels derived in Examples 1.1 and 1.3 in symmetric form.

1.6. Compute an overall kernel for the cascade connection shown below, and compare your result with Example 1.10.

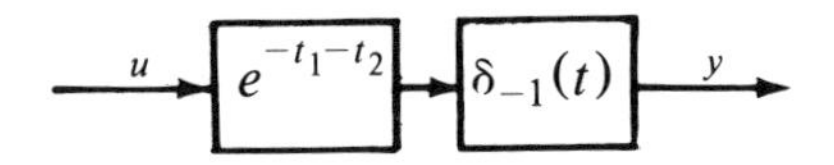

1.7. Show that the symmetric kernel for the interconnection structured system shown below is

$$h_{sym}(t_1,t_2) = 1 - e^{-min[t_1,t_2]}$$

1.8. Show that the kernel corresponding to the cascade connection of a degree-n system with symmetric kernel followed by a linear system is "automatically" symmetric.

1.9. Show that the symmetric form of a kernel is separable if and only if the triangular form is separable (neglecting the unit step functions, of course).

1.10. List the possible "direct transmission" terms for the degree-3 case. Then show how these can be represented by introducing impulses into the kernel.

1.11. Compute kernels for the two systems shown below.

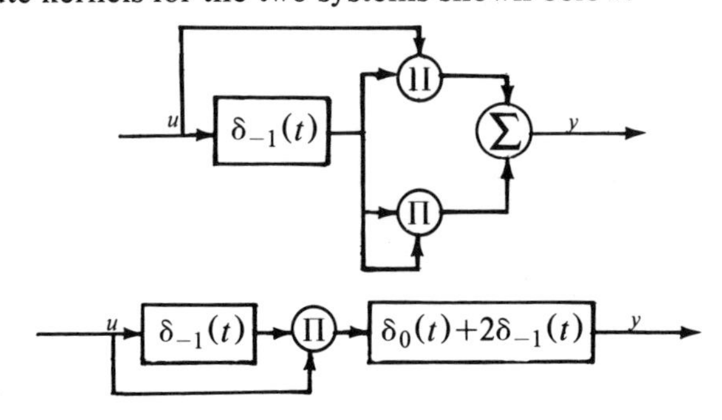

1.12. Find kernels for the polynomial system representation of the interconnection shown below.

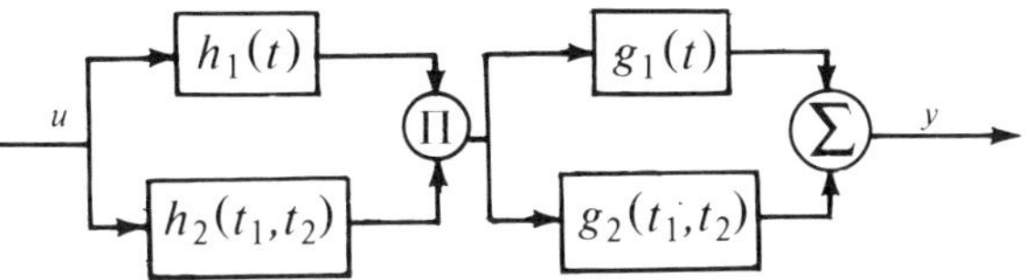

1.13. Suppose that in Example 1.3 each $h_i(t)$ can be written as

$$h_i(t) = \sum_{j_i=1}^{n_i} \sum_{k_i=1}^{\sigma_i} a_{j_i k_i} t^{k_i - 1} e^{\lambda_{j_i} t}, \quad t \geq 0$$

Show that the overall system kernel is separable.

1.14. Consider a Volterra system of the form

$$y(t) = \sum_{n=1}^{\infty} \int_0^t h_n(t-\sigma_1, \ldots, t-\sigma_n) u(\sigma_1) \cdots u(\sigma_n) \, d\sigma_n \cdots d\sigma_1$$

where $h_1(t) = \delta_{-1}(t)$, and for $n > 1$,

$$h_n(t_1, \ldots, t_n) = a_n \delta_0(t_1 - t_2) \delta_0(t_2 - t_3) \cdots \delta_0(t_{n-1} - t_n)$$

Give conditions under which this series converges. Can you devise a simple interconnection diagram for this system?

1.15. Given the symmetric kernel $h_{sym}(t_1, \ldots, t_n)$, show that the regular kernel on the first orthant is given by

$$h_{reg}(t_1, \ldots, t_n) = n! h_{sym}(t_1 + \cdots + t_n, t_2 + \cdots + t_n, \ldots, t_n)$$

1.16. Show that

$$G_2 * (H_1 + H_2) - G_2 * H_1 - G_2 * H_2$$

represents a degree-3 homogeneous system.

1.17. Analyze the feedback system diagramed below, and show that awful things happen for a unit step input.

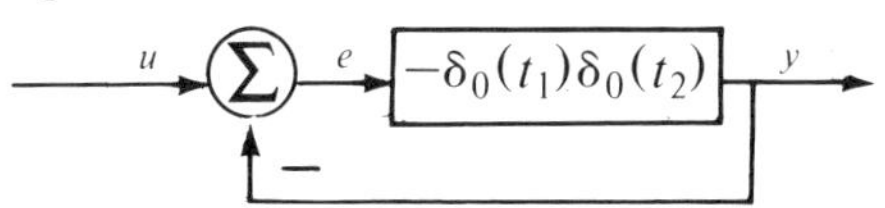

1.18. Use the heuristic justification of the Volterra series representation given in Section 1.5 to substantiate the following claim. Any (suitably smooth) nonlinear system can be approximated by a linear system followed by a polynomial nonlinearity.

1.19. Define a pth-degree preinverse for a nonlinear system, show how to calculate the homogeneous terms through degree 3, and show that these terms are identical to those in the pth-degree postinverse.

APPENDIX 1.1 Convergence Conditions for Interconnections of Volterra Systems

When Volterra systems are interconnected in the manners discussed in Section 1.4, some care should be exercised with regard to convergence issues. An analysis of the situation is most conveniently performed in terms of the convergence argument in Example 1.9. Although the various bounds that are obtained are very conservative, the bounds are less important than the fact of convergence.

Suppose $y = G[u]$ and $y = H[u]$ are two stationary Volterra systems, where the respective kernels are bounded according to

$$\int_{-\infty}^{\infty} |g_n(t_1,\ldots,t_n)|\, dt_1 \cdots dt_n = g_n$$

$$\int_{-\infty}^{\infty} |h_n(t_1,\ldots,t_n)|\, dt_1 \cdots dt_n = h_n$$

where g_n and h_n are finite nonnegative numbers, and $n = 1,2,\ldots$. (I might as well be conservative in the matter of technicalities, and assume that the kernels are continuous for nonnegative arguments, and that the input and output signals are continuous.) Permitting degree-0 terms with absolute values g_0 and h_0 to be included in the Volterra systems, suppose that the power series

$$b_G(x) = \sum_{n=0}^{\infty} g_n x^n\,,\quad b_H(x) = \sum_{n=0}^{\infty} h_n x^n$$

have positive radii of convergence r_G and r_H, respectively. Then the Volterra system representation for $y = G[u]$ converges if the input satisfies $|u(t)| < r_G$ for all t. Furthermore, if $|u(t)| \leqslant r < r_G$ for all t, then as discussed in Example 1.9, $|y(t)| \leqslant b_G(r)$ for all t. For this reason $b_G(x)$ is called the *bound function* for the system, and the values of $b_G(x)$ for $x \geqslant 0$ are of interest.

From this formulation a number of conclusions about convergence of interconnections of Volterra systems and bounds on responses follow from elementary facts about power series. For the Volterra system representation for $F = H + G$ in (55), the kernels satisfy

$$\int_0^{\infty} |f_n(t_1,\ldots,t_n)|\, dt_1 \cdots dt_n = \int_0^{\infty} |h_n(t_1,\ldots,t_n) + g_n(t_1,\ldots,t_n)|\, dt_1 \cdots dt_n$$

$$\leqslant h_n + g_n$$

so that the bound function for $H + G$ satisfies

$$b_{H+G}(x) \leqslant b_H(x) + b_G(x),\quad x \geqslant 0$$

and the radius of convergence satisfies $r_{H+G} \geqslant min[r_H, r_G]$. Also, the kernels for the Volterra system representation for $F = HG$ in (56) satisfy

$$\int_0^\infty |f_n(t_1,\ldots,t_n)|\,dt_1\cdots dt_n = \int_0^\infty |\sum_{j=0}^{n} h_j(t_1,\ldots,t_j)g_{n-j}(t_{j+1},\ldots,t_n)|\,dt_1\cdots dt_n$$
$$\leqslant \sum_{j=0}^{n} h_j g_{n-j}$$

Thus,

$$b_{HG}(x) \leqslant b_H(x)\,b_G(x), \quad x \geqslant 0$$

and $r_{HG} \geqslant min[r_H,r_G]$. Stating these results informally: if H and G are Volterra systems that converge for sufficiently small inputs, then $H + G$ and HG are Volterra systems that converge for sufficiently small inputs.

The cascade connection $F = G*H$ is less easy to handle in an explicit manner because the formula for the kernels of $G*H$ in terms of the kernels for G and H is quite messy. However, an indirect approach can be used to conclude convergence. Suppose $|u(t)| \leqslant r < r_H$, and $r > 0$ is such that

$$\sum_{n=0}^{\infty} h_n r^n < r_G$$

Then the Volterra system representation for H converges, and the output signal from H has a bound that is within the radius of convergence of the system G. Thus the Volterra system representation for $G*H$ will have a positive radius of convergence, specifically,

$$r_{G*H} \geqslant r$$

Notice that if the degree-0 term h_0 is too large, it can be impossible to conclude convergence of the Volterra system representation for $G*H$ from this argument. On the other hand, if $h_0 = 0$, then the argument insures that the Volterra system representation of the cascade connection $G*H$ will converge for sufficiently small inputs.

For use in the sequel, it is important to consider the bound function for the cascade system. Assuming that no degree-0 terms are present for simplicity, the development that leads to (70) in Section 1.4 shows that the degree-k term in the cascade representation is given by

$$\sum_{m=1}^{\infty} \underbrace{\sum_{n_1=1}^{\infty}\cdots\sum_{n_m=1}^{\infty}}_{n_1+\cdots+n_m=k} \int_{-\infty}^{\infty} g_{sym}(\sigma_1,\ldots,\sigma_m)\,w_{n_1}(t-\sigma_1)\cdots w_{n_m}(t-\sigma_m)\,d\sigma_1\cdots d\sigma_m$$

where

$$w_n(t) = \int_{-\infty}^{\infty} h_{sym}(\tau_1,\ldots,\tau_n)\,u(t-\tau_1)\cdots u(t-\tau_n)\,d\tau_1\cdots d\tau_n$$

From this expression the kernel of the degree-k term in the cascade connection satisfies

$$\int_{-\infty}^{\infty} |f_k(t_1,\ldots,t_k)|\,dt_1\cdots dt_k \leqslant \sum_{m=1}^{\infty} \underbrace{\sum_{n_1=1}^{\infty}\cdots\sum_{n_m=1}^{\infty}}_{n_1+\cdots+n_m=k} g_m h_{n_1}\cdots h_{n_m}$$

and thus the bound function for the cascade satisfies

$$\sum_{k=1}^{\infty} f_k x^k \leqslant \sum_{m=1}^{\infty} \sum_{n_1=1}^{\infty} \cdots \sum_{n_m=1}^{\infty} g_m h_{n_1} \cdots h_{n_m} x^{n_1+\cdots+n_m}$$

$$\leqslant \sum_{m=1}^{\infty} g_m \Big(\sum_{n=1}^{\infty} h_n x^n \Big), \quad x \geqslant 0$$

That is,

$$b_{G*H}(x) \leqslant b_G(b_H(x)), \quad x \geqslant 0$$

When the feedback connection is considered, issues of well-posedness, convergence, and stability intertwine in a nasty way. The following discussion of convergence for feedback Volterra systems will illustrate the situation, and at least partially unravel it.

The basic feedback connection is shown in Figure A1.1. If the Volterra system K is the cascade connection $K = H*G$, and if this feedback system is followed in cascade by G, then the general feedback system shown in Figure 1.12 is obtained. Since cascade systems have been considered already, I will concentrate on the basic feedback system alone. From another point of view, the basic feedback system is much like the "error system" described by (81) with $K = H*G$.

The first assumption to be made is that K represents a Volterra system with no degree-0 term, and with radius of convergence $r_K > 0$. Writing $K = K_1 + (K - K_1)$ to exhibit the degree-1 term, the basic feedback system can be redrawn as shown in Figure A1.2. Here the degree-1 term of the closed-loop representation $y = F[u]$ is given by $F_1 = (I + K_1)^{-1}$, and it is assumed that this is a well defined, causal system so that the procedure discussed in Section 1.4 can be used to compute the higher-degree terms in F. (Thus, the kinds of problems indicated in Problem 1.17 are avoided.) Furthermore, it is assumed that the kernel for F_1 satisfies

$$\int_{-\infty}^{\infty} |f_1(t)|\, dt = f_1 < \infty$$

Figure A1.1. The basic feedback system.

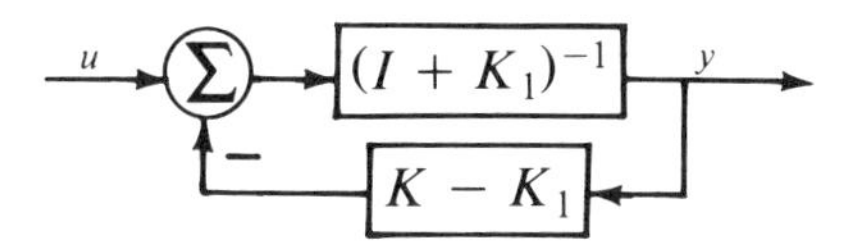

Figure A1.2. The basic feedback system with separated degree-1 terms.

where $f_1 > 0$ is assumed (without loss of generality) to avoid trivialities in the sequel. Notice that since $f_1(t)$ is the kernel for a linear feedback system, this boundedness assumption involves the stability properties of linear feedback systems in an essential way.

The remainder of the discussion is devoted to computing bounds on the higher-degree kernels and establishing convergence of the feedback system under the foregoing assumptions. Using (82) the feedback system in Figure A1.2 is described by the operator equation

$$F = F_1*[I - (K - K_1)*F], \quad F_1 = (I + K_1)^{-1}$$

Writing out the homogeneous terms for F and $(K - K_1)$ gives

$$\begin{aligned} F_1 + F_2 + F_3 + \cdots &= F_1 - F_1*(K_2 + K_3 + \cdots)*(F_1 + F_2 + F_3 + \cdots) \\ &= F_1 + Q*(F_1 + F_2 + F_3 + \cdots) \end{aligned}$$

where $Q = -F_1*(K_2 + K_3 + \cdots)$ is a cascade Volterra system that contains no degree-0 or degree-1 terms, and that by previous results has a bound function

$$b_Q(x) = \sum_{n=2}^{\infty} q_n x^n$$

with positive radius of convergence. Now suppose there is a bound function for the feedback system,

$$b_F(x) = \sum_{n=1}^{\infty} f_n x^n$$

where f_1 is known and $f_n \geq 0$, $n \geq 2$. Then, using the argument in the cascade case, $b_F(x)$ satisfies

$$b_F(x) \leq f_1 x + b_Q(b_F(x))$$

Writing this equation in power-series form gives

$$\sum_{n=1}^{\infty} f_n x^n \leq f_1 x + \sum_{n=2}^{\infty} q_n \Big(\sum_{m=1}^{\infty} f_m x^m\Big)^n, \quad x \geq 0$$

Thus

$$\sum_{n=2}^{\infty} f_n x^n \leqslant \sum_{n=2}^{\infty} q_n \Big(\sum_{m=1}^{\infty} f_m x^m\Big)^n$$
$$= q_2(f_1^2 x^2 + 2f_1 f_2 x^3 + \cdots) + q_3(f_1^3 x^3 + \cdots) + \cdots, \quad x \geqslant 0$$

and it is clear from this expression that bounds on the individual coefficients in $b_F(x)$ can be determined sequentially:

$$f_2 \leqslant q_2 f_1^2, \quad f_3 \leqslant 2q_2 f_1 f_2 + q_3 f_1^3, \quad \ldots$$

To ascertain the convergence properties of a power series that satisfies the coefficient bounds recursively constructed above, an indirect approach is needed. Suppose $y(x)$ is a solution of

$$y = f_1 x + b_Q(y)$$

Rearranging the equation, and substituting the power series representation for b_Q gives the convergent power series expression

$$x = \frac{1}{f_1} y - \frac{1}{f_1} \sum_{n=2}^{\infty} q_n y^n, \; |y| < r_Q$$

Now the theory of inversion of power series can be applied since the linear term on the right side has nonzero coefficient. This gives $y(x)$ as a power series in x that has a positive radius of convergence. Furthermore, a recursive computation similar to that above shows that the coefficients in the power series for $y(x)$ are given by

$$y(x) = f_1 x + (q_2 f_1^2) x^2 + (2q_2 f_1 f_2 + q_3 f_1^3) x^3 + \cdots$$

Therefore $b_F(x) \leqslant y(x)$ for $x \geqslant 0$, and it follows that $b_F(x)$ has a positive radius of convergence.

To summarize informally, the Volterra system representation for the feedback system in Figure A1.1 converges for inputs that are sufficiently small if the Volterra system representation for K converges for sufficiently small inputs. It should be emphasized again that this result depends crucially on the existence and boundedness assumptions on the degree-1 term in the closed-loop system.

APPENDIX 1.2 The Volterra Representation for Functionals

The Volterra series representation originated in the mathematics of functional analysis. A review of the mathematics therefore is appropriate, and moreover will indicate the rigorous approach that lies behind the purely symbolic treatment of generalized functions as impulse functions in the main text. A complete and self-contained exposition is beyond the scope of an appendix, so I will be more detailed than usual in citing reference material.

Recalling that a (real) functional is a real-valued function defined on a linear space, it will be helpful to show first how the representation of functionals is connected to the representation of systems. Of course, in this case the linear space is assumed to be a linear space of functions, namely, the input signals. At a particular time t, the output value $y(t)$ of a system can be viewed as depending on the past history of the input signal, $u(t-\sigma)$, $\sigma \geqslant 0$. In other words, the output at a particular time is a real functional of the input signal. If the system is stationary, then this functional will be the same regardless of the particular choice of t, and so the system is completely characterized. If the system is nonstationary, more care is required in the interpretation of the functional representation, and so that case will be ignored for simplicity.

It is natural to consider real functionals on the space $L_2(0,\infty)$ of real functions defined for $0 \leqslant t < \infty$ that satisfy

$$\|f\|^2 = \int_0^\infty f^2(\sigma)\, d\sigma < \infty$$

A functional will be denoted by $F{:}L_2(0,\infty) \rightarrow R$, and it will be assumed that F is $n+1$ times Frechet differentiable at the point of interest in $L_2(0,\infty)$, here taken to be zero for convenience. Let $F^{(k)}[w_1,\ldots,w_k]$ denote the kth Frechet differential of F at 0 with increments $w_1,\ldots,w_k$, and recall that $F^{(k)}[w_1,\ldots,w_k]$ is a symmetric, multilinear (k-linear), continuous mapping from $L_2(0,\infty)\ X \cdots X\ L_2(0,\infty)$ (k-tuples of functions) into R. Then Taylor's formula can be applied to yield the following result. For all $\phi \in L_2(0,\infty)$ in some neighborhood of 0,[1]

$$F[\phi] = \sum_{k=1}^{n} \frac{1}{k!} F^{(k)}[\phi,\ldots,\phi] + R_n[\phi]$$

where $\|R_n[\phi]\| < M\|\phi\|^{n+1}$, and $F[0] = 0$ is assumed for simplicity. (Alternatively, stronger hypotheses give the existence of an infinite series that converges in some neighborhood of 0.)

Since the kth term in this representation is determined by a k-linear functional, the Volterra representation involves the integral representation of such functionals. I will concentrate on the first two terms in the representation, namely, the integral representation of linear and bilinear functionals. The remaining terms are handled in a manner similar to the bilinear case. And before proceeding further, it is worthwhile to point out that some casually apparent approaches are not sufficiently general, or appropriate, for the class of functionals (systems) of interest.

Beginning with the first term in the Taylor expansion, it is tempting to apply the Riesz Theorem for linear functionals on $L_2(0,\infty)$ and write

$$F^{(1)}[\phi] = \int_0^\infty \psi(\sigma)\phi(\sigma)\, d\sigma$$

1 J. Dieudonne, *Foundations of Modern Analysis*, Academic Press, New York, 1960, Chapter 8.

where $\psi \in L_2(0,\infty)$ is fixed. Unfortunately, this approach to obtaining an integral representation excludes certain linear functionals that correspond to systems of interest. As an example, consider the linear functional

$$F^{(1)}[\phi] = \phi(0)$$

which corresponds to the identity system when $\phi(\sigma)$ is identified with $u(t-\sigma)$. In this case the Riesz Theorem cannot be applied, and moreover in the space $L_2(0,\infty)$ the value of a function at a point cannot be discussed.

Another example of an approach that fails to be sufficiently general can be constructed by using a slightly different setting. Let $L_2(0,T)$ denote the Hilbert space of square-integrable functions defined for $0 \leqslant t \leqslant T$. Consider a symmetric bilinear functional $F^{(2)}[\phi,\psi]:L_2(0,T) \rightarrow R$ such as the second term in the Taylor expansion given earlier. Suppose that for all $\phi,\psi \in L_2(0,T)$ for which the derivatives $\dot{\phi}$ and $\dot{\psi}$ are also elements of $L_2(0,T)$ the bilinear functional satisfies

$$|F^{(2)}[\dot{\phi},\psi]| \leqslant M\|\phi\|\,\|\psi\|$$

$$|F^{(2)}[\phi,\dot{\psi}]| \leqslant M\|\phi\|\,\|\psi\|$$

Then there exists $k(t_1,t_2) \in L_2((0,T)X(0,T))$ such that [2]

$$F^{(2)}[\phi,\psi] = \int_0^T\int_0^T k(\sigma_1,\sigma_2)\phi(\sigma_1)\psi(\sigma_2)\,d\sigma_1 d\sigma_2$$

Unfortunately, the hypotheses here are too restrictive to allow consideration of a bilinear functional of the form

$$F^{(2)}[\phi,\psi] = \int_0^T \phi(\sigma)\psi(\sigma)\,d\sigma$$

which, upon taking $\phi(t) = \psi(t)$, corresponds to a system composed of a squarer followed by an integrator:

$$F^{(2)}[\phi,\phi] = \int_0^T \phi^2(\sigma)\,d\sigma$$

Here $u(t-\sigma)$ is identified with $\phi(\sigma)$, and only finite-length input signals are considered. Of course, in the main text an integral representation for this kind of system involves an impulse in the kernel,

$$F^{(2)}[\phi,\phi] = \int_0^T\int_0^T \delta(\sigma_1-\sigma_2)\phi(\sigma_1)\phi(\sigma_2)\,d\sigma_1 d\sigma_2$$

and a few simple rules are used to manipulate impulses.

From these considerations it is clear that some of the more readily available integral representation theorems are not appropriate tools for investigating the Vol-

2 I. Gelfand, N. Vilenkin, *Generalized Functions*, Vol. 4, Academic Press, New York, 1964, pp. 11 - 13.

terra series representation for nonlinear systems, and that the issue of impulses arises in a basic way. To consider the matter further involves using a more restricted input space than L_2. Let K be the set of all real functions that are infinitely differentiable and that vanish outside some finite interval in $[0,\infty)$. Then K is a linear space, $K \subset L_2(0,\infty)$, and it can be shown that K is dense in $L_2(0,\infty)$. A sequence of functions in K, say $\phi_1(t), \phi_2(t), \ldots$ is said to *converge to zero in* K if (a) there is some finite interval such that all of the functions vanish outside the interval, and (b) the sequence converges uniformly to zero on the interval, and for any positive integer j the sequence of derivatives $\phi_1^{(j)}(t), \phi_2^{(j)}(t), \ldots$ converges uniformly to zero on the interval. A functional $F:K \rightarrow R$ is called a *continuous linear functional on* K if $F[\cdot]$ is linear and if for every $\phi_1(t), \phi_2(t), \ldots$ converging to zero on K, the sequence $F[\phi_1], F[\phi_2], \ldots$ converges to zero in R.[3] It should be noted that it is a purely technical exercise to show that a continuous linear functional on $L_2(0,\infty)$ also is a continuous linear functional on K. Thus a functional of the form

$$F[\phi] = \int_0^\infty f(\sigma)\phi(\sigma)\, d\sigma$$

where $f \in L_2(0,\infty)$, is a continuous linear functional on K. But

$$F[\phi] = \phi(0)$$

also is a continuous linear functional on K, although there is no corresponding way to write an integral representation.

The established, though very confusing, terminology is to call any continuous linear functional on K a *generalized function* or *distribution.* It is standard practice, however, to write generalized functions in integral form. Accordingly, the functional $F[\phi] = \phi(0)$ is written in the integral form

$$F[\phi] = \int_0^\infty \delta(\sigma)\phi(\sigma)\, d\sigma = \phi(0)$$

where $\delta(t)$ is the *impulse function,* though of course such a "function" doesn't exist, and the notation is purely symbolic. The fact remains that as long as a simple set of rules is followed, such a symbolic integral representation is quite convenient. The cited references show in detail the correspondence between the special rules for manipulating impulses, and the corresponding rigorous interpretation in terms of generalized functions (continuous linear functionals on K).

Now consider bilinear functionals on the space $K \times K$ of pairs of infinitely differentiable functions, each of which vanishes outside some finite interval. Such a functional $F^{(2)}[\phi,\psi]$ is called *continuous in each argument* if whenever either $\phi \in K$ or $\psi \in K$ is fixed, $F^{(2)}[\phi,\psi]$ is a continuous linear functional on K. It is also necessary to consider the linear space K_2 of infinitely differentiable functions of two variables

3 L. Zadeh, C. Desoer, *Linear System Theory*, McGraw-Hill, New York, 1963, Appendix A; or I. Gelfand, G. Shilov, *Generalized Functions*, Vol. 1, Academic Press, New York, 1964, Chapter 1.

$\phi(t_1,t_2)$ that vanish outside of a finite region in $R \ X \ R$. The notions of convergence to zero in K_2 and of continuity of a linear functional $F^{(1)}:K_2 \rightarrow R$ are the obvious extensions of the corresponding notions in K. And just as for generalized functions on K, an integral representation of the form

$$F^{(1)}[\phi(t_1,t_2)] = \int_0^\infty \int_0^\infty k(\sigma_1,\sigma_2)\phi(\sigma_1,\sigma_2)\ d\sigma_1 d\sigma_2$$

is used for generalized functions on K_2, although $k(t_1,t_2)$ is often of a purely symbolic nature. A celebrated result in the theory of generalized functions now can be stated.[4]

The Kernel Theorem Suppose $F^{(2)}:K \ X \ K \rightarrow R$ is a bilinear functional that is continuous in each of its arguments. Then there exists a generalized function $F^{(1)}:K_2 \rightarrow R$ such that

$$F^{(2)}[\phi,\psi] = F^{(1)}[\phi(t_1)\psi(t_2)]$$

Of course, this result immediately provides a symbolic integral representation for bilinear functionals that are continuous in each argument. For $\psi(t) = \phi(t)$, the integral representation becomes

$$F^{(2)}[\phi,\phi] = \int_0^\infty \int_0^\infty k(\sigma_1,\sigma_2)\phi(\sigma_1)\phi(\sigma_2)\ d\sigma_1 d\sigma_2$$

and for $\phi(\sigma) = u(t-\sigma)$ this takes the form of the degree-2 homogeneous system representation used in the main text.

The symbolic integral representation for higher-degree homogeneous systems follows in a similar way using a more general version of the kernel theorem.[5] Finally, it is a technical matter to extend the symbolic representation to $L_2(0,\infty)$ using the fact that K is dense in $L_2(0,\infty)$, and the fact that the notions of continuity used in the development are quite strong. Again, it should be emphasized that these symbolic representations are convenient because they permit writing a large class of homogeneous systems in a standard integral form.

4 I. Gelfand, N. Vilenkin, *Generalized Functions*, Vol. 4, Academic Press, New York, 1964, p. 18.
5 Ibid., p. 20.

CHAPTER 2

Input/Output Representations in the Transform Domain

The generalization of the Laplace transform to functions of several variables yields a tool of considerable importance in stationary nonlinear system theory. Just as for linear systems, the Laplace transform of a multivariable kernel is called a transfer function. This representation is useful both for characterizing system properties, and for describing system input/output behavior. Furthermore, many of the rules for describing interconnections of systems can be expressed most neatly in terms of transfer functions. A basic reason for all these features is that certain multivariable convolutions can be represented in terms of products of Laplace transforms, much as in the single-variable case.

Corresponding to each of the special forms for the kernel of a homogeneous system, there is a special form of the transfer function. Polynomial and Volterra systems can be described by the collection of transfer functions corresponding to the homogeneous subsystems. All of these representations will be used extensively in the sequel.

2.1 The Laplace Transform

I begin by reviewing for a moment the definition of the Laplace transform of a one-sided, real-valued function $f(t)$:

$$F(s) = L[f(t)] = \int_0^\infty f(t)e^{-st}\,dt \tag{1}$$

Of course, a comment should be included about the region of convergence of $F(s)$, that is, the range of values of the complex variable s for which the integral converges. However, the reader is probably accustomed to treating convergence considerations for the Laplace transform in an off-handed manner. The reason is that for the typical functions arising in linear system theory convergence regions always exist.

Usually, the functions encountered are exponential forms, in other words, finite linear combinations of terms of the form $t^m e^{\lambda t}$. (Of course λ may be complex, but the terms appear in conjugate pairs so the function is real.) It is easily verified that for functions of this type the integral in (1) converges for all s in some half-plane of the complex plane, and the resulting transform is a real-coefficient rational function of s that is strictly proper (numerator-polynomial degree less than denominator-polynomial degree). Also, the typical operations on these functions of time (addition, integration, convolution) yield functions of the same type. One result of these observations is that a strictly algebraic viewpoint is valid for the Laplace transform in the setting of the limited class of functions described above. But, for the purposes here, a more relaxed, informal treatment based on the integral definition will do. (What is called informal here, a mathematician would call formal!)

For the inverse Laplace transform, the computation of $f(t)$ from $F(s)$, the familiar line integration formula is

$$f(t) = L^{-1}[F(s)] = \frac{1}{2\pi i} \int_{\sigma - i\infty}^{\sigma + i\infty} F(s)e^{st}\, ds \tag{2}$$

where σ is chosen within the convergence region of $F(s)$. For rational Laplace transforms, of course, the partial fraction expansion method is used, and the calculations are worry-free as far as convergence issues are concerned.

Given a function of n variables $f(t_1,\ldots,t_n)$ that is one-sided in each variable, the Laplace transform is defined in a fashion analogous to (1):

$$\begin{aligned} F(s_1,\ldots,s_n) &= L[f(t_1,\ldots,t_n)] \\ &= \int_0^\infty f(t_1,\ldots,t_n)e^{-s_1 t_1} \cdots e^{-s_n t_n}\, dt_1 \cdots dt_n \end{aligned} \tag{3}$$

Of course, this definition also is subject to convergence considerations. However, for reasons similar to those given above, I will proceed in an informal way.

The (perhaps justifiably) nervous reader should investigate the situation further. In particular, it is easy to show that if $f(t_1,\ldots,t_n)$ is a linear combination of terms of the form

$$t_1^{m_1} e^{\lambda_1 t_1} t_2^{m_2} e^{\lambda_2 t_2} \cdots t_n^{m_n} e^{\lambda_n t_n}, \quad t_1,\ldots,t_n \geq 0$$

that is, an exponential form on the first orthant, then the integral in (2) can be written as a sum of products of integrals of the form in (1). This indicates that convergence regions always exist. Carrying out the integrations shows that Laplace transforms are obtained that are rational functions in more than one variable (ratios of multivariable polynomials). Similar investigations for exponential forms on a triangular domain, or for symmetric exponential forms, lead to similar conclusions, though the convergence regions have more complicated geometry in general. (See Problem 2.3.)

Example 2.1 To compute the Laplace transform of

$$f(t_1,t_2) = t_1 - t_1 e^{-t_2}, \quad t_1,t_2 \geq 0$$

the definition in (3) gives

$$
\begin{aligned}
F(s_1,s_2) &= \int_0^\infty \int_0^\infty (t_1 - t_1 e^{-t_2}) e^{-s_1 t_1} e^{-s_2 t_2}\, dt_1 dt_2 \\
&= \int_0^\infty \int_0^\infty t_1 e^{-s_1 t_1} e^{-s_2 t_2}\, dt_1 dt_2 - \int_0^\infty \int_0^\infty t_1 e^{-t_2} e^{-s_1 t_1} e^{-s_2 t_2}\, dt_1 dt_2 \\
&= \frac{1}{s_1^2} \int_0^\infty e^{-s_2 t_2}\, dt_2 - \frac{1}{s_1^2} \int_0^\infty e^{-t_2} e^{-s_2 t_2}\, dt_2 \\
&= \frac{1}{s_1^2 s_2 (s_2+1)}
\end{aligned}
$$

The properties of the multivariable Laplace transform that will be used in the sequel are rather simple to state, and straightforward to prove. In fact, the proofs of the properties are left to the reader with the hint that they are quite similar to the corresponding proofs in the single-variable case. This is not meant to imply that the calculations involved in using the multivariable transform are as simple as in the single-variable transform. To the contrary, it is easy to think of examples where the frequency-domain convolution in Theorem 2.6 is at best exhausting to carry out, and at worst frightening even to contemplate.

In the following list of theorems, and throughout the sequel, one-sidedness is assumed, and the capital letter notation is used for transforms.

Theorem 2.1 The Laplace transform operation is linear:

$$
\begin{aligned}
L[f(t_1,\ldots,t_n) + g(t_1,\ldots,t_n)] &= F(s_1,\ldots,s_n) + G(s_1,\ldots,s_n) \\
L[\alpha f(t_1,\ldots,t_n)] &= \alpha F(s_1,\ldots,s_n), \quad \text{for } scalar\ \alpha
\end{aligned}
\tag{4}
$$

Theorem 2.2 If $f(t_1,\ldots,t_n)$ can be written as a product of two factors,

$$f(t_1,\ldots,t_n) = h(t_1,\ldots,t_k)g(t_{k+1},\ldots,t_n) \tag{5}$$

then

$$F(s_1,\ldots,s_n) = H(s_1,\ldots,s_k)G(s_{k+1},\ldots,s_n) \tag{6}$$

Theorem 2.3 If $f(t_1,\ldots,t_n)$ can be written as a convolution of the form

$$f(t_1,\ldots,t_n) = \int_0^\infty h(\sigma)g(t_1-\sigma,\ldots,t_n-\sigma)\, d\sigma \tag{7}$$

then

$$F(s_1,\ldots,s_n) = H(s_1+\cdots+s_n)G(s_1,\ldots,s_n) \tag{8}$$

Theorem 2.4 If $f(t_1,\ldots,t_n)$ can be written as an n-fold convolution of the form

$$f(t_1,\ldots,t_n) = \int_0^\infty h(t_1-\sigma_1,\ldots,t_n-\sigma_n)g(\sigma_1,\ldots,\sigma_n)\, d\sigma_1 \cdots d\sigma_n \tag{9}$$

then

$$F(s_1,\ldots,s_n) = H(s_1,\ldots,s_n)G(s_1,\ldots,s_n) \tag{10}$$

Theorem 2.5 If $T_1,\ldots,T_n$ are nonnegative constants, then

$$L[f(t_1-T_1,\ldots,t_n-T_n)] = F(s_1,\ldots,s_n)e^{-s_1T_1-\cdots-s_nT_n} \tag{11}$$

Theorem 2.6 If $f(t_1,\ldots,t_n)$ is given by the product

$$f(t_1,\ldots,t_n) = h(t_1,\ldots,t_n)g(t_1,\ldots,t_n) \tag{12}$$

then

$$F(s_1,\ldots,s_n) = \frac{1}{(2\pi i)^n}\int_{\sigma-i\infty}^{\sigma+i\infty} H(s_1-w_1,\ldots,s_n-w_n)G(w_1,\ldots,w_n)\, dw_1 \cdots dw_n \tag{13}$$

Example 2.2 For the function

$$f(t_1,t_2) = e^{-t_1-2t_2} - e^{-t_1-3t_2},\quad t_1,\ t_2 \geq 0$$

the definition of the Laplace transform in (3) can be applied, or I can write

$$f(t_1,t_2) = e^{-t_1}(e^{-2t_2} - e^{-3t_2})$$

and apply Theorem 2.2. Choosing the latter approach, results from the single-variable case imply that

$$F(s_1,s_2) = \frac{1}{s_1+1}\left[\frac{1}{s_2+2} - \frac{1}{s_2+3}\right] = \frac{1}{s_1s_2^2+s_2^2+5s_1s_2+6s_1+5s_2+6}$$

It is natural to call a rational, multivariable Laplace transform *strictly proper* if the degree of the numerator polynomial in s_j is less than the denominator-polynomial degree in s_j for each j. The discussion so far might give the impression that in the multivariable case Laplace transforms that are strictly proper rational functions correspond to exponential forms. Unfortunately, such a degree of similarity to the single-variable case is too much for which to hope. I will give two examples to show this. The first involves the unit step function, and the second involves the unit impulse function. Note that the second example indicates that the treatment of generalized functions in the multivariable Laplace transform follows in a natural way from the single-variable case using the sifting property of the impulse.

Example 2.3 Consider the function

$$f(t_1,t_2) = \delta_{-1}(t_2-2t_1), \quad t_1,\ t_2 \geqslant 0$$

which clearly is discontinuous along $2t_1 = t_2$. The corresponding Laplace transform is

$$F(s_1,s_2) = \int_0^\infty \int_0^\infty \delta_{-1}(t_2-2t_1) e^{-s_1t_1} e^{-s_2t_2}\, dt_1 dt_2$$

$$= \int_0^\infty \int_0^{t_2/2} e^{-s_1t_1} e^{-s_2t_2}\, dt_1 dt_2$$

$$= \frac{1}{s_1} \int_0^\infty e^{-s_2t_2}\, dt_2 - \frac{1}{s_1} \int_0^\infty e^{-[(1/2)s_1+s_2]t_2}\, dt_2$$

$$= \frac{1}{s_2(s_1+2s_2)}$$

Example 2.4 For the impulse function

$$f(t_1,t_2) = \delta_0(t_1-t_2)$$

the Laplace transform can be computed directly from the definition.

$$F(s_1,s_2) = \int_0^\infty \int_0^\infty \delta_0(t_1-t_2) e^{-s_1t_1} e^{-s_2t_2}\, dt_1 dt_2$$

$$= \int_0^\infty [\int_0^\infty \delta_0(t_1-t_2) e^{-s_1t_1} dt_1] e^{-s_2t_2}\, dt_2$$

$$= \int_0^\infty e^{-s_1t_2} e^{-s_2t_2}\, dt_2$$

$$= \frac{1}{s_1+s_2}$$

The basic relationship used to determine thc one-sided function $f(t_1,\ldots,t_n)$ corresponding to a given $F(s_1,\ldots,s_n)$ is a multiple line integration of the form

$$f(t_1,\ldots,t_n) = L^{-1}[F(s_1,\ldots,s_n)]$$

$$= \frac{1}{(2\pi i)^n} \int_{\sigma-i\infty}^{\sigma+i\infty} F(s_1,\ldots,s_n) e^{s_1t_1} \cdots e^{s_nt_n}\, ds_1 \cdots ds_n \tag{14}$$

The value of σ is different for each integral, in general, and must be suitably chosen to avoid convergence difficulties. Under appropriate technical hypotheses, the line integrals can be replaced by Bromwich contour integrals, and the calculus of residues applied. Often this will be done, but without explicit mention of the technicalities.

Again, for $n = 1$ this is precisely the inverse-transform formula mentioned just before the method of partial fraction expansion is discussed in detail. That is to say, for rational, single-variable Laplace transforms, the line integration need never be performed. Unfortunately, such a nice alternative inversion procedure for multivariable Laplace transforms is not available. Naive generalization of partial fraction expansion is doomed since a multivariable polynomial in general cannot be written as a product of simple factors, each in a single variable. Thus, except for certain special cases to be discussed later, the line integrals indicated in (14) must be evaluated. In effect, the inverse transform must be found one variable at a time. But even a simple example should indicate why (14) can be a lot more fun to talk about than to use.

Example 2.5 To compute the inverse Laplace transform of

$$F(s_1,s_2) = \frac{1}{s_1s_2(s_1+s_2)}$$

write

$$f(t_1,t_2) = \frac{1}{(2\pi i)^2}\int_{\sigma-i\infty}^{\sigma+i\infty} \frac{1}{s_1s_2(s_1+s_2)} e^{s_1t_1}e^{s_2t_2}\, ds_1ds_2$$

$$= \frac{1}{2\pi i}\int_{\sigma-i\infty}^{\sigma+i\infty} \frac{1}{s_2} e^{s_2t_2} [\frac{1}{2\pi i}\int_{\sigma-i\infty}^{\sigma+i\infty} \frac{1}{s_1(s_1+s_2)} e^{s_1t_1}\, ds_1]\, ds_2$$

The term in brackets can be regarded as a single-variable inverse Laplace transform in s_1 with s_2 a constant. This gives

$$f(t_1,t_2) = \frac{1}{2\pi i}\int_{\sigma-i\infty}^{\sigma+i\infty} \frac{1}{s_2} e^{s_2t_2}(\frac{1}{s_2} - \frac{1}{s_2} e^{-s_2t_1})\, ds_2$$

$$= \frac{1}{2\pi i}\int_{\sigma-i\infty}^{\sigma+i\infty} \frac{1}{s_2^2} e^{s_2t_2}\, ds_2 - \frac{1}{2\pi i}\int_{\sigma-i\infty}^{\sigma+i\infty} \frac{1}{s_2^2} e^{-s_2t_1}e^{s_2t_2}\, ds_2$$

The first term is the single-variable inverse Laplace transform of $1/(s_2^2)$, namely, t_2, while the second term is similar with a time delay of t_1 units indicated. Thus, inserting step functions to make the one-sided nature of the function explicit,

$$f(t_1,t_2) = t_2\delta_{-1}(t_2) - (t_2-t_1)\delta_{-1}(t_2-t_1)$$

or, being a bit more clever,

$$f(t_1,t_2) = min\,[t_1,t_2],\quad t_1,\ t_2 \geqslant 0$$

In fact I could have been clever at the beginning of this example by noting that Theorem 2.3 can be applied. Taking

$$H(s_1+s_2) = \frac{1}{s_1+s_2}$$

$$G(s_1,s_2) = \frac{1}{s_1s_2}$$

corresponding to

$$h(t) = \delta_{-1}(t)$$
$$g(t_1,t_2) = \delta_{-1}(t_1)\delta_{-1}(t_2)$$

gives

$$f(t_1,t_2) = \int_0^\infty h(\sigma)g(t_1-\sigma,t_2-\sigma)\,d\sigma$$
$$= \int_0^\infty \delta_{-1}(\sigma)\delta_{-1}(t_1-\sigma)\delta_{-1}(t_2-\sigma)\,d\sigma$$
$$= \int_0^\infty \delta_{-1}(t_1-\sigma)\delta_{-1}(t_2-\sigma)\,d\sigma = \int_0^{min[t_1,t_2]} d\sigma$$
$$= min\,[t_1,t_2]$$

The reader might reflect on just how simple this example is. The main feature is that the denominator of $F(s_1,s_2)$ is given in terms of a product of simple factors. Thus, the line integrals are easily evaluated by residue calculations, or partial fraction expansions. Without this factored form, it generally is impossible to perform the sequence of single-variable inverse transforms that leads to the multivariable inverse transform. Apparently, there is no easy way around this dilemma. The good news is that the inverse-transform operation is required only rarely in the sequel. Even so, the factoring problem will arise in other contexts, and in the next section I will discuss it further.

2.2 Laplace-Transform Representation of Homogeneous Systems

For a stationary linear system with kernel $h(t)$, the *transfer function* of the system is the Laplace transform of $h(t)$:

$$H(s) = \int_0^\infty h(t)e^{-st}\,dt \tag{15}$$

Restricting attention to one-sided input signals, and using the convolution property of the Laplace transform, the input/output relation

$$y(t) = \int_{-\infty}^\infty h(\sigma)u(t-\sigma)\,d\sigma = \int_0^t h(\sigma)u(t-\sigma)\,d\sigma \tag{16}$$

can be written in the form

$$Y(s) = H(s)U(s) \tag{17}$$

where $Y(s)$ and $U(s)$ are the transforms of $y(t)$ and $u(t)$.

If a system transfer function is known, and the input signal of interest has a simple transform $U(s)$, then the utility of this representation for computing the corresponding output signal is clear. Another reason for the importance of the transfer function is that many system properties can be expressed rather simply as properties of $H(s)$. Also the transfer function of a "linear" interconnection of linear systems is easily computed from the subsystem transfer functions. In developing a transform representation for homogeneous systems, these features of the degree-1 case give a preview of the goals.

A degree-n homogeneous system with one-sided input signals can be represented by

$$y(t) = \int_{-\infty}^{\infty} h(\sigma_1,\ldots,\sigma_n)u(t-\sigma_1)\cdots u(t-\sigma_n)\,d\sigma_1\cdots d\sigma_n$$

$$= \int_0^t h(\sigma_1,\ldots,\sigma_n)u(t-\sigma_1)\cdots u(t-\sigma_n)\,d\sigma_1\cdots d\sigma_n \tag{18}$$

Inspection of the list of properties of the multivariable Laplace transform yields no direct way to write this in a form similar to (17). Therefore, an indirect approach is adopted by writing (18) as the pair of equations

$$y_n(t_1,\ldots,t_n) = \int_0^{t_1}\cdots\int_0^{t_n} h(\sigma_1,\ldots,\sigma_n)u(t_1-\sigma_1)\cdots u(t_n-\sigma_n)\,d\sigma_1\cdots d\sigma_n$$

$$y(t) = y_n(t_1,\ldots,t_n)\Big|_{t_1=\cdots=t_n=t} = y_n(t,\ldots,t) \tag{19}$$

Now, the first equation in (19) can be written as a relationship between Laplace transforms by using Theorems 2.4 and 2.2:

$$Y_n(s_1,\ldots,s_n) = H(s_1,\ldots,s_n)U(s_1)\cdots U(s_n) \tag{20}$$

I call

$$H(s_1,\ldots,s_n) = L[h(t_1,\ldots,t_n)] \tag{21}$$

a (multivariable) *transfer function* of the homogeneous system.

At this point the utility of the multivariable transfer function for response computation is far from clear. Given $H(s_1,\ldots,s_n)$ and $U(s)$, it is easy to compute $Y_n(s_1,\ldots,s_n)$. However, the inverse Laplace transform must be computed before $y(t)$ can be found from the second equation in (19), and often this is not easy.

Before proceeding to a further investigation of response computation, I will discuss some simple properties of the multivariable transfer function representation with regard to the interconnection of systems. In doing so, the transform-domain system diagram shown in Figure 2.1 will be used.

Perhaps the most obvious feature of the multivariable transfer function representation involves the additive parallel connection of homogeneous systems of the same degree. The overall transfer function is the sum of the subsystem transfer functions

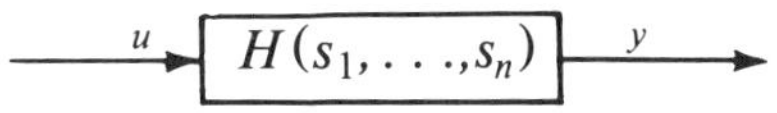

Figure 2.1. A degree-n homogeneous system.

by the linearity property of the Laplace transform. On the other hand, the kernel of a multiplicative parallel connection of homogeneous systems is factorable in the sense of Theorem 2.2. Therefore, the overall transfer function of this connection also can be written by inspection of the subsystem transfer functions.

When the cascade connection of two homogeneous systems is considered, the going quickly gets rougher. Apparently, there is no neat way to write the multivariable Laplace transform of the overall cascade-system kernel derived in Section 1.4. However, there are two special cases that can be handled because they correspond precisely to Theorems 2.3 and 2.4.

Consider the cascade connection of a degree-n homogeneous system followed by a linear system as shown in Figure 2.2. From the analysis in Section 1.4, a kernel for the overall system is

$$f_n(t_1,\ldots,t_n) = \int_{-\infty}^{\infty} g_1(\sigma)h_n(t_1-\sigma,\ldots,t_n-\sigma)\,d\sigma \tag{22}$$

Thus, from Theorem 2.3, a system transfer function is

$$F_n(s_1,\ldots,s_n) = H_n(s_1,\ldots,s_n)G_1(s_1+\cdots+s_n) \tag{23}$$

Shown in Figure 2.3 is a cascade connection of a linear system followed by a degree-n homogeneous system. From Section 1.4, an overall system kernel is given by

$$f_n(t_1,\ldots,t_n) = \int_{-\infty}^{\infty} g_n(\sigma_1,\ldots,\sigma_n)h_1(t_1-\sigma_1)\cdots h_1(t_n-\sigma_n)\,d\sigma_1\cdots d\sigma_n \tag{24}$$

Application of Theorem 2.4 to this expression shows that an overall system transfer function is

$$F_n(s_1,\ldots,s_n) = H_1(s_1)\cdots H_1(s_n)G_n(s_1,\ldots,s_n) \tag{25}$$

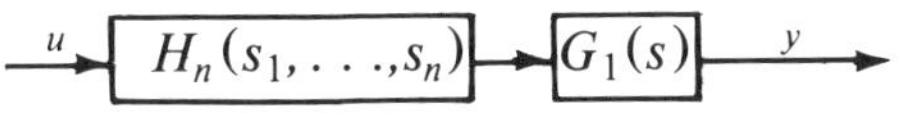

Figure 2.2. A cascade connection.

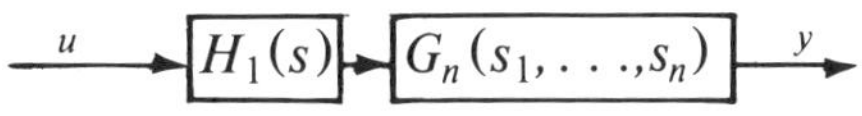

Figure 2.3. A cascade connection.

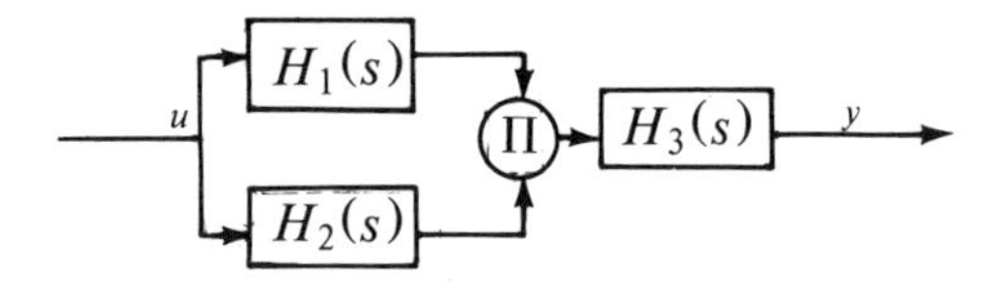

Figure 2.4. The system discussed in Example 2.6.

Example 2.6 Application of these results to the interconnection structured system shown in Figure 2.4 gives the transfer function

$$H(s_1,s_2) = H_1(s_1)H_2(s_2)H_3(s_1+s_2)$$

(In system diagrams such as that in Figure 2.4, I retain the meaning of the Π symbol as a time-domain multiplication, even though the subsystems are represented in the transform domain. Also, the notational collision involved in using subscripts to denote different single-variable transfer functions, rather than the number of variables, will be ignored.)

Example 2.7 Computing an overall transfer function of the system shown in Figure 2.5 is a bit more subtle because it must be remembered that a distinct variable should be reserved for the unity transfer functions as well as the others. The relationship between the intermediate signal and the input signal can be written from Example 2.4 as

$$W(s_1,s_2) = H_3(s_1)H_2(s_1+s_2)U(s_1)U(s_2)$$

Then

$$Y(s_1,s_2,s_3) = H_3(s_1)H_2(s_1+s_2)H_1(s_1+s_2+s_3)U(s_1)U(s_2)U(s_3)$$

and an overall transfer function is

$$H(s_1,s_2,s_3) = H_1(s_1+s_2+s_3)H_2(s_1+s_2)H_3(s_1)$$

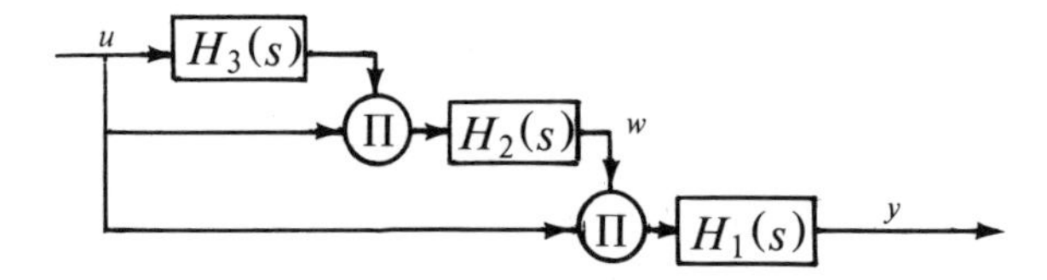

Figure 2.5. The system discussed in Example 2.7.

Consideration of less elementary interconnections will be postponed until another tool for working with transfer functions is developed in Section 2.4. For the moment, I will return briefly to the problem of response computation using the multivariable transfer function. The point to be made is that, the general formulas notwithstanding, sometimes it is quite easy to compute the response of a nonlinear system.

Example 2.8 Suppose that for the system shown in Figure 2.4 the subsystem transfer functions are

$$H_1(s) = \frac{1}{s}, \quad H_2(s) = \frac{1}{s+1}, \quad H_3(s) = 1$$

Then compute the system response to the input signal $u(t) = \delta_{-1}(t)$ as follows. First write

$$Y(s_1,s_2) = H(s_1,s_2)U(s_1)U(s_2) = \frac{1}{s_1^2 s_2(s_2+1)}$$

Taking the inverse Laplace transform—easy in this case—gives

$$y(t_1,t_2) = t_1(1 - e^{-t_2})$$

Thus, the output signal is

$$y(t) = t(1 - e^{-t})$$

Of course, in a simple case such as this it is just as easy to trace the input signal through the system. The response of the first subsystem is $y_1(t) = L^{-1}[1/s^2] = t$, and the response of the second is $y_2(t) = L^{-1}[1/s(s+1)] = 1 - e^{-t}$. Multiplying these together gives $y(t)$.

The transform representation and the rules for describing interconnections and computing the response to given inputs are valid regardless of any special form that the kernel might take. That is, it doesn't matter whether the transform of the symmetric kernel, triangular kernel, or just any everyday kernel is used (leaving aside the regular kernel representation). However, the symmetric transfer function and triangular transfer function, corresponding, respectively, to the symmetric and triangular kernels, will have particular applications in the sequel. Thus, it is useful to derive relationships between these special forms.

It is clear from the definition of the Laplace transform that the symmetric transfer function is a symmetric function in the variables $s_1,\ldots,s_n$. The triangular transfer function, on the other hand, possesses no particular feature that by inspection distinguishes it from any other asymmetric transfer function.

Computing the symmetric transfer function corresponding to a given triangular transfer function is straightforward, in principle. Choosing the obvious notation, the time-domain formula

$$h_{sym}(t_1,\ldots,t_n) = \frac{1}{n!} \sum_{\pi(.)} h_{tri}(t_{\pi(1)},\ldots,t_{\pi(n)}) \tag{26}$$

gives, by linearity of the Laplace transform,

$$H_{sym}(s_1,\ldots,s_n) = \frac{1}{n!} \sum_{\pi(.)} H_{tri}(s_{\pi(1)},\ldots,s_{\pi(n)}) \tag{27}$$

Obtaining $H_{tri}(s_1,\ldots,s_n)$ from a given $H_{sym}(s_1,\ldots,s_n)$ is more difficult because the time-domain formula

$$h_{tri}(t_1,\ldots,t_n) = n!\, h_{sym}(t_1,\ldots,t_n)\delta_{-1}(t_1-t_2)\delta_{-1}(t_2-t_3)\cdots\delta_{-1}(t_{n-1}-t_n) \tag{28}$$

must be expressed in the transform domain. I should pause to emphasize that the triangular domain $t_1 \geqslant t_2 \geqslant \cdots \geqslant t_n \geqslant 0$ is being used here. This particular choice is nonessential, but a different choice of triangular domain will give a different triangular transfer function.

Using Theorem 2.6 and the Laplace transform

$$L[\delta_{-1}(t_1-t_2)\delta_{-1}(t_2-t_3)\cdots\delta_{-1}(t_{n-1}-t_n)] = \frac{1}{s_1(s_1+s_2)(s_1+s_2+s_3)\cdots(s_1+s_2+\cdots+s_n)} \tag{29}$$

gives

$$H_{tri}(s_1,\ldots,s_n) = \frac{n!}{(2\pi i)^n} \int_{\sigma-i\infty}^{\sigma+i\infty} \frac{H_{sym}(s_1-w_1,\ldots,s_n-w_n)}{w_1(w_1+w_2)\cdots(w_1+\cdots+w_n)}\, dw_1\cdots dw_n \tag{30}$$

Example 2.9 For the system shown in Figure 2.6, it is clear from the interconnection rules that

$$H_{sym}(s_1,s_2) = \frac{1}{s_1s_2(s_1+s_2+1)}$$

To compute the triangular transfer function representation, (30) gives

$$H_{tri}(s_1,s_2) = \frac{2}{(2\pi i)^2} \int_{\sigma-i\infty}^{\sigma+i\infty}\int_{\sigma-i\infty}^{\sigma+i\infty} \frac{H_{sym}(s_1-w_1,s_2-w_2)}{w_1(w_1+w_2)}\, dw_2dw_1$$

$$= \frac{2}{(2\pi i)^2} \int_{\sigma-i\infty}^{\sigma+i\infty}\int_{\sigma-i\infty}^{\sigma+i\infty} \frac{1}{(s_1-w_1)(s_2-w_2)(s_1+s_2-w_1-w_2+1)(w_1)(w_1+w_2)}\, dw_2dw_1$$

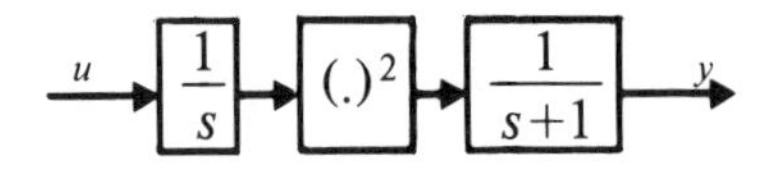

Figure 2.6. The system discussed in Example 2.9.

It is convenient to rearrange this expression in the following way to apply the residue calculus:

$$H_{tri}(s_1,s_2) =$$

$$\frac{2}{2\pi i}\int_{\sigma-i\infty}^{\sigma+i\infty}\frac{1}{(s_1-w_1)w_1}\frac{1}{2\pi i}\int_{\sigma-i\infty}^{\sigma+i\infty}\frac{1}{(s_2-w_2)(s_2-w_2+s_1-w_1+1)(w_2+w_1)}\,dw_2dw_1$$

Now the inner integral can be evaluated by calculating the residue of the integrand corresponding to the pole $w_2 = -w_1$. This residue is

$$\frac{1}{(s_2-w_2)(s_2-w_2+s_1-w_1+1)}\bigg|_{w_2=-w_1} = \frac{1}{(s_2+w_1)(s_1+s_2+1)}$$

so that

$$H_{tri}(s_1,s_2) = \frac{2}{s_1+s_2+1}\frac{1}{2\pi i}\int_{\sigma-i\infty}^{\sigma+i\infty}\frac{1}{(s_1-w_1)w_1(w_1+s_2)}\,dw_1$$

To evaluate the integral using the residue calculus, the sum of the residues of the integrand at the poles $w_1 = 0$ and $w_1 = -s_2$ must be calculated. This gives

$$\frac{1}{(s_1-w_1)(w_1+s_2)}\bigg|_{w_1=0} + \frac{1}{(s_1-w_1)w_1}\bigg|_{w_1=-s_2} = \frac{1}{s_1(s_1+s_2)}$$

and thus

$$H_{tri}(s_1,s_2) = \frac{2}{s_1(s_1+s_2)(s_1+s_2+1)}$$

The alert reader probably has noted that these residue calculations depend upon knowing the denominator polynomial of the integrand in a factored form. This begs again the question of factoring multivariable polynomials, and the question of just what the term "factor" means in this context. These involve nontrivial problems in mathematics, and a general discussion would be at once lengthy, and not very beneficial for the purposes here. However, there are a few special classes of polynomials that can be handled more or less effectively, and these classes suffice at least to provide examples of typical calculations. Thus, I will avoid generalities and concentrate on two of these classes, both of which involve symmetric polynomials.

First, consider the situation where the polynomial $P(s_1,\ldots,s_n)$ is known to have the form

$$P(s_1,\ldots,s_n) = P_1(s_1)\cdots P_1(s_n)P_2(s_1+\cdots+s_n)$$

where $P_1(s)$ and $P_2(s)$ are (single-variable) polynomials. Furthermore, for simplicity, assume that $P_1(s)$ and $P_2(s)$ each have distinct (multiplicity-one) roots. Clearly these are severe restrictions on the form of $P(s_1,\ldots,s_n)$ when compared with all possible n-variable polynomials. Some measure of the system-theoretic importance of this class of polynomials can be obtained by reviewing the examples so far presented. These polynomials arise in the symmetric transfer function of a cascade connection of a linear system followed by an nth-power nonlinearity followed by another linear system.

The nice feature of the special form of $P(s_1,\ldots,s_n)$ is that the factoring problem can be solved by factoring the single-variable polynomial $P(s,\ldots,s) = P_1^n(s)P_2(ns)$. If $s = \alpha$ is a root of $P(s,\ldots,s)$ of multiplicity one, then $(s_1+\cdots+s_n-n\alpha)$ is a factor of $P_2(s_1+\cdots+s_n)$, and thus of $P(s_1,\ldots,s_n)$. If $s = \alpha$ is a root of $P(s,\ldots,s)$ of multiplicity n, then $(s-\alpha)$ is a factor of $P_1(s)$, and thus $(s_1-\alpha)\cdots(s_n-\alpha)$ is a factor of $P(s_1,\ldots,s_n)$. If $s = \alpha$ is a root of $P(s,\ldots,s)$ of multiplicity $n+1$, then $(s_1-\alpha)\cdots(s_n-\alpha)$ and $(s_1+\cdots+s_n-n\alpha)$ are factors of $P(s_1,\ldots,s_n)$. That these are the only possibilities should be clear because of the restricted form of the polynomial. A similar procedure can be developed for the case where $P_1(s)$ and $P_2(s)$ are permitted to have multiple roots, but this is left to Section 2.6.

I will also discuss briefly a factoring approach for somewhat more general symmetric polynomials. However, for the degree-2 case, this form agrees precisely with that considered above: $P(s_1,s_2) = P_1(s_1)P_1(s_2)P_2(s_1+s_2)$. For various numerical values of s_2, compute the roots of $P(s,s_2)$. A root $s = \lambda(s_2)$ that is fixed with respect to different values of s_2 gives a candidate for a factor $(s_1-\lambda)(s_2-\lambda)$ of $P(s_1,s_2)$. This candidate is readily checked by performing long division. A root $s = \lambda(s_2)$ for which $\lambda(s_2) + s_2$ is constant for various values of s_2 gives a candidate factor $(s_1+s_2-\lambda)$ for $P(s_1,s_2)$. Again, long division can be used to check and remove such a factor. In this way the factors are found one at a time.

Now consider 3-variable, symmetric polynomials of the general form

$$P(s_1,s_2,s_3) = P_1(s_1)P_1(s_2)P_1(s_3)P_2(s_1+s_2)P_2(s_1+s_3)P_2(s_2+s_3)P_3(s_1+s_2+s_3)$$

For various numerical values of s_2 $(= s_3)$, compute the roots of the single-variable polynomial

$$P(s,s_2,s_2) = P_1(s)P_1^2(s_2)P_2^2(s+s_2)P_2(2s_2)P_3(s+2s_2)$$

A root $s = \lambda(s_2)$ that is fixed with respect to s_2 gives a candidate factor of the form $(s_1-\lambda)(s_2-\lambda)(s_3-\lambda)$ for $P(s_1,s_2,s_3)$. A root $s = \lambda(s_2)$ for which $\lambda(s_2) + s_2$ is constant gives a candidate factor of the form $(s_1+s_2-\lambda)(s_1+s_3-\lambda)(s_2+s_3-\lambda)$. Finally, a root for which $\lambda(s_2) + 2s_2$ is constant gives a candidate factor $(s_1+s_2+s_3-\lambda)$. In all cases, the factor candidates can be checked and removed by long division.

In the general case, symmetric polynomials of the form

$$P(s_1,\ldots,s_n) =$$

$$\prod_{j=1}^{n} P_1(s_j) \prod_{\pi} P_2(s_{\pi(1)}+s_{\pi(2)}) \prod_{\pi} P_3(s_{\pi(1)}+s_{\pi(2)}+s_{\pi(3)}) \cdots P_n(s_1+\cdots+s_n)$$

where the products are over all permutations π of the integers $1,\ldots,n$, can be attacked using this procedure. However, the complications rapidly increase, and the attack involves a great deal of artful guessing and single-variable factoring.

Of course, there are many other forms for symmetric polynomials. Even in the 2-variable case, factors of the form (s_1s_2+1) or $(s_1-s_2)(s_2-s_1)$ have not been considered. There are a couple of reasons for ignoring these situations: procedures are unknown, and, more importantly, for the system theory of interest here such factors do not arise. Further explanation must be postponed until Chapter 4.

2.3 Response Computation and the Associated Transform

Because of the need to perform the multivariable inverse Laplace transform, the response computation procedure used in Example 2.8 often is unwieldy. There is an alternative method that is based on the idea of computing the Laplace transform of the output signal, $Y(s)$, directly from $Y_n(s_1,\ldots,s_n)$. Then only a single-variable inverse Laplace transform is required to find $y(t)$. The procedure for computing $Y(s)$ from $Y_n(s_1,\ldots,s_n)$ is called *association of variables,* and $Y(s)$ is called the *associated transform.* The notation

$$Y(s) = A_n[Y_n(s_1,\ldots,s_n)] \tag{31}$$

is used to denote the association of variables operation.

I begin with the $n = 2$ case, for then the general case is an easy extension.

Theorem 2.7 For the Laplace transform $Y_2(s_1,s_2)$, the associated transform $Y(s) = A_2[Y_2(s_1,s_2)]$ is given by the line integral

$$Y(s) = \frac{1}{2\pi i}\int_{\sigma-i\infty}^{\sigma+i\infty} Y_2(s-s_2,s_2)\, ds_2 \tag{32}$$

Proof Writing the inverse Laplace transform

$$y_2(t_1,t_2) = \frac{1}{(2\pi i)^2}\int_{\sigma-i\infty}^{\sigma+i\infty} Y_2(s_1,s_2)e^{s_1t_1}e^{s_2t_2}\, ds_1ds_2$$

and setting $t_1 = t_2 = t$ yields

$$y(t) = \frac{1}{2\pi i}\int_{\sigma-i\infty}^{\sigma+i\infty} [\frac{1}{2\pi i}\int_{\sigma-i\infty}^{\sigma+i\infty} Y_2(s_1,s_2)e^{s_1t}\, ds_1]e^{s_2t}\, ds_2$$

Changing the variable of integration s_1 to $s = s_1+s_2$ gives

$$y(t) = \frac{1}{2\pi i}\int_{\sigma-i\infty}^{\sigma+i\infty} [\frac{1}{2\pi i}\int_{\sigma-i\infty}^{\sigma+i\infty} Y_2(s-s_2,s_2)e^{(s-s_2)t}\, ds]e^{s_2t}\, ds_2$$

Finally, interchanging the order of integration to obtain

$$y(t) = \frac{1}{2\pi i}\int_{\sigma-i\infty}^{\sigma+i\infty} [\frac{1}{2\pi i}\int_{\sigma-i\infty}^{\sigma+i\infty} Y_2(s-s_2,s_2)\, ds_2]e^{st}\, ds$$

shows that the bracketed term must be $Y(s) = L[y(t)]$, and the proof is complete.

Note that by reordering the integrations in this proof, another formula for the association operation is obtained. Namely

$$Y(s) = \frac{1}{2\pi i} \int_{\sigma - i\infty}^{\sigma + i\infty} Y_2(s_1, s - s_1)\, ds_1 \tag{33}$$

In the degree-n case, computations similar to those in the proof of Theorem 2.7 lead to a number of different formulas for association of variables. These different formulas arise from different orderings of the integrations when manipulating the inverse-Laplace-transform expression for $y_n(t,\ldots,t)$, (not to mention different variable labelings). Two forms are shown below, with verification left to Problem 2.12:

$$Y(s) = \frac{1}{(2\pi i)^{n-1}} \int_{\sigma - i\infty}^{\sigma + i\infty} Y_n(s - s_2, s_2 - s_3, \ldots, s_{n-1} - s_n, s_n)\, ds_n \cdots ds_2 \tag{34}$$

$$Y(s) = \frac{1}{(2\pi i)^{n-1}} \int_{\sigma - i\infty}^{\sigma + i\infty} Y_n(s - s_1 - s_2 - \cdots - s_{n-1}, s_1, s_2, \ldots, s_{n-1})\, ds_1 \cdots ds_{n-1} \tag{35}$$

But the details of the calculations that establish (34) and (35) also show that the association operation can be regarded as a sequence of pairwise associations. Such a sequence can be written as shown below, where semicolons are used to set off the two variables to be associated.

$$\begin{aligned} Y_{n-1}(s, s_3, \ldots, s_n) &= A_2[Y_n(s_1, s_2; s_3, \ldots, s_n)] \\ Y_{n-2}(s, s_4, \ldots, s_n) &= A_2[Y_{n-1}(s, s_3; s_4, \ldots, s_n)] \\ &\vdots \\ Y(s) &= A_2[Y_2(s, s_n)] \end{aligned} \tag{36}$$

It is readily demonstrated that this particular sequence of computing the pairwise associations corresponds to the formula (35) if (32) is used at each step.

Before going further, I should point out one situation in which the associated transform has a clear system-theoretic interpretation. If $H_n(s_1,\ldots,s_n)$ is the multivariable transfer function of a degree-n homogeneous system, and if $u(t) = \delta_0(t)$, then from (10), $Y_n(s_1,\ldots,s_n) = H_n(s_1,\ldots,s_n)$. Therefore, the Laplace transform of the impulse response of the system is $Y(s) = H(s) = A_n[H_n(s_1,\ldots,s_n)]$.

Of course, some caution should be exercised because, as noted in Chapter 1, the impulse response of a homogeneous system may not exist. This implies that difficulties can arise with the association operation. As an example, consider the system shown in Figure 2.4 with $H_1(s) = H_2(s) = 1$. In this case, the input signal $u(t) = \delta_0(t)$ results in the undefined signal $\delta_0^2(t)$ at the input to $H_3(s)$. The reader should set up the association operation for the transfer function of this system, namely $H_3(s_1+s_2)$, and attempt to evaluate the line integral to see how this difficulty appears.

To perform the association operation in (32), the line integral usually is replaced by a Bromwich contour integration and the residue calculus is applied (assuming the requisite technical hypotheses). However, simpler approaches should not be overlooked. For instance, there are cases where the time-domain route yields the answer easily.

Example 2.10 If

$$F_2(s_1,s_2) = \frac{1}{(s_1-\alpha)^k(s_2-\beta)^j},$$

where α and β are real numbers, and k and j are positive integers, then

$$f_2(t_1,t_2) = \frac{1}{(k-1)!}t_1^{k-1}e^{\alpha t_1}\frac{1}{(j-1)!}t_2^{j-1}e^{\beta t_2}$$

Therefore,

$$f(t) = f_2(t,t) = \frac{1}{(k-1)!(j-1)!}t^{k+j-2}\, e^{(\alpha+\beta)t}$$

and from the single-variable Laplace transform,

$$F(s) = A_2[F(s_1,s_2)] = \frac{1}{(k-1)!(j-1)!}(k+j-2)!\,\frac{1}{(s-\alpha-\beta)^{k+j-1}}$$

$$= \binom{k+j-2}{k-1}\frac{1}{(s-\alpha-\beta)^{k+j-1}}$$

For many commonly occurring types of rational functions, the association operation can be accomplished by the application of a few simple properties to a small group of tabulated associations. The proofs of these involve familiar integral manipulations, so I leave them to the reader. Note that the notation indicated in (31) is used throughout. Two tables of simple associated transforms are given at the end of this section. These are not extensive; just enough transforms are given to facilitate computations in simple examples.

Theorem 2.8 If $F(s_1,\ldots,s_n)$ can be written in the factored form

$$F(s_1,\ldots,s_n) = H(s_1,\ldots,s_k)G(s_{k+1},\ldots,s_n) \tag{37}$$

then

$$F(s) = A_n[F(s_1,\ldots,s_n)] = \frac{1}{2\pi i}\int_{\sigma-i\infty}^{\sigma+i\infty} H(s-s_1)G(s_1)\,ds_1 \tag{38}$$

where $H(s) = A_k[H(s_1,\ldots,s_k)]$ and $G(s) = A_{n-k}[G(s_{k+1},\ldots,s_n)]$.

For a number of special cases, (38) can be worked out explicitly, as the following results indicate.

Corollary 2.1 If $F(s_1,\ldots,s_n)$ can be written in the factored form

$$F(s_1,\ldots,s_n) = \frac{1}{(s_k+\alpha)^{q+1}}\,G(s_1,\ldots,s_{k-1},s_{k+1},\ldots,s_n) \tag{39}$$

where α is a scalar, then

$$F(s) = \frac{(-1)^q}{q!} \frac{d^q}{ds^q} G(s+\alpha) \tag{40}$$

Corollary 2.2 If $F(s_1,\ldots,s_n)$ can be written in the form

$$F(s_1,\ldots,s_n) = \frac{\alpha}{s_k^2+\alpha^2} G(s_1,\ldots,s_{k-1},s_{k+1},\ldots,s_n) \tag{41}$$

where α is a scalar, then

$$F(s) = \frac{1}{2i} [G(s-i\alpha) - G(s+i\alpha)] \tag{42}$$

Corollary 2.3 If $F(s_1,\ldots,s_n)$ can be written in the form

$$F(s_1,\ldots,s_n) = \frac{s_k}{s_k^2+\alpha^2} G(s_1,\ldots,s_{k-1},s_{k+1},\ldots,s_n) \tag{43}$$

then

$$F(s) = \frac{1}{2}[G(s-i\alpha) + G(s+i\alpha)] \tag{44}$$

Theorem 2.9 If $F(s_1,\ldots,s_n)$ can be written in the form

$$F(s_1,\ldots,s_n) = H(s_1+\cdots+s_n)\, G(s_1,\ldots,s_n) \tag{45}$$

then

$$F(s) = H(s)G(s) \tag{46}$$

Example 2.11 The impulse response of the degree-3 system described by

$$H(s_1,s_2,s_3) = \frac{1}{(s_1+1)(s_2^2+3s_2+2)(s_3+2)}$$

can be computed as follows. Using Corollary 2.1 to associate the variables s_1 and s_2 gives

$$A_2[\frac{1}{(s_1+1)(s_2^2+3s_2+2)}] = \frac{1}{(s+1)^2+3(s+1)+2} = \frac{1}{s^2+5s+6}$$

The second step of the procedure again involves Corollary 2.1:

$$A_3[H(s_1,s_2,s_3)] = A_2[\frac{1}{(s^2+5s+6)(s_3+2)}]$$

$$= \frac{1}{(s+2)^2+5(s+2)+6} = \frac{1}{s^2+9s+20}$$

Thus, by partial fraction expansion, the impulse response of the system is

$$y(t) = e^{-4t} - e^{-5t}$$

A considerably different input/output representation can be obtained in the transform domain when the regular kernel representation is the starting point in the time domain. If a system is described by

$$y(t) = \int_0^\infty h_{reg}(\sigma_1,\ldots,\sigma_n)u(t-\sigma_1-\cdots-\sigma_n)u(t-\sigma_2-\cdots-\sigma_n)$$
$$\cdots u(t-\sigma_n)\, d\sigma_1 \cdots d\sigma_n \tag{47}$$

I will call

$$H_{reg}(s_1,\ldots,s_n) = L[h_{reg}(t_1,\ldots,t_n)] \tag{48}$$

the *regular transfer function* for the system. It is a simple matter to relate the regular transfer function to the triangular transfer function since

$$h_{reg}(t_1,\ldots,t_n) = h_{tri}(t_1+\cdots+t_n,t_2+\cdots+t_n,\ldots,t_n)$$

for $t_1,\ldots,t_n \geqslant 0$. A change of variables in the definition of the Laplace transform of $h_{reg}(t_1,\ldots,t_n)$ gives

$$H_{reg}(s_1,\ldots,s_n) = H_{tri}(s_1,s_2-s_1,s_3-s_2,\ldots,s_n-s_{n-1}) \tag{49}$$

And it is clear from this expression that

$$H_{tri}(s_1,\ldots,s_n) = H_{reg}(s_1,s_1+s_2,s_1+s_2+s_3,\ldots,s_1+\cdots+s_n) \tag{50}$$

It is unfortunate, although perhaps no surprise, that the relationships between the regular and symmetric transfer functions are much more difficult. In fact, the connection is made through the triangular transfer function by using (27) and (30) in conjunction with (49) and (50).

To derive interconnection rules for the regular transfer function is a quite tedious process using the theory so far developed. Therefore, I will postpone this topic until Appendix 4.1, at which point a different approach to the problem is available. But the computation of input/output behavior in terms of $H_{reg}(s_1,\ldots,s_n)$ need not be delayed.

The transform-domain input/output representation in terms of the regular transfer function is most easily derived from the association of variables formula involving the triangular transfer function. Namely, from (35),

$$Y(s) = \frac{1}{(2\pi i)^{n-1}} \int_{\sigma-i\infty}^{\sigma+i\infty} H_{tri}(s-s_1-\cdots-s_{n-1},s_1,\ldots,s_{n-1})$$
$$U(s-s_1-\cdots-s_{n-1})U(s_1)\cdots U(s_{n-1})\, ds_{n-1}\cdots ds_1 \tag{51}$$

Using (50) gives

$$Y(s)=\frac{1}{(2\pi i)^{n-1}} \int_{\sigma-i\infty}^{\sigma+i\infty} H_{reg}(s-s_1-\cdots-s_{n-1},s-s_2-\cdots-s_{n-1},\ldots,s-s_{n-1},s)$$
$$U(s-s_1-\cdots-s_{n-1})U(s_1)\cdots U(s_{n-1})\, ds_{n-1}\cdots ds_1 \tag{52}$$

The important fact about this formula is that there is one situation of wide interest wherein the line integrations can be evaluated in a general fashion. The restrictions on the forms of $H_{reg}(s_1,\ldots,s_n)$ and $U(s)$ in the following result are to permit application of the residue calculus. The form of the regular transfer function may seem a bit strange, but in Chapter 4 it will become very familiar.

Theorem 2.10 Suppose that a degree-n homogeneous system is described by a strictly proper, rational regular transfer function of the form

$$H_{reg}(s_1,\ldots,s_n) = \frac{P(s_1,\ldots,s_n)}{Q_1(s_1)\cdots Q_n(s_n)} \tag{53}$$

where each $Q_j(s_j)$ is a single-variable polynomial. If the input signal is described by

$$U(s) = \sum_{i=1}^{r} \frac{a_i}{s+\gamma_i} \tag{54}$$

where $\gamma_1,\ldots,\gamma_r$ are distinct, then the response of the system is given by

$$Y(s) = \sum_{i_1=1}^{r} \cdots \sum_{i_{n-1}=1}^{r} a_{i_1} \cdots a_{i_{n-1}} H_{reg}(s+\gamma_{i_1}+\cdots+\gamma_{i_{n-1}},$$
$$s+\gamma_{i_2}+\cdots+\gamma_{i_{n-1}},\ldots,s+\gamma_{i_{n-1}},s)\,U(s+\gamma_{i_1}+\cdots+\gamma_{i_{n-1}}) \tag{55}$$

Proof This result is proved by evaluating the integrals in (52) one at a time using the residue calculus. The first integration to be performed is

$$\frac{1}{2\pi i}\int_{\sigma-i\infty}^{\sigma+i\infty} H_{reg}(s-s_1-\cdots-s_{n-1},s-s_2-\cdots-s_{n-1},\ldots,s-s_{n-1},s)$$
$$U(s-s_1-\cdots-s_{n-1})\,U(s_{n-1})\,ds_{n-1}$$
$$= \frac{1}{2\pi i}\int_{\sigma-i\infty}^{\sigma+i\infty} \frac{P(s-s_1-\cdots-s_{n-1},\ldots,s-s_{n-1},s)}{Q_1(s-s_1-\cdots-s_{n-1})\cdots Q_{n-1}(s-s_{n-1})Q_n(s)}$$
$$U(s-s_1-\cdots-s_{n-1})\,U(s_{n-1})\,ds_{n-1}$$

A key point here is that $1/Q_n(s)$ can be factored out of the integral, leaving the denominator of the integrand in the form $F(s-s_{n-1})G(s_{n-1})$. Thus the residue calculus can be applied, and the sum of the residues of the integrand at the poles of $U(s_{n-1})$, namely, the poles $-\gamma_1,\ldots,-\gamma_r$, gives the result

$$\sum_{i_{n-1}=1}^{r} a_{i_{n-1}} H_{reg}(s-s_1-\cdots-s_{n-2}+\gamma_{i_{n-1}},s-s_2-\cdots-s_{n-2}+\gamma_{i_{n-1}},$$
$$\ldots,s+\gamma_{i_{n-1}},s)\,U(s-s_1-\cdots-s_{n-2}+\gamma_{i_{n-1}})$$

Now (52) can be written in the form

$$Y(s) =$$

$$\sum_{i_{n-1}=1}^{r} a_{i_{n-1}} \frac{1}{(2\pi i)^{n-2}} \int_{\sigma-i\infty}^{\sigma+i\infty} H_{reg}(s-s_1-\cdots-s_{n-2}+\gamma_{i_{n-1}}, s-s_2-\cdots-s_{n-2}+\gamma_{i_{n-1}},$$

$$\ldots, s+\gamma_{i_{n-1}}, s)\, U(s-s_1-\cdots-s_{n-2}+\gamma_{i_{n-1}})\, U(s_1)\cdots U(s_{n-2})\, ds_{n-2}\cdots ds_1$$

Performing the integrations with respect to s_{n-2} using the residue calculus just as before,

$$Y(s) = \sum_{i_{n-2}=1}^{r} \sum_{i_{n-1}=1}^{r} a_{i_{n-2}} a_{i_{n-1}} \frac{1}{(2\pi i)^{n-3}} \int_{\sigma-i\infty}^{\sigma+i\infty} H_{reg}(s-s_1-\cdots+\gamma_{i_{n-2}}+\gamma_{i_{n-1}},$$

$$\ldots, s+\gamma_{i_{n-1}}, s)\, U(s-s_1-\cdots-s_{n-3}+\gamma_{i_{n-2}}+\gamma_{i_{n-1}})\, U(s_1)\cdots U(s_{n-3})\, ds_{n-3}\cdots ds_1$$

Continuing to work through the integrations will lead directly to the input/output relation in (54).

Example 2.12 The triangular transfer function corresponding to the system in Figure 2.6 is

$$H_{tri}(s_1,s_2) = \frac{2}{s_1(s_1+s_2)(s_1+s_2+1)}$$

Applying (50) gives the regular transfer function for the system:

$$H_{reg}(s_1,s_2) = H_{tri}(s_1,s_2-s_1) = \frac{2}{s_1 s_2(s_2+1)}$$

Now it is a straightforward matter to compute the unit-step response of the system via (55). In this case

$$Y(s) = H_{reg}(s,s)\,U(s) = \frac{2}{s^3(s+1)}$$

and a simple partial fraction expansion yields

$$y(t) = 2 - 2t + t^2 - 2e^{-t}, \quad t \geqslant 0$$

Of course, it is easy to verify this answer by tracing the input signal through the system diagram.

2.4 The Growing Exponential Approach

Another viewpoint towards the transform representation of homogeneous systems is based upon the following property of linear systems. Consider a system described by

$$y(t) = \int_{-\infty}^{\infty} h(\sigma)u(t-\sigma)\,d\sigma \tag{56}$$

with the growing exponential input signal $u(t) = e^{\lambda t}$, $\lambda > 0$, defined for all t. The response is

$$y(t) = \int_{-\infty}^{\infty} h(\sigma)\,e^{\lambda(t-\sigma)}\,d\sigma$$

$$= \int_{0}^{\infty} h(\sigma)e^{-\lambda\sigma}\,d\sigma\; e^{\lambda t} \tag{57}$$

where the lower limit has been raised to 0 in view of the one-sidedness of $h(t)$. Thus, if $H(s)$ is the system transfer function, and if λ is in the region of convergence of the transform,

$$y(t) = H(\lambda)e^{\lambda t},\quad t \in (-\infty,\infty) \tag{58}$$

In particular, if the linear system is stable, then λ will be in the region of convergence since $\lambda > 0$.

The fact that a growing exponential input signal is simply scaled by a linear system to produce the output signal is sometimes called the eigenfunction property of linear systems. In addition, the response of the linear system to a linear combination of growing exponentials

$$u(t) = \sum_{i=1}^{p} \alpha_i\, e^{\lambda_i t},\quad \lambda_1,\ldots,\lambda_p > 0 \tag{59}$$

is given by

$$y(t) = \sum_{i=1}^{p} \alpha_1 H(\lambda_i)\, e^{\lambda_i t} \tag{60}$$

This is sometimes useful since arbitrary two-sided input signals can be approximated by linear combinations of growing exponentials. In the present context the transfer function $H(s)$ of a linear system is regarded as that function that characterizes the response to growing exponentials via (58) or (60).

For a homogeneous system of degree $n > 1$, I will proceed in an analogous fashion. Using the representation

$$y(t) = \int_{-\infty}^{\infty} h_n(\sigma_1,\ldots,\sigma_n)u(t-\sigma_1)\cdots u(t-\sigma_n)\,d\sigma_1\cdots d\sigma_n \tag{61}$$

the reader can see readily that the response to $u(t) = e^{\lambda t}$, $\lambda > 0$, $t \in (-\infty, \infty)$, is

$$y(t) = H_n(\lambda, \ldots, \lambda) e^{n\lambda t} \tag{62}$$

where $H_n(s_1, \ldots, s_n)$ is a system transfer function. But it should come as no surprise that the response to a single growing exponential indicates little about the behavior of the system for other inputs. In other words, it is clear that a complete characterization of the multivariable function $H_n(s_1, \ldots s_n)$ is not obtained from the single-variable function $H_n(\lambda, \ldots, \lambda)$.

Before treating input signals that are arbitrary linear combinations of growing exponentials, consider the case of "double exponential" inputs

$$u(t) = \alpha_1 e^{\lambda_1 t} + \alpha_2 e^{\lambda_2 t}, \quad \lambda_1, \lambda_2 > 0, \quad t \in (-\infty, \infty) \tag{63}$$

These will be of particular importance in the sequel. From (63) and (61), write

$$\begin{aligned} y(t) &= \int_{-\infty}^{\infty} h_n(\sigma_1, \ldots, \sigma_n) \prod_{j=1}^{n} [\alpha_1 e^{\lambda_1 (t-\sigma_j)} + \alpha_2 e^{\lambda_2 (t-\sigma_j)}] \, d\sigma_1 \cdots d\sigma_n \\ &= \int_{-\infty}^{\infty} h_n(\sigma_1, \ldots, \sigma_n) \sum_{k_1=1}^{2} \cdots \sum_{k_n=1}^{2} (\prod_{j=1}^{n} \alpha_{k_j}) \exp[\sum_{j=1}^{n} \lambda_{k_j}(t-\sigma_j)] \, d\sigma_1 \cdots d\sigma_n \\ &= \sum_{k_1=1}^{2} \cdots \sum_{k_n=1}^{2} (\prod_{j=1}^{n} \alpha_{k_j}) \\ &\quad \int_{0}^{\infty} h_n(\sigma_1, \ldots, \sigma_n) \exp(-\sum_{j=1}^{n} \lambda_{k_j} \sigma_j) \, d\sigma_1 \cdots d\sigma_n \exp(\sum_{j=1}^{n} \lambda_{k_j} t) \\ &= \sum_{k_1=1}^{2} \cdots \sum_{k_n=1}^{2} (\prod_{j=1}^{n} \alpha_{k_j}) H_n(\lambda_{k_1}, \ldots, \lambda_{k_n}) \exp(\sum_{j=1}^{n} \lambda_{k_j} t) \end{aligned} \tag{64}$$

Note that many of the terms in this output expression have identical exponents $(\lambda_{k_1} + \cdots + \lambda_{k_n} t)$. I can write all terms with exponent $k\lambda_1 + (n-k)\lambda_2) t$ as the single term

$$\alpha_1^k \alpha_2^{n-k} G_{k,n-k}(\lambda_1, \lambda_2) \exp[(k\lambda_1 + (n-k)\lambda_2)t] \tag{65}$$

where

$$G_{k,n-k}(\lambda_1, \lambda_2) = \underbrace{\sum_{k_1=1}^{2} \cdots \sum_{k_n=1}^{2}}_{k_1 + \cdots + k_n = 2n-k} H_n(\lambda_{k_1}, \ldots, \lambda_{k_n}) \tag{66}$$

Thus, the response of the system to the input (63) can be written in the form

$$y(t) = \sum_{k=0}^{n} \alpha_1^k \alpha_2^{n-k} G_{k,n-k}(\lambda_1, \lambda_2) \, e^{[k\lambda_1 + (n-k)\lambda_2]t} \tag{67}$$

There are two observations to be made in the context of the double-exponential case. First, note that if $H_n(s_1, \ldots, s_n)$ is the symmetric transfer function of the system, then (66) can be written as

$$G_{k,n-k}(\lambda_1, \lambda_2) = \binom{n}{k} H_{nsym}(\underbrace{\lambda_1, \ldots, \lambda_1}_{k}; \underbrace{\lambda_2, \ldots, \lambda_2}_{n-k}) \tag{68}$$

Second, if the system is of degree $n = 2$, then (68) implies that

$$H_{2sym}(\lambda_1, \lambda_2) = \frac{1}{2}\, G_{1,1}(\lambda_1, \lambda_2) \tag{69}$$

That is, for a double-exponential input to a degree-2 system, the symmetric transfer function is determined by the coefficient of the $e^{(\lambda_1 + \lambda_2)t}$ term in the output.

Example 2.12 To determine the multivariable transfer function of the system shown in Figure 2.4, the double exponential input method can be applied since the system is of degree 2. Denoting the input to $H_3(s)$ by $v(t)$, and choosing the coefficients $\alpha_1 = \alpha_2 = 1$ in (63),

$$\begin{aligned} v(t) &= [H_1(\lambda_1)e^{\lambda_1 t} + H_1(\lambda_2)e^{\lambda_2 t}][H_2(\lambda_1)e^{\lambda_1 t} + H_2(\lambda_2)e^{\lambda_2 t}] \\ &= H_1(\lambda_1)H_2(\lambda_1)e^{2\lambda_1 t} + [H_1(\lambda_1)H_2(\lambda_2) + H_1(\lambda_2)H_2(\lambda_1)]e^{(\lambda_1+\lambda_2)t} \\ &\quad + H_1(\lambda_2)H_2(\lambda_2)e^{2\lambda_2 t} \end{aligned}$$

Thus,

$$\begin{aligned} y(t) &= H_1(\lambda_1)H_2(\lambda_1)H_3(2\lambda_1)e^{2\lambda_1 t} \\ &\quad + [H_1(\lambda_1)H_2(\lambda_2) + H_1(\lambda_2)H_2(\lambda_1)]H_3(\lambda_1+\lambda_2)e^{(\lambda_1+\lambda_2)t} \\ &\quad + H_1(\lambda_2)H_2(\lambda_2)H_3(2\lambda_2)e^{2\lambda_2 t} \end{aligned}$$

and the symmetric transfer function of the system is

$$H_{sym}(s_1,s_2) = \frac{1}{2}\, [H_1(s_1)H_2(s_2) + H_1(s_2)H_2(s_1)]H_3(s_1+s_2)$$

It is clear that this is the symmetric version of the transfer function derived in Example 2.6.

Now consider the response of a degree-n homogeneous system to an arbitrary linear combination of growing exponentials such as in (59). My purpose is again twofold: to present a method for determining the symmetric transfer function of such a system, and to develop a representation for the system response to such an input signal.

Substituting (59) into (61) gives

$$y(t) = \int_{-\infty}^{\infty} h_n(\sigma_1,\ldots,\sigma_n) \prod_{j=1}^{n} [\alpha_1 e^{\lambda_1(t-\sigma_j)} + \cdots + \alpha_p e^{\lambda_p(t-\sigma_j)}]\, d\sigma_1 \cdots d\sigma_n$$

$$= \int_{-\infty}^{\infty} h_n(\sigma_1,\ldots,\sigma_n) \sum_{k_1=1}^{p} \cdots \sum_{k_n=1}^{p} [\prod_{j=1}^{n} \alpha_{k_j}] \exp[\sum_{j=1}^{n} \lambda_{k_j}(t-\sigma_j)]\, d\sigma_1 \cdots d\sigma_n$$

$$= \sum_{k_1=1}^{p} \cdots \sum_{k_n=1}^{p} [\prod_{j=1}^{n} \alpha_{k_j}]$$

$$\int_{-\infty}^{\infty} h_n(\sigma_1,\ldots,\sigma_n) \exp(-\sum_{j=1}^{n} \lambda_{k_j}\sigma_j)\, d\sigma_1 \cdots d\sigma_n \exp(\sum_{j=1}^{n} \lambda_{k_j} t)$$

$$= \sum_{k_1=1}^{p} \cdots \sum_{k_n=1}^{p} [\pi_{j=1}^{n} \alpha_{k_j}] H_n(\lambda_{k_1},\ldots,\lambda_{k_n}) \exp(\sum_{j=1}^{n} \lambda_{k_j} t) \tag{70}$$

where, as before, certain assumptions have been made concerning the regions of convergence. Of course, many of the terms in this expression contain identical exponents. By collecting all those terms with like exponents, a simplified expression can be obtained.

In particular, consider all those terms in (70) with the exponent $(m_1\lambda_1+m_2\lambda_2 + \cdots + m_p\lambda_p)t$, where each m_i is an integer satisfying $0 \leqslant m_i \leqslant n$ with $\sum_{i=1}^{p} m_i = n$. I can write this collection of terms as

$$\alpha_1^{m_1} \cdots \alpha_p^{m_p} G_{m_1,\ldots,m_p}(\lambda_1,\ldots,\lambda_p) \exp[(m_1\lambda_1 + \cdots + m_p\lambda_p)t] \tag{71}$$

where, using an implicit summation in which m_j of the n indices take the value j, $j = 1,\ldots,p$,

$$G_{m_1,\ldots,m_p}(\lambda_1,\ldots,\lambda_p) = \underbrace{\sum_{k_1=1}^{p} \cdots \sum_{k_n=1}^{p}}_{m_j \text{ indices } = j} H_n(\lambda_{k_1},\ldots,\lambda_{k_n}) \tag{72}$$

Thus, the output signal is expressed in the form

$$y(t) = \sum_{m} \alpha_1^{m_1} \cdots \alpha_p^{m_p} G_{m_1,\ldots,m_p}(\lambda_1,\ldots,\lambda_p) \exp[(m_1\lambda_1 + \cdots + m_p\lambda_p)t] \tag{73}$$

where $\sum_{m}$ indicates a p-fold sum over all integer indices $m_1,\ldots, m_p$ such that $0 \leqslant m_i \leqslant n$, and $m_1 + \cdots + m_p = n$.

If the transfer function $H_n(s_1,\ldots,s_n)$ is symmetric, then the implicit sum in (72) can be made somewhat more explicit. Since there will be $\binom{n}{m_1}$ identical terms with m_1 of the indices 1, then $\binom{n-m_1}{m_2}$ identical terms with m_2 of the indices 2, and so forth;

and since

$$\binom{n}{m_1}\binom{n-m_1}{m_2}\binom{n-m_1-m_2}{m_3}\cdots\binom{n-m_1-\cdots-m_{p-2}}{m_{p-1}} = \frac{n!}{m_1!m_2!\cdots m_p!} \tag{74}$$

(72) can be written in the form

$$G_{m_1,\ldots,m_p}(\lambda_1,\ldots,\lambda_p) = \frac{n!}{m_1!m_2!\cdots m_p!} H_{nsym}(\underbrace{\lambda_1,\ldots,\lambda_1}_{m_1};\cdots;\underbrace{\lambda_p,\ldots,\lambda_p}_{m_p}) \tag{75}$$

Aside from being a neater way to write the exponential coefficients, when $p = n$ the relationship

$$G_{1,\ldots,1}(\lambda_1,\ldots,\lambda_n) = n!H_{nsym}(\lambda_1,\ldots,\lambda_n) \tag{76}$$

is obtained. To rephrase: when a linear combination of n growing exponentials is applied to a degree-n homogeneous system, the symmetric transfer function of the system is given by $1/n!$ times the coefficient of $\exp[(\lambda_1+\cdots+\lambda_n)t]$. The utility of this approach will be demonstrated by the following not quite trivial example. I invite the reader to compute the transfer function by using the interconnection rules, just to compare the two approaches.

Example 2.13 The system shown in Figure 2.7 is assumed to have stable linear subsystems (for convergence purposes) and integer-power nonlinearities, with $m_1m_2 = m$. To find the transfer function, consider the input

$$u(t) = \sum_{i=1}^{m} e^{\lambda_i t},\quad \lambda_1,\ldots,\lambda_m > 0$$

Tracing this signal through the system diagram gives

$$v(t) = [\sum_{i=1}^{m} H_0(\lambda_i)e^{\lambda_i t}]^{m_1}$$

$$= \sum_{i_1=1}^{m}\cdots\sum_{i_{m_1}=1}^{m} [\prod_{j=1}^{m_1} H_0(\lambda_{i_j})]\exp(\sum_{k=1}^{m_1}\lambda_{i_k}t)$$

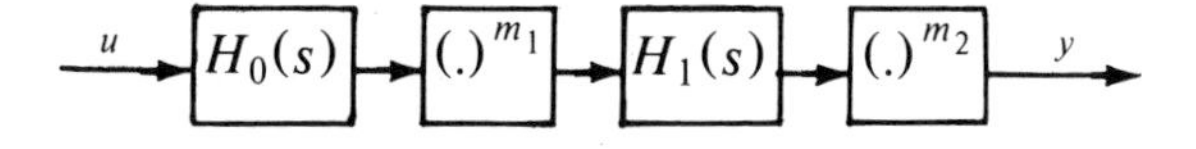

Figure 2.7. A cascade connection of linear systems and integer-power nonlinearities.

Then

$$y(t) = [\sum_{i_1=1}^{m} \cdots \sum_{i_{m_1}=1}^{m} [\prod_{j=1}^{m_1} H_0(\lambda_{i_j})] H_1(\sum_{k=1}^{m_1} \lambda_{i_k}) \exp(\sum_{k=1}^{m_1} \lambda_{i_k} t)\ ]^{m_2}$$

$$= [\sum_{i_1=1}^{m} \cdots \sum_{i_{m_1}=1}^{m} [\prod_{j=1}^{m_1} H_0(\lambda_{i_j})] H_1(\sum_{k=1}^{m_1} \lambda_{i_k}) \exp(\sum_{k=1}^{m_1} \lambda_{i_k} t)\ ]$$

$$[\sum_{i_{m_1+1}=1}^{m} \cdots \sum_{i_{2m_1}=1}^{m} [\prod_{j=m_1+1}^{2m_1} H_0(\lambda_{i_j})] H_1(\sum_{k=m_1+1}^{2m_1} \lambda_{i_k}) \exp(\sum_{k=m_1+1}^{2m_1} \lambda_{i_k} t)\ ] \cdots$$

$$= \sum_{i_1=1}^{m} \cdots \sum_{i_m=1}^{m} [\prod_{j=1}^{m} H_0(\lambda_{i_j})][\prod_{j=1}^{m_2} H_1(\sum_{k=1}^{m_1} \lambda_{i_{k+(j-1)m_1}})] \exp(\sum_{k=1}^{m} \lambda_{i_k} t)$$

where in the middle equality only the first two factors have been written for simplicity. The important point is that distinct variables must be maintained in each factor.

Now collect all those terms that correspond to the exponent $(\lambda_1 + \cdots + \lambda_m)t$ in order to obtain the symmetric transfer function. There are $m!$ such terms, m choices of which index is 1, $m-1$ choices of which index is 2, and so on. However, each of these $m!$ terms is itself an asymmetric transfer function; the various choices of indices correspond to the various permutations of the variables. Taking, for instance, $i_1=1,\ i_2=2, \ldots, i_m=m$, one system transfer function is obtained:

$$H_m(s_1, \ldots, s_m) = \prod_{j=1}^{m} H_0(s_j) \prod_{j=1}^{m_2} H_1(\sum_{k=1}^{m_1} s_{k+(j-1)m_1})$$

The properties of growing exponentials just discussed provide another viewpoint from which the transfer function can be considered. Recall that in the linear case the transfer function $H(s)$ can be viewed as that function that characterizes the response to a linear combination of growing exponentials (59), as shown in (60). For a degree-n homogeneous system, a transfer function $H_n(s_1, \ldots, s_n)$ can be viewed as a function that characterizes the response to (59) via the expression

$$y(t) = \sum_{k_1=1}^{p} \cdots \sum_{k_m=1}^{p} (\prod_{j=1}^{n} \alpha_{k_j}) H_n(\lambda_{k_1}, \ldots, \lambda_{k_n}) \exp(\sum_{j=1}^{n} \lambda_{k_j} t) \tag{77}$$

I emphasize that this characterization of the transfer function is appropriate for both symmetric and asymmetric forms of the transfer function, although the special features of (77) in the case of symmetric transfer functions will prove most useful.

2.5 Polynomial and Volterra Systems

The transform representation of stationary polynomial or Volterra systems basically involves the collection of homogeneous subsystem transfer functions. Thus, for example, response calculations are performed by summing the responses of the subsystems as calculated individually by association of variables or from Theorem 2.10. For Volterra systems this summation requires consideration of the convergence properties of an infinite series of time functions. Often, convergence is crucially dependent on properties of the input, for example, bounds on the amplitude of the input signal.

The growing exponential approach to transfer-function determination can be a very convenient tool for stationary polynomial and Volterra systems. To determine the first N symmetric transfer functions, assume that the input signal is a sum of N distinct growing exponentials. Then if the output is a sum of growing exponentials, either by calculation or by assumption, the coefficient of $e^{(\lambda_1+\cdots+\lambda_n)t}$ is $n!H_{nsym}(\lambda_1,\ldots,\lambda_n)$, $n = 1,2,\ldots,N$. For interconnections of homogeneous systems, this approach obviates the need to explicitly unravel the various homogeneous subsystems from the structure.

Example 2.14 A quick trace of the input αu through the system in Figure 2.8 shows that it is polynomial of degree 2. Application of the input signal

$$u(t) = e^{\lambda_1 t} + e^{\lambda_2 t}, \quad \lambda_1,\lambda_2 > 0$$

yields the output signal

$$\begin{aligned} y(t) &= H_1(\lambda_1)H_2(\lambda_1)e^{\lambda_1 t} + H_1(\lambda_2)H_2(\lambda_2)e^{\lambda_2 t} + H_1(\lambda_1)H_3(\lambda_1)e^{2\lambda_1 t} \\ &\quad + [H_1(\lambda_1)H_3(\lambda_2)+H_1(\lambda_2)H_3(\lambda_1)]e^{(\lambda_1+\lambda_2)t} + H_1(\lambda_2)H_3(\lambda_2)e^{2\lambda_2 t} \\ &\quad + H_4(\lambda_1)e^{\lambda_1 t} + H_4(\lambda_2)e^{\lambda_2 t} \\ &= [H_1(\lambda_1)H_2(\lambda_1)+H_4(\lambda_1)]e^{\lambda_1 t} + [H_1(\lambda_1)H_3(\lambda_2)+H_1(\lambda_2)H_3(\lambda_1)]e^{(\lambda_1+\lambda_2)t} + \cdots \end{aligned}$$

Thus, the transfer function of the degree-1 subsystem is

$$H(s_1) = H_1(s_1)H_2(s_1) + H_4(s_1)$$

and the symmetric transfer function of the degree-2 subsystem is

$$H_{2sym}(s_1,s_2) = \frac{1}{2}[H_1(s_1)H_3(s_2) + H_1(s_2)H_3(s_1)]$$

Of course, in this simple case the transfer-function interconnection rules produce the same result more efficiently.

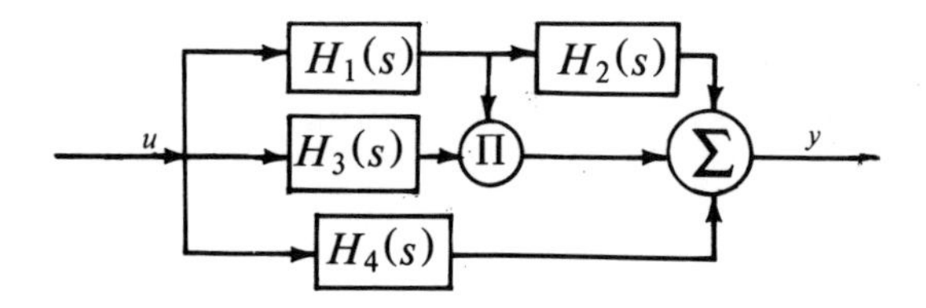

Figure 2.8. A polynomial system of degree 2.

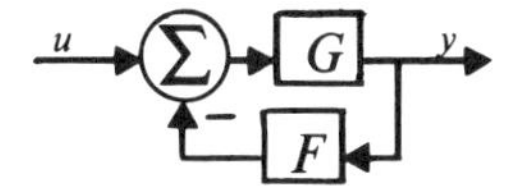

Figure 2.9. A feedback system in operator notation.

When a linear or nonlinear feedback loop is closed around a homogeneous, polynomial, or Volterra system, the result is a Volterra system, in general. This situation was discussed in Chapter 1 only in terms of a general operator notation, principally because time-domain analysis of the feedback connection is prohibitively complex. Now that transform-domain tools are available, I will illustrate the feedback computations for the first three terms for the general feedback system shown in Figure 2.9. Considerable simplification is achieved if the system is redrawn as in Figure 2.10. This configuration shows that the problem of computing the closed-loop representation can be viewed as two cascade-connection problems, and a feedback connection of the relatively simple form shown in Figure 2.11. The cascade connections are left to the reader (Problem 2.16), while the feedback connection in Figure 2.11 now will be discussed in detail.

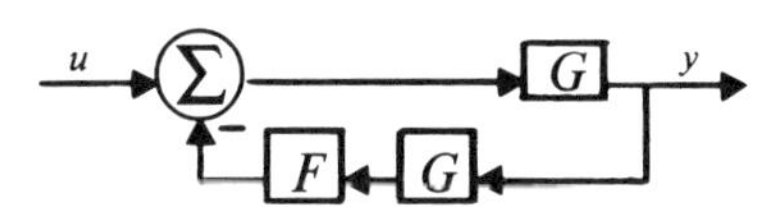

Figure 2.10. The system in Figure 2.9, redrawn.

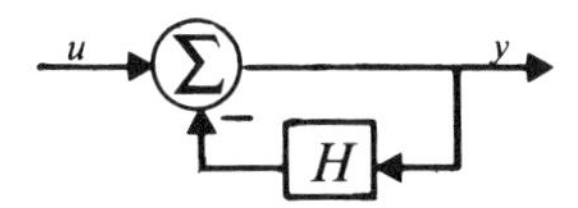

Figure 2.11. The basic feedback system.

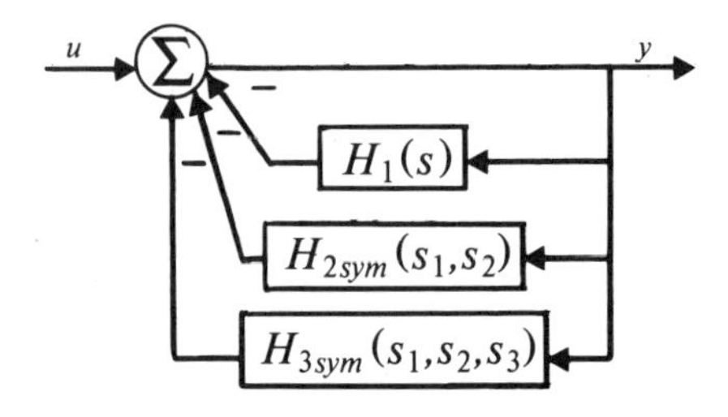

Figure 2.12. Symmetric transfer function representations for the basic feedback system.

To compute the first three closed-loop symmetric transfer functions, the feedback system in Figure 2.11 is redrawn using transform-domain notation as shown in Figure 2.12. It is assumed that the feedback transfer functions are symmetric, and only the first three are shown, as the higher-degree transfer functions will not enter the calculation.

The procedure for determining the first three symmetric transfer functions is to assume an input of the form

$$u(t) = e^{\lambda_1 t} + e^{\lambda_2 t} + e^{\lambda_3 t}$$

and a response in terms of the unknown closed-loop symmetric transfer functions of the form

$$\begin{aligned} y(t) = {} & F_1(\lambda_1)e^{\lambda_1 t} + F_1(\lambda_2)e^{\lambda_2 t} + F_1(\lambda_3)e^{\lambda_3 t} + 2!F_{2sym}(\lambda_1,\lambda_2)e^{(\lambda_1+\lambda_2)t} \\ & + 2!F_{2sym}(\lambda_1,\lambda_3)e^{(\lambda_1+\lambda_3)t} + 2!F_{2sym}(\lambda_2,\lambda_3)e^{(\lambda_2+\lambda_3)t} \\ & + 3!F_{3sym}(\lambda_1,\lambda_2,\lambda_3)e^{(\lambda_1+\lambda_2+\lambda_3)t} + \cdots \end{aligned}$$

The economy of notation obtained by hiding many terms behind the ellipsis in this expression is required to avoid writer's cramp. At the same time extreme care is required, for the right answer can be avoided if the economical notation causes contributing terms to be neglected. Notice also that here is where the assumption that the system can be described by a Volterra series expression enters the calculation.

Now the strategy is to trace the signals through the system diagram in Figure 2.11 to obtain another expression for the output signal. In particular, tracing the assumed $y(t)$ through the homogeneous systems in the feedback path involves the previously derived formulas for the response of homogeneous systems to sums of growing exponentials. Of course, the majority of terms in these formulas are discarded because they do not contribute to the end result. Finally, equating coefficients of like exponentials in the two output expressions yields a set of equations for the closed-loop symmetric transfer functions. Proceeding in this manner gives

$$y(t) = e^{\lambda_1 t} + e^{\lambda_2 t} + e^{\lambda_3 t} - H_1(\lambda_1)F_1(\lambda_1)e^{\lambda_1 t} - H_1(\lambda_2)F_1(\lambda_2)e^{\lambda_2 t}$$
$$- H_1(\lambda_3)F_1(\lambda_3)e^{\lambda_3 t} - 2H_1(\lambda_1+\lambda_2)F_{2sym}(\lambda_1,\lambda_2)e^{(\lambda_1+\lambda_2)t} - \cdots$$
$$- 6H_1(\lambda_1+\lambda_2+\lambda_3)F_{3sym}(\lambda_1,\lambda_2,\lambda_3)e^{(\lambda_1+\lambda_2+\lambda_3)t} - \cdots$$
$$- 2H_{2sym}(\lambda_1,\lambda_2)F_1(\lambda_1)F_1(\lambda_2)e^{(\lambda_1+\lambda_2)t} - \cdots$$
$$- 4[F_1(\lambda_1)F_{2sym}(\lambda_2,\lambda_3)H_{2sym}(\lambda_1,\lambda_2+\lambda_3) + F_1(\lambda_2)F_{2sym}(\lambda_1,\lambda_3)$$
$$H_{2sym}(\lambda_2,\lambda_1+\lambda_3) + F_1(\lambda_3)F_2(\lambda_1,\lambda_2)H_{2sym}(\lambda_3,\lambda_1+\lambda_2)]e^{(\lambda_1+\lambda_2+\lambda_3)t}$$
$$- 6F_1(\lambda_1)F_1(\lambda_2)F_1(\lambda_3)H_{3sym}(\lambda_1,\lambda_2,\lambda_3)e^{(\lambda_1+\lambda_2+\lambda_3)t} + \cdots$$

Equating the coefficients of $e^{\lambda_1 t}$ in the two expressions for $y(t)$ gives

$$F_1(\lambda_1) = 1 - H_1(\lambda_1)F_1(\lambda_1)$$

Solving yields the degree-1 closed-loop transfer function

$$F_1(s) = \frac{1}{1 + H_1(s)}$$

Equating coefficients of $e^{(\lambda_1+\lambda_2)t}$ gives the equation

$$2F_{2sym}(\lambda_1,\lambda_2) = -2H_1(\lambda_1+\lambda_2)F_{2sym}(\lambda_1,\lambda_2) - 2H_{2sym}(\lambda_1,\lambda_2)F_1(\lambda_1)F_1(\lambda_2)$$

the solution of which gives

$$F_{2sym}(s_1,s_2) = \frac{-H_{2sym}(s_1,s_2)}{[1+H_1(s_1+s_2)][1+H_1(s_1)][1+H_1(s_2)]}$$

Equating coefficients of $e^{(\lambda_1+\lambda_2+\lambda_3)t}$ gives

$$6F_{3sym}(\lambda_1,\lambda_2,\lambda_3) = -6H_1(\lambda_1+\lambda_2+\lambda_3)F_{3sym}(\lambda_1,\lambda_2,\lambda_3)$$
$$- 6H_{3sym}(\lambda_1,\lambda_2,\lambda_3)F_1(\lambda_1)F_1(\lambda_2)F_1(\lambda_3) - 4[F_1(\lambda_1)F_{2sym}(\lambda_2,\lambda_3)$$
$$H_{2sym}(\lambda_1,\lambda_2+\lambda_3) + F_1(\lambda_2)F_{2sym}(\lambda_1,\lambda_3)H_{2sym}(\lambda_2,\lambda_1+\lambda_3)$$
$$+ F_1(\lambda_3)F_{2sym}(\lambda_1,\lambda_2)H_{2sym}(\lambda_3,\lambda_1+\lambda_2)]$$

Thus,

$$F_{3sym}(s_1,s_2,s_3) = \frac{1}{[1+H_1(s_1+s_2+s_3)][1+H_1(s_1)][1+H_1(s_2)][1+H_1(s_3)]}$$
$$[-H_{3sym}(s_1,s_2,s_3) + \frac{(2/3)H_{2sym}(s_1,s_2+s_3)H_{2sym}(s_2,s_3)}{1+H_1(s_2+s_3)}$$
$$+ \frac{(2/3)H_{2sym}(s_2,s_1+s_3)H_{2sym}(s_1,s_3)}{1+H_1(s_1+s_3)}$$
$$+ \frac{(2/3)H_{2sym}(s_3,s_1+s_2)H_{2sym}(s_1,s_2)}{1+H_1(s_1+s_2)}]$$

2.6 Remarks and References

Remark 2.1 An introduction to the multivariable Laplace transform as a mathematical tool can be found in

V. Ditkin, A. Prudnikov, *Operational Calculus in Two Variables and Its Applications,* Pergamon Press, New York, 1962.

This is a translation of the original volume in Russian published by Fizmatgiz, in Moscow, in 1958. A very readable treatment of convergence issues, numerous examples, properties, and extensive tables of the 2-variable transform are provided. These go well beyond the introduction in Section 2.1. The multivariable Fourier transform is closely related to the multivariable Laplace transform, and expositions of the Fourier transform that include the multivariable case are much more numerous. For example, see

D. Champeney, *Fourier Transforms and Their Physical Applications,* Academic Press, New York, 1973.

(The multivariable Fourier transform will be used extensively in chapters 5 and 7.)

Remark 2.2 The use of the multivariable Laplace transform in system theory, the association operation, and interconnection rules are discussed in

D. George, "Continuous Nonlinear Systems," MIT RLE Technical Report No. 355, 1959 (AD246-281).

More readily available expositions include

Y. Ku, A. Wolff, "Volterra-Wiener Functionals for the Analysis of Nonlinear Systems," *Journal of The Franklin Institute,* Vol. 281, pp. 9-26, 1966.

R. Parente, "Nonlinear Differential Equations and Analytic System Theory," *SIAM Journal on Applied Mathematics,* Vol. 18, pp. 41-66, 1970.

L. Chua, C. Ng, "Frequency-Domain Analysis of Nonlinear Systems: General Theory, Formulation of Transfer Functions," *IEE Journal on Electronic Circuits and Systems,* Vol. 3, pp. 165-185, 257-269, 1979.

Remark 2.3 Simple properties of the association operation and a table of associated transforms are given in

C. Chen, R. Chiu, "New Theorems of Association of Variables in Multidimensional Laplace Transforms," *International Journal of Systems Science,* Vol. 4, pp. 647-664, 1973.

General formulas for performing the association operation in a wide class of Laplace transforms (with factored denominators) are derived in

L. Crum, J. Heinen, "Simultaneous Reduction and Expansion of Multidimensional Laplace Transform Kernels," *SIAM Journal on Applied Mathematics,* Vol. 26, pp. 753-771, 1974.

A basic review of the residue calculations that I have used several times to evaluate complex convolution integrals can be found in

R. Schwartz, B. Friedland, *Linear System Theory,* McGraw-Hill, New York, 1965.

The proof of Theorem 2.10 touches upon some relatively subtle issues in the residue method. In particular, the need for the factored form for the denominator of the regular transfer function can be appreciated more fully by analyzing the residue calculation in detail for the example

$$H_{reg}(s_1,s_2) = \frac{1}{s_1 s_2 + 1}$$

Remark 2.4 Many authors have discussed the properties of the response of a homogeneous system to sums of exponentials. Usually the exponents are considered to be purely imaginary so there are close ties to the frequency response of the system—a topic to be considered in Chapter 5. However, the growing exponential viewpoint adopted in Section 2.4 seems to lead more economically to a characterization of the symmetric transfer function. A more general version of Example 2.13 is considered in

W. Smith, W. Rugh, "On the Structure of a Class of Nonlinear Systems," *IEEE Transactions on Automatic Control,* Vol. AC-19, pp. 701-706, 1974.

In particular, it is shown that the response to two growing exponentials suffices to characterize homogeneous systems that are cascades of power nonlinearities and linear systems.

Remark 2.5 The regular transfer function representation, its relationship to the triangular and symmetric transfer functions, and the input/output formula in Theorem 2.10 were introduced in

G. Mitzel, S. Clancy, W. Rugh, "On Transfer Function Representations for Homogeneous Nonlinear Systems," *IEEE Transactions on Automatic Control,* Vol. AC-24, pp. 242-249, 1979.

Remark 2.6 Further discussion of the transform-domain analysis of the feedback connection for nonlinear systems can be found in the report by George cited in Remark 2.1. See also

M. Brilliant, "Theory of the Analysis of Nonlinear Systems," MIT RLE Technical Report No. 345, 1958 (AD216-209).

and

J. Barrett, "The Use of Functionals in the Analysis of Nonlinear Physical Systems," *Journal of Electronics and Control,* Vol. 15, pp. 567-615, 1963.

Recurrence relations for the closed-loop symmetric transfer function of certain feedback systems are derived in

E. Bedrosian, S. Rice, "The Output Properties of Volterra Systems (Nonlinear Systems with Memory) Driven by Harmonic and Gaussian Inputs," *Proceedings of the IEEE,* Vol. 59, pp. 1688-1707, 1971.

Remark 2.7 The suggested factoring procedure for symmetric polynomials of the form $P_1(s_1) \cdots P_1(s_n)P_2(s_1+ \cdots +s_n)$ is discussed in

K. Shanmugam, M. Lal, "Analysis and Synthesis of a Class of Nonlinear Systems," *IEEE Transactions on Circuits and Systems,* Vol. CAS-23, pp. 17-25, 1976.

The factoring procedure I have outlined for more general symmetric polynomials was suggested by E. G. Gilbert.

2.7 Problems

2.1. Compute the Laplace transforms of

$$f(t_1,t_2) = \delta_0(t_1)\delta_0(t_2)$$

$$f(t_1,t_2) = \delta_0(t_1)\delta_0(t_1-t_2)$$

2.2. Compute the Laplace transforms of the symmetric functions

$$f(t_1,t_2) = \frac{1}{2}t_1 + \frac{1}{2}t_2 - \frac{1}{2}t_1e^{-t_2} - \frac{1}{2}t_2e^{-t_1}, \quad t_1,t_2 \geqslant 0$$

$$g(t_1,t_2) = \begin{cases} t_1 - t_1e^{-t_2}, & t_1 \geqslant t_2 \geqslant 0 \\ t_2 - t_2e^{-t_1}, & t_2 > t_1 > 0 \end{cases}$$

2.3. Find convergence regions for the Laplace transforms of the one-sided functions

$$f(t_1,t_2) = e^{min[t_1,t_2]}, \quad f(t_1,t_2) = \begin{cases} e^{t_2}, & t_1 \geqslant t_2 \geqslant 0 \\ 0, & otherwise \end{cases}$$

2.4. Find the inverse Laplace transform of

$$F(s_1,s_2) = \frac{1}{(s_1+1)(s_2+1)(s_1+s_2+2)}$$

using the line integration formula. Then check your answer using the cascade formula and cleverness.

2.5. Prove Theorem 2.3.

2.6. Prove Theorem 2.4.

2.7. Prove Theorem 2.5.

2.8. Show that if

$$L[f(t_1,\ldots,t_n)] = F(s_1,\ldots,s_n)$$

then

$$L[e^{-a_1t_1-\cdots-a_nt_n}f(t_1,\ldots,t_n)] = F(s_1+a_1,\ldots,s_n+a_n)$$

2.9. State and prove a final-value theorem for multivariable Laplace transforms.

2.10. Suppose $L[f(t)] = F(s)$. Find a formula for the 2-variable Laplace transforms $L\{f(min[t_1,t_2])\}$ and $L\{f(max[t_1,t_2])\}$.

2.11. If $L[f(t)] = F(s)$, find the 2-variable transform $L[f(t_1+t_2)]$.

2.12. Show that the association formulas (34) and (35) are equivalent.

2.13. Compute a transfer function for the cascade connection shown below.

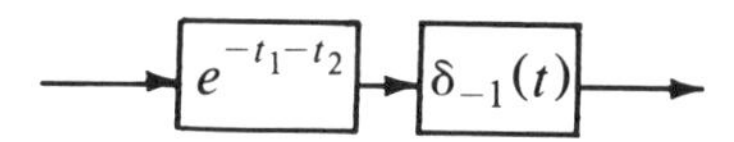

2.14. Verify the symmetric transfer function given in Example 2.9 by computing the Laplace transform of the symmetric kernel given in Problem 1.7.

2.15. Using the kernels found in Example 1.4, compute the symmetric transfer functions for the integral-square and square-integral computers. Then calculate these transfer functions using the interconnection rules in Section 2.2.

2.16. Consider the system $y(t) = h(u(t))$, where $h(u)$ is an analytic function given by

$$h(u) = \sum_{j=0}^{\infty} h_j u^j$$

Show that the degree-n symmetric transfer function for the system is given by

$$H_{nsym}(s_1,\ldots,s_n) = h_n, \quad n = 1,2,\ldots$$

2.17. Find transfer functions through degree 3 for the polynomial system shown below.

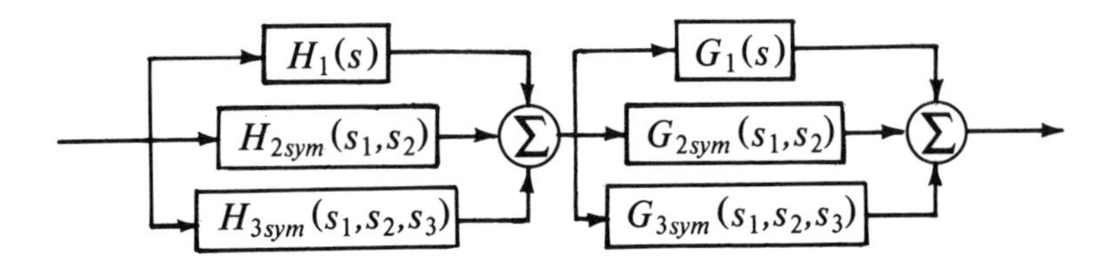

2.18. Compute the symmetric transfer function for the system shown below.

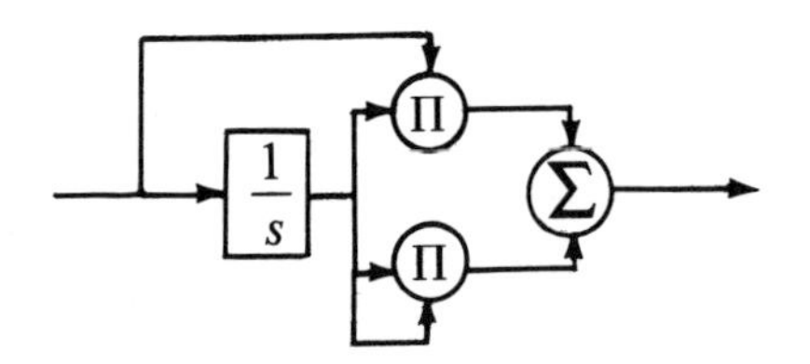

2.19. Show that the two systems below have identical responses to all single growing exponentials. Is this so for double growing exponential inputs?

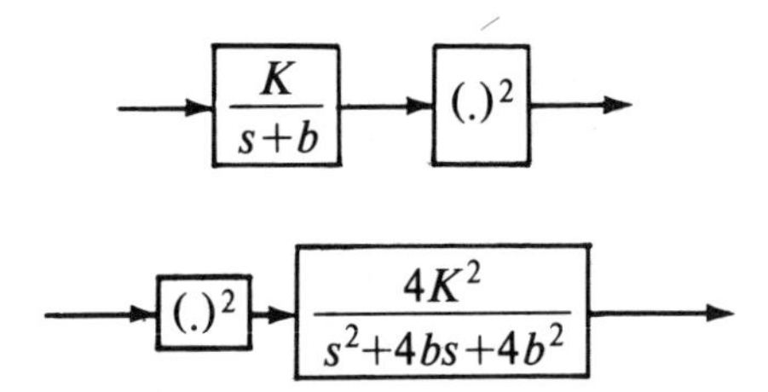

2.20. If

$$F(s_1,s_2) = \frac{1}{(s_1+s_2)^2+2(s_1+s_2)+3}$$

compute $F(s) = A_2[F(s_1,s_2)]$. (Be careful!) Can you compute the association for $F(s_1,s_2)/s_1s_2$?

2.21. Consider the system described by

$$y(t) = u(t) + \epsilon[\dot{u}(t)]^2\ddot{u}(t)$$

where the dot notation is used for differentiation. Show that the first three symmetric transfer functions for this system are

$$H_1(s) = 1, \quad H_2(s_1,s_2) = 0, \quad H_3(s_1,s_2,s_3) = 2\epsilon s_1s_2s_3(s_1+s_2+s_3)$$

2.22. Show that

$$L[\delta_{-1}(t_1-t_2)\delta_{-1}(t_2-t_3)\cdots\delta_{-1}(t_{n-1}-t_n)] = \frac{1}{s_1(s_1+s_2)\cdots(s_1+s_2+\cdots+s_n)}$$

2.23. Given a polynomial $P(s_1,\ldots,s_n)$, devise a simple test to determine if it can be written as a product of single-variable polynomials

$$P(s_1,\ldots,s_n) = P_1(s_1)\cdots P_n(s_n)$$

How would you determine the $P_j(s_j)$'s?

2.24. For the feedback system shown below, show that the first three closed-loop symmetric transfer functions are

$$F_1(s) = G(s)$$

$$F_{2sym}(s_1,s_2) = -h_2G(s_1)G(s_2)G(s_1+s_2)$$

$$F_{3sym}(s_1,s_2,s_3) = [\frac{2h_2^2}{3}[G(s_1+s_2) + G(s_1+s_3) + G(s_2+s_3)]$$

$$- h_3]G(s_1)G(s_2)G(s_3)G(s_1+s_2+s_3)$$

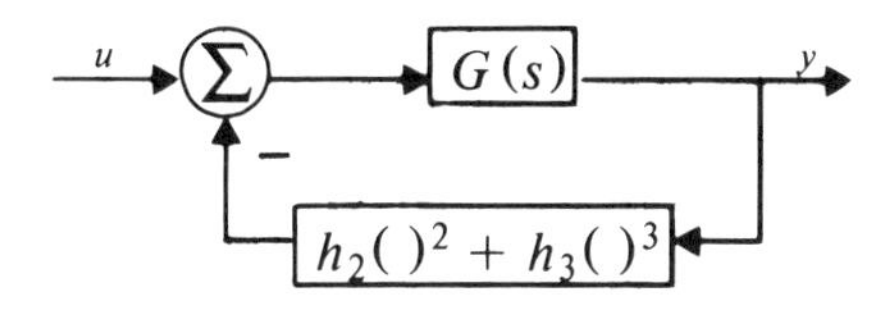

2.25. For the feedback system shown below, find the first three closed-loop transfer functions.

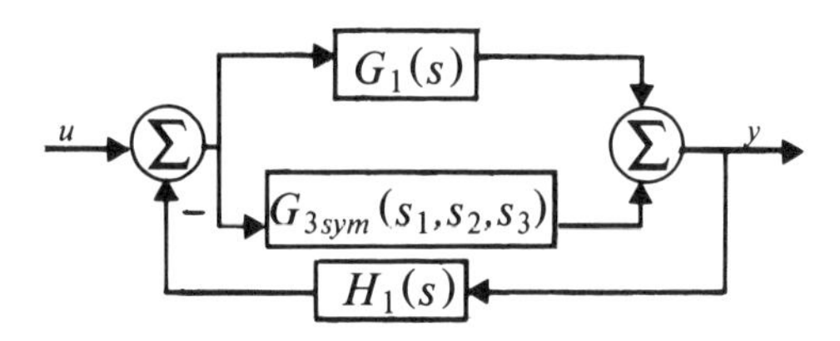

Table 2.1
Simple 2-Variable Associated Transforms

$$A_2[\frac{K}{(s_1+a)(s_1+b)} = \frac{K}{s+a+b}$$

$$A_2[\frac{K}{(s_1+a)(s_1+b)(s_2+a)(s_2+b)}] = \frac{2K}{(s+2a)(s+2b)(s+a+b)}$$

$$A_2\frac{[K}{(s_1+a)(s_1+b)(s_2+a)(s_2+b)[c(s_1+s_2)^2+d(s_1+s_2)+e]}] = \frac{2K}{(cs^2+ds+e)(s+a+b)(s+2a)(s+2b)}$$

$$A_2[\frac{K(s_1+c)(s_2+c)}{(s_1+a)(s_1+b)(s_2+a)(s_2+b)}] = \frac{K}{(b-a)^2}[\frac{(c-a)^2}{s+2a} - \frac{2(c-a)(c-b)}{s+a+b} + \frac{(c-b)^2}{s+2b}]$$

$$A_2[\frac{K(s_1+c)(s_2+c)}{(s_1+a)(s_1+b)(s_2+a)(s_2+b)[d(s_1+s_2)^2+e(s_1+s_2)+f]}] = \frac{K/(b-a)^2}{ds^2+es+f}[\frac{(c-a)^2}{s+2a} - \frac{2(c-a)(c-b)}{s+a+b} + \frac{(c-b)^2}{s+2b}]$$

Table 2.2
Simple 3-Variable Associated Transforms

$$A_3[\frac{K}{(s_1+a)(s_2+b)(s_3+c)}] = \frac{K}{s+a+b+c}$$

$$A_3[\frac{K}{(s_1+a)(s_1+b)(s_2+a)(s_2+b)(s_3+a)(s_3+b)}] = \frac{K}{(a-b)^3}[\frac{1}{s+3b} - \frac{3}{s+2a+b} + \frac{3}{s+a+2b} - \frac{1}{s+3a}]$$

$$A_3[\frac{K(s_1+c)(s_2+c)(s_3+c)}{(s_1+a)(s_1+b)(s_2+a)(s_2+b)(s_3+a)(s_3+b)}] = \frac{K}{(a-b)^3}[\frac{(a-c)^3}{s+3a} - \frac{3(a-c)^2(b-c)}{s+2a+b} + \frac{3(a-c)(b-c)^2}{s+a+2b} - \frac{(c-b)^3}{s+3b}]$$

$$A_3[\frac{K}{(s_1+a)^2(s_2+a)^2(s_3+a)^2}] = \frac{6K}{(s+a)^4}$$

CHAPTER 3

Obtaining Input/Output Representations From Differential Equation Descriptions

Systems often are described in terms of a vector, first-order differential equation called the *state equation*. When the input/output behavior of a system described in this way is of interest, a representation for the solution of the state equation is needed. In this chapter, several procedures for determining the kernels or transfer functions in a Volterra/Wiener representation corresponding to a given state equation will be discussed. In general, an infinite Volterra series is required, and this raises again the issue of convergence. Although general convergence results will be mentioned, most of the discussion will be phrased in terms of finding degree-N polynomial-system truncations of the full Volterra system. (A proof of one general convergence result is given in Appendix 3.1.)

A major difficulty in dealing with nonlinear differential equations is that existence and/or uniqueness of solutions, even in a local sense, cannot be taken for granted. The nasty things that sometimes occur can be demonstrated with very simple, innocent-appearing examples, and I presume the reader is well aware of the situation. To avoid all this, it will be assumed that the differential equations under study all have unique solutions on the time interval of interest, regardless of the particular initial state or (nominally assumed to be piecewise-continuous) input signal. This means that well known conditions on the growth and smoothness properties of the nonlinear functions in a given differential equation should be checked before methods based on the Volterra/Wiener representation are used. In fact, they should be checked before any methods are used.

Much of the development in the following pages is in terms of differential equations with time-variable parameters, that is, the nonstationary case. The reader uninterested in this case can specialize the development readily. Indeed, not much more is required than to drop arguments in the right places and replace $\Phi(t,\tau)$ by $e^{A(t-\tau)}$.

3.1 Introduction

To ease into the subject, I begin with a review of one technique for determining an input/output representation corresponding to the linear state equation

$$\dot{x}(t) = A(t)x(t) + b(t)u(t), \quad t \geq 0$$
$$y(t) = c(t)x(t), \quad x(0) = x_0 \tag{1}$$

In this expression $x(t)$ is the n-dimensional state vector, $u(t)$ is the scalar input, and $y(t)$ is the scalar output. Typical assumptions would be that on some finite time interval $[0,T]$, $A(t)$, $b(t)$, and $c(t)$ are continuous, and the input signal is bounded and piecewise continuous. Such assumptions are sufficient to guarantee the existence of a unique solution of (1) for all $t \epsilon [0,T]$. This standard result usually is derived from successive approximations, though that will not be demonstrated here. My principal interest is to get the form of the solution to (1) in a suggestive way with respect to an approach to nonlinear state equations.

First consider the solution of (1) with $u(t) = 0$ for all $t \geq 0$. In that case, both sides of the differential equation can be integrated to obtain

$$x(t) = x_0 + \int_0^t A(\sigma_1)x(\sigma_1)\, d\sigma_1 \tag{2}$$

Based upon this expression, repeated substitutions can be performed. More specifically, write

$$x(\sigma_1) = x_0 + \int_0^{\sigma_1} A(\sigma_2)x(\sigma_2)\, d\sigma_2 \tag{3}$$

and substitute into (2) to obtain

$$x(t) = x_0 + \int_0^t A(\sigma_1)\, d\sigma_1 x_0 + \int_0^t A(\sigma_1)\int_0^{\sigma_1} A(\sigma_2)x(\sigma_2)\, d\sigma_2 d\sigma_1 \tag{4}$$

Continuing by substituting for $x(\sigma_2)$ in (4) using an expression of the form (3), gives

$$x(t) = [I + \int_0^t A(\sigma_1)\, d\sigma_1 + \int_0^t A(\sigma_1)\int_0^{\sigma_1} A(\sigma_2)\, d\sigma_2 d\sigma_1]x_0$$
$$+ \int_0^t A(\sigma_1)\int_0^{\sigma_1} A(\sigma_2)\int_0^{\sigma_2} A(\sigma_3)x(\sigma_3)\, d\sigma_3 d\sigma_2 d\sigma_1$$

Repeating this process indefinitely, and showing that the last term approaches 0 (in norm) in a uniform way, gives a solution in the form

$$x(t) = \Phi(t,0)\, x_0 \tag{5}$$

where the *transition matrix* $\Phi(t,\tau)$ is defined on any finite square $[0,T]\ X\ [0,T]$ by the uniformly convergent series

$$\Phi(t,\tau) = I + \int_\tau^t A(\sigma_1)\, d\sigma_1 + \int_\tau^t A(\sigma_1) \int_\tau^{\sigma_1} A(\sigma_2)\, d\sigma_2 d\sigma_1$$

$$+ \cdots + \int_\tau^t A(\sigma_1) \int_\tau^{\sigma_1} A(\sigma_2) \cdots \int_\tau^{\sigma_{k-1}} A(\sigma_k)\, d\sigma_k \cdots d\sigma_1 + \cdots \tag{6}$$

known as the *Peano-Baker* series.

An important property of the transition matrix that will be used in the sequel without benefit of derivation is the multiplication formula

$$\Phi(t,\sigma)\Phi(\sigma,\tau) = \Phi(t,\tau) \tag{7}$$

This formula in conjunction with the fact that $\Phi(t,\tau)$ is invertible at each t and τ gives $\Phi^{-1}(t,\tau) = \Phi(\tau,t)$. Finally, when $A(t)$ is actually a constant matrix A, it is not difficult to show that $\Phi(t,\tau)$ is precisely the matrix exponential $e^{A(t-\tau)}$.

The solution of (1) for zero input can be used to obtain a representation for the solution of (1) with an arbitrary input signal $u(t)$. Since $\Phi(t,\tau)$ is invertible for all t and τ, change variables to $z(t) = \Phi^{-1}(t,0)x(t)$ and rewrite (1) as

$$\begin{aligned} \dot{z}(t) &= \hat{b}(t)u(t), \quad t \geqslant 0 \\ y(t) &= \hat{c}(t)z(t), \quad z(0) = x_0 \end{aligned} \tag{8}$$

where

$$\begin{aligned} \hat{b}(t) &= \Phi^{-1}(t,0)b(t) \\ \hat{c}(t) &= c(t)\Phi(t,0) \end{aligned} \tag{9}$$

In the state equation (8), there is no term of the form $A(t)z(t)$, which was the objective of the variable change. Integrating both sides of the differential equation in (8) gives

$$z(t) = x_0 + \int_0^t \hat{b}(\sigma)u(\sigma)\, d\sigma \tag{10}$$

which becomes, in terms of the original variables,

$$x(t) = \Phi(t,0)x_0 + \int_0^t \Phi(t,\sigma)b(\sigma)u(\sigma)\, d\sigma \tag{11}$$

Thus,

$$y(t) = c(t)\Phi(t,0)x_0 + \int_0^t c(t)\Phi(t,\sigma)b(\sigma)u(\sigma)\, d\sigma \tag{12}$$

For the case where $x_0 = 0$, the degree-1 homogeneous input/output representation

$$y(t) = \int_0^t h(t,\sigma)u(\sigma)\, d\sigma \tag{13}$$

with kernel

$$h(t,\sigma) = c(t)\Phi(t,\sigma)b(\sigma) \tag{14}$$

has been obtained. Furthermore, if $A(t)$, $b(t)$, and $c(t)$ actually are constant matrices, then $\Phi(t,\sigma) = e^{A(t-\sigma)}$, and (13) becomes a convolution integral with

$$h(t,\sigma) = h(t-\sigma) = ce^{A(t-\sigma)}b \tag{15}$$

I will commence consideration of the nonlinear case by taking this same resubstitution approach to bilinear state equations. This starting point is appropriate in part because the class of bilinear state equations was the first wide class of nonlinear equations for which a general form for the kernels was obtained—and in part because the general form is such a splendid example of mathematical pulchritude. Moreover, it will become clear in later sections that the treatment of the bilinear case is a precursor to more general developments.

A bilinear state equation is a vector differential equation of the form

$$\begin{aligned} \dot{x}(t) &= A(t)x(t) + D(t)x(t)u(t) + b(t)u(t) \\ y(t) &= c(t)x(t),\ t \geqslant 0,\ x(0) = x_0 \end{aligned} \tag{16}$$

where, as before, $x(t)$ is $n \times 1$, while $u(t)$ and $y(t)$ are scalars. Typical assumptions for (16) are the same as in the linear case. In Problem 3.9, the reader is invited to mimic a standard successive approximation proof to show that these assumptions guarantee existence of a unique solution on any finite time interval.

Using the variable change $z(t) = \Phi^{-1}(t,0)x(t)$, where $\Phi(t,\tau)$ is the transition matrix corresponding to $A(t)$, yields a simplified form of (16):

$$\begin{aligned} \dot{z}(t) &= \hat{D}(t)z(t)u(t) + \hat{b}(t)u(t) \\ y(t) &= \hat{c}(t)z(t),\quad t \geqslant 0,\quad z(0) = z_0 \end{aligned} \tag{17}$$

where

$$\begin{aligned} \hat{b}(t) &= \Phi^{-1}(t,0)b(t) \\ \hat{D}(t) &= \Phi^{-1}(t,0)D(t)\Phi(t,0) \\ \hat{c}(t) &= c(t)\Phi(t,0) \end{aligned} \tag{18}$$

Just as in the linear case, the technique to find the form of the input/output representation is to integrate both sides of the differential equation in (17), and then resubstitute for $z(t)$. The first step of the procedure gives

$$z(t) = z_0 + \int_0^t \hat{D}(\sigma_1)z(\sigma_1)u(\sigma_1)\, d\sigma_1 + \int_0^t \hat{b}(\sigma_1)u(\sigma_1)\, d\sigma_1 \tag{19}$$

Substituting for $z(\sigma_1)$ using an expression of this same form,

$$z(t) = z_0 + \int_0^t \hat{D}(\sigma_1)z_0u(\sigma_1)\, d\sigma_1$$

$$+\int_0^t \int_0^{\sigma_1} \hat{D}(\sigma_1)\hat{D}(\sigma_2)z(\sigma_2)u(\sigma_1)u(\sigma_2)\, d\sigma_2 d\sigma_1$$

$$+\int_0^t \int_0^{\sigma_1} \hat{D}(\sigma_1)\hat{b}(\sigma_2)u(\sigma_1)u(\sigma_2)\, d\sigma_2 d\sigma_1 + \int_0^t \hat{b}(\sigma_1)u(\sigma_1)\, d\sigma_1 \tag{20}$$

Substituting for $z(\sigma_2)$ in (20) using an expression of the form (19), and continuing in this manner yields, after $N-1$ steps,

$$z(t) = z_0 + \sum_{k=1}^{N} \int_0^t \int_0^{\sigma_1} \cdots \int_0^{\sigma_{k-1}} \hat{D}(\sigma_1) \cdots \hat{D}(\sigma_k)z_0 u(\sigma_1) \cdots u(\sigma_k)\, d\sigma_k \cdots d\sigma_1$$

$$+ \sum_{k=1}^{N} \int_0^t \int_0^{\sigma_1} \cdots \int_0^{\sigma_{k-1}} \hat{D}(\sigma_1) \cdots \hat{D}(\sigma_{k-1})\hat{b}(\sigma_k)u(\sigma_1) \cdots u(\sigma_k)\, d\sigma_k \cdots d\sigma_1$$

$$+ \int_0^t \int_0^{\sigma_1} \cdots \int_0^{\sigma_{N-1}} \hat{D}(\sigma_1) \cdots \hat{D}(\sigma_N)z(\sigma_N)u(\sigma_1) \cdots u(\sigma_N)\, d\sigma_N \cdots d\sigma_1 \tag{21}$$

Actually the notation in (21) is rather poor for the $k = 1$ terms in the summations. A clearer expression would be

$$z(t) = z_0 + \int_0^t \hat{D}(\sigma_1)z_0 u(\sigma_1)\, d\sigma_1$$

$$+ \sum_{k=2}^{N} \int_0^t \int_0^{\sigma_1} \cdots \int_0^{\sigma_{k-1}} \hat{D}(\sigma_1) \cdots \hat{D}(\sigma_k)z_0 u(\sigma_1) \cdots u(\sigma_k)\, d\sigma_k \cdots d\sigma_1$$

$$+ \int_0^t b(\sigma_1)u(\sigma_1)\, d\sigma_1$$

$$+ \sum_{k=2}^{N} \int_0^t \int_0^{\sigma_1} \cdots \int_0^{\sigma_{k-1}} \hat{D}(\sigma_1) \cdots \hat{D}(\sigma_{k-1})\hat{b}(\sigma_k)u(\sigma_1) \cdots u(\sigma_k)\, d\sigma_k \cdots d\sigma_1$$

$$+ \int_0^t \int_0^{\sigma_1} \cdots \int_0^{\sigma_{N-1}} \hat{D}(\sigma_1) \cdots \hat{D}(\sigma_N)z(\sigma_N)u(\sigma_1) \cdots u(\sigma_N)\, d\sigma_N \cdots d\sigma_1$$

However, for reasons of economy I will continue to use the collapsed version in (21).

Equation (21) is in many ways analogous to (5) in the linear case, and it can be shown that the last term in (21) approaches 0 in a uniform way on any finite time interval. Therefore on any finite time interval the solution of the bilinear state equation can be represented by the uniformly convergent (vector) Volterra series:

$$z(t) = z_0 + \sum_{k=1}^{\infty} \int_0^t \int_0^{\sigma_1} \cdots \int_0^{\sigma_{k-1}} \hat{D}(\sigma_1) \cdots \hat{D}(\sigma_k)z_0 u(\sigma_1) \cdots u(\sigma_k)\, d\sigma_k \cdots d\sigma_1$$

$$+\sum_{k=1}^{\infty}\int_0^t\int_0^{\sigma_1}\cdots\int_0^{\sigma_{k-1}}\hat{D}(\sigma_1)\cdots\hat{D}(\sigma_{k-1})\hat{b}(\sigma_k)u(\sigma_1)\cdots u(\sigma_k)\,d\sigma_k\cdots d\sigma_1 \qquad (22)$$

(Problems 3.12 and 3.13 show cleverly the convergence property of (22).)

Incorporating the output equation and changing back to the original variables gives the Volterra system representation

$$y(t) = c(t)\Phi(t,0)x_0 + \sum_{k=1}^{\infty}\int_0^t\int_0^{\sigma_1}\cdots\int_0^{\sigma_{k-1}} c(t)\Phi(t,\sigma_1)D(\sigma_1)\Phi(\sigma_1,\sigma_2)D(\sigma_2)$$
$$\cdots D(\sigma_k)\Phi(\sigma_k,0)x_0u(\sigma_1)\cdots u(\sigma_k)\,d\sigma_k\cdots d\sigma_1$$
$$+\sum_{k=1}^{\infty}\int_0^t\int_0^{\sigma_1}\cdots\int_0^{\sigma_{k-1}} c(t)\Phi(t,\sigma_1)D(\sigma_1)\Phi(\sigma_1,\sigma_2)D(\sigma_2)$$
$$\cdots D(\sigma_{k-1})\Phi(\sigma_{k-1},\sigma_k)b(\sigma_k)u(\sigma_1)\cdots u(\sigma_k)\,d\sigma_k\cdots d\sigma_1 \qquad (23)$$

which also converges uniformly on any finite time interval.

There are three kinds of terms in (23): those that depend on the initial state alone, those that depend on the input alone, and those that depend on both. That is, unlike the linear case, the response is not simply the sum of the forced and unforced responses. If $u(t) = 0$ for all $t \geqslant 0$, the bilinear state equation looks like a linear state equation, and the response has the corresponding familiar form. If $x_0 = 0$, the input/output behavior is described by a reasonably simple Volterra system. Finally, if $x_0 \neq 0$ is fixed, the input/output behavior is again described by a Volterra system, but the kernels depend on the specific value of x_0.

It should not be too surprising that the input/output behavior of a nonlinear system depends in a somewhat complicated way on the initial state of the system. With fixed initial state, (23) is a Volterra system representation with a degree-0 term that is a specified time function, and with the kernels of the higher-degree terms completely specified. However, it usually is most convenient to introduce a variable change in the bilinear state equation to allow the choice of zero initial state in the new variables. This will be discussed more generally in due course, but for now simple examples will show how variable-change ideas can be implemented.

Example 3.1 The direct method for generating frequency modulated (FM) signals is to use a voltage controlled oscillator. That is, the frequency of a harmonic oscillator is changed in accordance with a message signal $u(t)$. The basic differential equation model is

$$\ddot{y}(t) + [\omega^2 + u(t)]y(t) = 0,\quad t \geqslant 0$$
$$y(0) = 0,\quad \dot{y}(0) = 1$$

where $y(t)$ is the generated FM signal. This model can be written in the state equation form by setting

$$z(t) = \begin{bmatrix} y(t) \\ \dot{y}(t) \end{bmatrix}$$

to obtain

$$\dot{z}(t) = \begin{bmatrix} 0 & 1 \\ -\omega^2 & 0 \end{bmatrix} z(t) + \begin{bmatrix} 0 & 0 \\ -1 & 0 \end{bmatrix} z(t)u(t)$$

$$y(t) = [1 \quad 0]\, z(t), \quad z(0) = \begin{bmatrix} 0 \\ 1 \end{bmatrix}$$

Now introduce a new state equation description for which the initial state is 0 by subtracting the zero input response. For $u(t) = 0$,

$$z(t) = e^{At}z_0 = \begin{bmatrix} \cos(\omega t) & \frac{1}{\omega}\sin(\omega t) \\ -\omega\sin(\omega t) & \cos(\omega t) \end{bmatrix} \begin{bmatrix} 0 \\ 1 \end{bmatrix}$$

$$= \begin{bmatrix} \frac{1}{\omega}\sin(\omega t) \\ \cos(\omega t) \end{bmatrix}$$

so let

$$x(t) = z(t) - \begin{bmatrix} \frac{1}{\omega}\sin(\omega t) \\ \cos(\omega t) \end{bmatrix}$$

Writing the differential equation in terms of $x(t)$ gives

$$\dot{x}(t) = \begin{bmatrix} 0 & 1 \\ -\omega^2 & 0 \end{bmatrix} x(t) + \begin{bmatrix} 0 & 0 \\ -1 & 0 \end{bmatrix} x(t)u(t) + \begin{bmatrix} 0 \\ \frac{-1}{\omega}\sin(\omega t) \end{bmatrix} u(t)$$

$$y(t) = [\,1 \quad 0\,]\, x(t) + \frac{1}{\omega}\sin(\omega t), \quad x(0) = 0$$

Applying the result in (23) to this bilinear state equation yields

$$y(t) = \frac{1}{\omega}\sin(\omega t) + \int_0^t h(t,\sigma_1)u(\sigma_1)\, d\sigma_1$$

$$+ \int_0^t \int_0^{\sigma_1} h(t,\sigma_1,\sigma_2)u(\sigma_1)u(\sigma_2)\, d\sigma_2 d\sigma_1 + \cdots$$

where the first two triangular kernels are

$$h(t,\sigma_1) = \frac{-1}{\omega^2}\sin[\omega(t-\sigma_1)]\sin(\omega\sigma_1)\delta_{-1}(t-\sigma_1)$$

$$h(t,\sigma_1,\sigma_2) = \frac{1}{\omega^3}\sin[\omega(t-\sigma_1)]\sin[\omega(\sigma_1-\sigma_2)]\sin(\omega\sigma_2)\delta_{-1}(t-\sigma_1)\delta_{-1}(\sigma_1-\sigma_2)$$

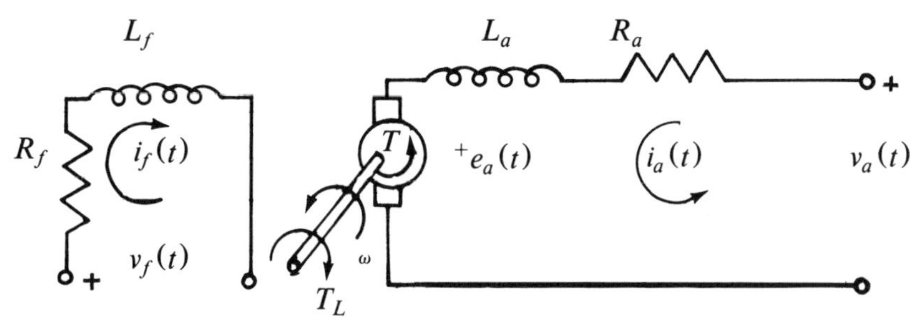

Figure 3.1. An ideal DC motor.

Example 3.2 As another illustration of the formulation of bilinear state equation models and the calculation of kernels, consider the ideal, separately excited, direct-current motor diagramed in Figure 3.1. The differential equation description of the field circuit is

$$\frac{d}{dt} i_f(t) = -\frac{R_f}{L_f} i_f(t) + \frac{1}{L_f} v_f(t)$$

The basic characteristics of the armature circuit require further explanation. The so-called generated voltage $e_a(t)$ is proportional to the product of the field current and the motor speed:

$$e_a(t) = K i_f(t) \omega(t)$$

The magnetic torque generated by the motor is similarly proportional to the product of the field and armature currents:

$$T(t) = K i_f(t) i_a(t)$$

Thus, the armature circuit is described by

$$\frac{d}{dt} i_a(t) = -\frac{R_a}{L_a} i_a(t) - \frac{K}{L_a} i_f(t)\omega(t) + \frac{1}{L_a} v_a(t)$$

and the mechanical load system is described by

$$\frac{d}{dt}\omega(t) = \frac{K}{J} i_f(t) i_a(t) - \frac{1}{J} T_L$$

where T_L is the mechanical load torque, and J is the moment of inertia.

A simple method for speed control in a DC motor is to keep the armature voltage constant, $v_a(t) = V_a$, and control the field current $i_f(t)$ by means of a variable resistor in the field circuit. To represent this scheme in a particular case, suppose the motor load acts as a damping device. That is, suppose $T_L = B\omega(t)$, where B is the viscous damping coefficient. (For example, the motor might be stirring a fluid.) Then with the input $u(t) = i_f(t)$, output $y(t) = \omega(t)$, and state vector

$$x(t) = \begin{bmatrix} i_a(t) \\ \omega(t) \end{bmatrix}$$

the system is described by

$$\dot{x}(t) = \begin{bmatrix} -R_a/L_a & 0 \\ 0 & -B/J \end{bmatrix} x(t) + \begin{bmatrix} 0 & -K/L_a \\ K/J & 0 \end{bmatrix} x(t)u(t) + \begin{bmatrix} V_a/L_a \\ 0 \end{bmatrix}$$

$$y(t) = [0 \quad 1]x(t), \quad x(0) = \begin{bmatrix} i_a(0) \\ \omega(0) \end{bmatrix}$$

This bilinear state equation is not quite in the form of (16) because of the constant term on the right side. To remove this term, let $x_c(t)$ be the solution of the differential equation with $x(0) = 0$ and $u(t) = 0$. Then it is readily verified that

$$x_c(t) = \begin{bmatrix} \frac{V_a}{R_a}(1 - e^{-\frac{R_a}{L_a}t}) \\ 0 \end{bmatrix}$$

Now let $z(t) = x(t) - x_c(t)$, and compute a differential equation for $z(t)$:

$$\begin{aligned} \dot{z}(t) &= \dot{x}(t) - \dot{x}_c(t) \\ &= \begin{bmatrix} -R_a/L_a & 0 \\ 0 & -B/J \end{bmatrix} z(t) + \begin{bmatrix} 0 & -K/L_a \\ K/J & 0 \end{bmatrix} z(t)u(t) \\ &+ \begin{bmatrix} 0 \\ (KV_a/JR_a)(1-e^{-(R_a/L_a)t}) \end{bmatrix} u(t) \end{aligned}$$

$$y(t) = [0 \quad 1]\, z(t), \quad z(0) = \begin{bmatrix} i_a(0) \\ \omega(0) \end{bmatrix}$$

This is a bilinear state equation description in the standard form (16), and the calculation of the solution via (23) is straightforward. For example, if the initial conditions are 0, then the first three triangular kernels are

$$h_1(t,\sigma_1) = \frac{KV_a}{JR_a} e^{-\frac{B}{J}(t-\sigma_1)}(1 - e^{-\frac{R_a}{L_a}\sigma_1})\delta_{-1}(t-\sigma_1)$$

$$h_2(t,\sigma_1,\sigma_2) = 0$$

$$\begin{aligned} h_3(t,\sigma_1,\sigma_2,\sigma_3) = -\frac{K^3V_a}{J^2L_aR_a} & e^{-\frac{B}{J}(t-\sigma_1)} e^{-\frac{R_a}{L_a}(\sigma_1-\sigma_2)} e^{-\frac{B}{J}(\sigma_2-\sigma_3)} \\ & (1 - e^{-\frac{R_a}{L_a}\sigma_3})\delta_{-1}(t-\sigma_1)\delta_{-1}(\sigma_1-\sigma_2)\delta_{-1}(\sigma_2-\sigma_3) \end{aligned}$$

3.2 A Digression on Notation

As more general nonlinear differential equations are considered, notational complexities begin to appear. These have to do with functions of several variables and their power series expansions. The difficulties probably are not unfamiliar, but their resolution here in terms of Kronecker (tensor) products is somewhat uncommon, hence this digression.

For matrices $A = (a_{ij})$ and $B = (b_{ij})$, of dimension $n_a \ x \ m_a$ and $n_b \ x \ m_b$ respectively, the *Kronecker product* is defined by

$$A \otimes B = \begin{bmatrix} a_{11}B & \cdots & a_{1m_a}B \\ \vdots & \vdots & \vdots \\ a_{n_a1}B & \cdots & a_{n_am_a}B \end{bmatrix} \tag{24}$$

It is clear that $A \otimes B$ has dimension $n_a n_b \ x \ m_a m_b$, and that any two matrices are conformable with respect to this product. The Kronecker product is associative so that $A \otimes B \otimes C$ is written without ambiguity. The following relationships are easily proved, assuming conformability with respect to ordinary matrix addition and multiplication.

$$(A + B) \otimes (C + D) = (A \otimes C) + (A \otimes D) + (B \otimes C) + (B \otimes D) \tag{25}$$

$$(AB) \otimes (CD) = (A \otimes C)(B \otimes D) \tag{26}$$

In fact, these properties can be written in simpler forms since the Kronecker product is given a higher precedence than matrix addition and multiplication:

$$(A + B) \otimes (C + D) = A \otimes C + A \otimes D + B \otimes C + B \otimes D \tag{27}$$

$$(AB) \otimes (CD) = A \otimes C \ B \otimes D \tag{28}$$

Additional properties that are not hard to prove are listed below.

Property 1 The product $A \otimes B=0$ if and only if $A = 0$ or $B = 0$.

Property 2 If A and B are invertible, then $A \otimes B$ is invertible and $(A \otimes B)^{-1} = A^{-1} \otimes B^{-1}$.

Property 3 If $rank \ A = r_a$ and $rank \ B = r_b$, then $rank \ A \otimes B = r_a r_b$.

The Kronecker product notation will be used for polynomials or power series in several variables. For example, if $f{:}R^n \rightarrow R^m$, then the power-series expansion of $f(x)$ about $x = 0$ is written

$$f(x) = F_0 + F_1 \, x + F_2 \, x \otimes x + F_3 \, x \otimes x \otimes x + \cdots \tag{29}$$

where each F_i is a coefficient matrix of appropriate dimension, to be specific, $m \times n^i$. Usually I will simplify the notation somewhat by setting $x^{(i)} = x \otimes \cdots \otimes x$ (i terms) and writing

$$f(x) = \sum_{i=0}^{\infty} F_i x^{(i)} \tag{30}$$

Taking a closer look reveals that there are redundancies hidden in this notation. In particular $x^{(i)}$ is an $n^i \times 1$ vector, but only $\binom{n+i-1}{i}$ entries are distinct. For example, writing transposes to save space, if

$$x = [\, x_1 \; x_2 \; x_3 \,]$$

then

$$x^{(2)} = [\, x_1^2 \;\; x_1x_2 \;\; x_1x_3 \;\; x_2x_1 \;\; x_2^2 \;\; x_2x_3 \;\; x_3x_1 \;\; x_3x_2 \;\; x_3^2 \,]$$

The redundancy could be eliminated by deleting the repeated entries and using, say, a lexicographic ordering for the rest. Adopting a square-bracket notation for the result, this procedure gives

$$x^{[2]} = [\, x_1^2 \;\; x_1x_2 \;\; x_1x_3 \;\; x_2^2 \;\; x_2x_3 \;\; x_3^2 \,]$$

For many purposes, this reduced Kronecker product is preferable because the dimensions are smaller. However, some explicitness is sacrificed for the economy of dimension when general calculations are performed. For example, suppose A is $n \times n$ and $y = Ax$. Then

$$\begin{aligned} y^{(2)} &= y \otimes y = (Ax) \otimes (Ax) \\ &= A \otimes A \; x \otimes x \\ &= A^{(2)} x^{(2)} \end{aligned}$$

While it is apparent that there exists a (smaller dimension) matrix $A^{[2]}$ such that

$$y^{[2]} = A^{[2]} \, x^{[2]}$$

it is difficult to write $A^{[2]}$ in explicit terms of A.

As another example, consider the linear differential equation

$$\dot{x}(t) = Ax(t), \quad x(0) = x_0$$

again with A an $n \times n$ matrix. After verifying the product rule

$$\frac{d}{dt}\,[x^{(2)}(t)] = \dot{x}(t) \otimes x(t) + x(t) \otimes \dot{x}(t)$$

a differential equation for $x^{(2)}(t)$ can be written in the form

$$\frac{d}{dt}\, x^{(2)}(t) = [A \otimes I_n + I_n \otimes A]x^{(2)}(t), \quad x^{(2)}(0) = x_0^{(2)}$$

where I_n is the $n \times n$ identity. Although it can be shown that $x^{[2]}$ also satisfies a linear differential equation, and one of lower dimension, there is no apparent way to write the coefficient matrix explicitly in terms of A. (Incidentally, the notation $\dot{x}^{(2)}(t)$ is being avoided for good reason. Notice that $(d/dt)[x^{(2)}(t)]$ is much different from $[(d/dt)x(t)]^{(2)}$, and thus the dot notation tends to ambiguity.)

This differential equation example is of interest for more than just notational reasons. What has been shown is that if $x(t)$ satisfies a linear differential equation, then so does $x^{(2)}(t)$. Clearly, this argument can be continued to show that $x^{(k)}(t)$ satisfies a linear differential equation, $k = 3, 4, \ldots$. A very similar observation provides the key for the methods to be discussed in Section 3.3.

The result of these considerations is that I will use the Kronecker product notation for the general developments in this chapter. However, it is clear that the more economical notation can be substituted with a concomitant loss in explicitness. Going further, in simple examples it probably is profitable to abandon both these special notations and work freestyle.

3.3 The Carleman Linearization Approach

The Carleman linearization method for computing kernels will be considered first in the context of state equations of the form

$$\dot{x}(t) = a(x(t),t) + b(x(t),t)u(t), \quad t \geqslant 0$$
$$y(t) = c(x(t),t), \quad x(0) = x_0 \tag{40}$$

where $x(t)$ is the $n \times 1$ state vector and the input $u(t)$ and output $y(t)$ are scalar signals. One reason for starting with this particular form is that the corresponding kernels do not contain impulses. This is a direct result of the fact that the input in (40) appears linearly. Toward the end of the section I will remove this restriction and briefly discuss a more general case.

Another reason for the form in (40) is that the existence of a convergent Volterra system representation can be guaranteed under general hypotheses. Suppose the functions $a(x,t)$, $b(x,t)$, and $c(x,t)$ are analytic in x and continuous in t, in which case (40) is called a *linear-analytic* state equation. Then various methods can be used to establish the following, somewhat loosely stated, result. (A proof using the techniques discussed in Section 3.4 is given in Appendix 3.1.)

Theorem 3.1 Suppose a solution to the unforced linear-analytic state equation exists for $t \in [0,T]$. Then there exists an $\epsilon > 0$ such that for all inputs satisfying $|u(t)| < \epsilon$ there is a Volterra system representation for the state equation that converges on $[0,T]$.

It is interesting to compare this with the corresponding result for bilinear state equations. For linear-analytic state equations, the existence of a convergent Volterra system representation is guaranteed only for sufficiently small input signals, while for bilinear state equations the input signals need only be bounded.

The first step in actually computing the kernels will be to perform some variable changes to put the state equation into a simpler form. These are not necessary, but they do make the subsequent derivation less fussy. Incidently, it is not clear that such variable changes are always a great idea. When dealing with particular problems or examples, significant features can be obscured. But I yield to maintaining simplicity of derivations, with the remark that the form of the Volterra system representation can be derived without the variable changes.

The first simplification is that the function $c(x,t)$ in (40) can be taken to be linear in x with little loss of generality. Differentiating the output equation under the assumption that $c(x,t)$ is continuously differentiable in t gives a differential equation for $y(t)$,

$$\begin{aligned} \dot{y}(t) &= [\frac{\partial}{\partial x} c(x,t)]\dot{x}(t) + \frac{\partial}{\partial t} c(x,t) \\ &= [\frac{\partial}{\partial x} c(x,t)][a(x,t) + b(x,t)u(t)] + \frac{\partial}{\partial t} c(x,t) \end{aligned} \tag{41}$$

with $y(0) = c(x_0, 0)$. Since the right side of (41) has the linear-analytic form, $y(t)$ can be adjoined to the bottom of $x(t)$ to form a new vector $\hat{x}(t)$. Then the state equation can be written in the form

$$\begin{aligned} \dot{\hat{x}}(t) &= \hat{a}(\hat{x}(t),t) + \hat{b}(\hat{x}(t),t)u(t), \quad \hat{x}(0) = \hat{x}_0 \\ y(t) &= \hat{c}(t)\hat{x}(t), \quad t \geqslant 0 \end{aligned} \tag{42}$$

where $\hat{x}(t)$ is an $(n+1)$ x 1 vector. In this case $\hat{c}(t) = [\,0 \ \cdots \ 0\ 1\,]$.

I also will assume that the solution of the differential equation in (42) with $u(t)=0$ is $\hat{x}(t) = 0$. To show this entails no loss of generality, suppose that the response for $u(t) = 0$ is $x_0(t)$. Then setting $\bar{x}(t) = \hat{x}(t) - x_0(t)$, (42) can be written in the form

$$\begin{aligned} \dot{\bar{x}}(t) &= \dot{\hat{x}}(t) - \dot{x}_0(t) \\ &= \hat{a}(\hat{x}(t),t) + \hat{b}(\hat{x}(t),t)u(t) - \hat{a}(x_0(t),t) \\ &= \hat{a}(\bar{x}(t)+x_0(t),t) + \hat{b}(\bar{x}(t)+x_0(t),t)u(t) - \hat{a}(x_0(t),t) \\ &= \bar{a}(\bar{x}(t),t) + \bar{b}(\bar{x}(t),t)u(t) \\ y(t) &= \hat{c}(t)\bar{x}(t) + \hat{c}(t)x_0(t), \ \bar{x}(0) = 0, \ t \geqslant 0 \end{aligned}$$

with the appropriate definitions of $\bar{a}(\bar{x},t)$ and $\bar{b}(\bar{x},t)$. Thus, simplifying the notation, state equations of the form

$$\begin{aligned} \dot{x}(t) &= a(x(t),t) + b(x(t),t)u(t), \quad t \geqslant 0 \\ y(t) &= c(t)x(t) + y_0(t), \ x(0) = 0 \end{aligned} \tag{43}$$

will be considered. Here $x(t)$ is an n x 1 state vector, $u(t) = 0$ implies $x(t) = 0$ and $y(t) = y_0(t)$, and $a(0,t) = 0$ because of the variable change employed.

It should be noted that there is a price to pay for this last variable change. Namely, the unforced solution $x_0(t)$ must be computed in order to obtain the right side of the simplified differential equation in (43). While this might not be a severe problem when the unforced system is linear in $x(t)$, clearly the computation of $x_0(t)$ in a more general situation can be arbitrarily difficult.

The goal now is to determine the terms through degree N of a polynomial input/output expression for (43). That is, to determine an input/output representation of the form

$$y(t) = y_0(t) + \sum_{k=1}^{N} \int_{-\infty}^{\infty} h(t,\sigma_1,\ldots,\sigma_k)\, u(\sigma_1) \cdots u(\sigma_k)\, d\sigma_1 \cdots d\sigma_k \tag{44}$$

Of course, in general there will be terms of higher degree that have been ignored in (44). Since the state equation (43) can be represented as a convergent Volterra system (under the conditions stated earlier), a polynomial truncation of the series will be an accurate approximation for inputs that are sufficiently small.

Actually, the method to be considered for determining the polynomial system representation generates a polynomial representation for $x(t)$. That is, a set of vector kernels is determined for an expression of the form

$$x(t) = \sum_{k=1}^{N} \int_{-\infty}^{\infty} w(t,\sigma_1,\ldots,\sigma_k) u(\sigma_1) \cdots u(\sigma_k)\, d\sigma_1 \cdots d\sigma_k$$

Then, since $y(t)$ is a linear function of $x(t)$, the kernels for the input/output representation are readily computed.

The Carleman linearization method begins with the replacement of the right side of the state equation (43) by power series representations. Adopting the Kronecker-product notation, write

$$\begin{aligned} a(x,t) &= A_1(t)x + A_2(t)x^{(2)} + \cdots + A_N(t)x^{(N)} + \cdots \\ b(x,t) &= B_0(t) + B_1(t)x + \cdots + B_{N-1}(t)x^{(N-1)} + \cdots \end{aligned} \tag{45}$$

where the terms not shown are of higher degree in x than the terms that are shown. Thus, (43) is written in the form

$$\begin{aligned} \dot{x}(t) &= \sum_{k=1}^{N} A_k(t)x^{(k)}(t) + \sum_{k=0}^{N-1} B_k(t)x^{(k)}(t)u(t) + \cdots \\ y(t) &= c(t)x(t) + y_0(t),\ \ x(0) = 0,\ t \geqslant 0 \end{aligned} \tag{46}$$

where I have explicitly retained terms through degree N in the expansion of $a(x,t)$, and terms through degree $N-1$ in the expansion of $b(x,t)$. That higher-degree terms in these expansions will not contribute to the first N kernels will be seen in due course.

The representation in (46) is an appropriate first step because it can be shown that the output of (46) with the higher-degree terms *deleted,* call it $\hat{y}(t)$, for any input $u(t)$, when compared to the response $y(t)$ of (43) to this same input, satisfies

$$|\hat{y}(t) - y(t)| \leqslant K\beta^{N+1},\ \ t \geqslant 0 \tag{47}$$

where K is a constant, and

$$\beta = \max_{t \geq 0} |u(t)|$$

Now consider the responses $\hat{y}(t)$ and $y(t)$ for inputs of the form $\alpha u(t)$, where α is any real number. In this situation,

$$|\hat{y}(t) - y(t)| \leq K |\alpha|^{N+1} \beta^{N+1}$$

so that the polynomial representations (truncations) for $y(t)$ and $\hat{y}(t)$ must be identical through degree N.

To determine the first N kernels corresponding to (46), a differential equation is developed for $x^{(2)}(t)$, dropping from explicit consideration terms of degree greater than N along the way.

$$\begin{aligned}
\frac{d}{dt}[x^{(2)}(t)] &= \frac{d}{dt}[x^{(1)}(t) \otimes x^{(1)}(t)] = \dot{x}^{(1)}(t) \otimes x^{(1)}(t) + x^{(1)}(t) \otimes \dot{x}^{(1)}(t) \\
&= [\sum_{k=1}^{N} A_k(t) x^{(k)}(t) + \sum_{k=0}^{N-1} B_k(t) x^{(k)}(t) u(t)] \otimes x^{(1)}(t) \\
&\quad + x^{(1)}(t) \otimes [\sum_{k=1}^{N} A_k(t) x^{(k)}(t) + \sum_{k=0}^{N-1} B_k(t) x^{(k)}(t) u(t)] + \cdots \\
&= \sum_{k=1}^{N-1} [A_k(t) \otimes I_n + I_n \otimes A_k(t)]\, x^{(k+1)}(t) \\
&\quad + \sum_{k=0}^{N-2} [B_k(t) \otimes I_n + I_n \otimes B_k(t)]\, x^{(k+1)}(t) u(t) + \cdots, \quad x^{(2)}(0) = 0 \qquad (48)
\end{aligned}$$

Thus $x^{(2)}(t)$ satisfies a differential equation that has the same general form as the differential equation for $x^{(1)}(t)$ in (46).

Continuing in this fashion yields a differential equation for $x^{(j)}(t)$ to degree N of the form

$$\frac{d}{dt}[x^{(j)}(t)] = \sum_{k=1}^{N-j+1} A_{j,k}(t) x^{(k+j-1)}(t) + \sum_{k=0}^{N-j} B_{j,k}(t) x^{(k+j-1)}(t) u(t) + \cdots,$$

$$x^{(j)}(0) = 0, \; j = 1, \ldots, N \qquad (49)$$

with the notation defined by $A_{1,k} = A_k$, and for $j > 1$,

$$\begin{aligned}
A_{j,k}(t) &= A_k(t) \otimes I_n \otimes \cdots \otimes I_n + I_n \otimes A_k(t) \otimes I_n \otimes \cdots \otimes I_n \\
&\quad + \cdots + I_n \otimes \cdots \otimes I_n \otimes A_k(t)
\end{aligned}$$

(There are $j-1$ Kronecker products in each term, and j terms.) A similar notation is used for $B_{j,k}(t)$. Now, the crucial observation is that by setting

$$x^{\otimes}(t) = \begin{bmatrix} x^{(1)}(t) \\ x^{(2)}(t) \\ \vdots \\ x^{(N)}(t) \end{bmatrix}$$

I can write the collection of differential equations in (49) as the big bilinear state equation (dropping some arguments) plus higher-degree terms:

$$\frac{d}{dt}x^{\otimes} = \begin{bmatrix} A_{11} & A_{12} & \cdots & A_{1N} \\ 0 & A_{21} & \cdots & A_{2,N-1} \\ 0 & 0 & \cdots & A_{3,N-2} \\ \vdots & \vdots & \vdots & \vdots \\ 0 & 0 & \cdots & A_{N1} \end{bmatrix} x^{\otimes}$$

$$+ \begin{bmatrix} B_{11} & B_{12} & \cdots & B_{1,N-1} & 0 \\ B_{20} & B_{21} & \cdots & B_{2,N-2} & 0 \\ 0 & B_{30} & \cdots & B_{3,N-3} & 0 \\ \vdots & \vdots & \vdots & & \vdots \\ 0 & 0 & \cdots & B_{N0} & 0 \end{bmatrix} x^{\otimes} u + \begin{bmatrix} B_{10} \\ 0 \\ 0 \\ \vdots \\ 0 \end{bmatrix} u + \cdots$$

$$y(t) = [c(t)\ 0\ \cdots\ 0]\, x^{\otimes}(t) + y_0(t) + \cdots, \quad x^{\otimes}(0) = 0 \tag{50}$$

Upon deleting all the higher-degree terms represented by the ellipses, this state equation is called a truncated *Carleman linearization* of the linear-analytic state equation in (43). (It might also be appropriate to call (50) a *bilinearization* of (43).)

It is straightforward in principle to find the degree-N polynomial representation for the input/output behavior of the bilinear state equation (50). Since the input/output behavior of (50) agrees with that of (43) through terms of degree N, the polynomial representation of degree N for (50) is precisely the same as that for (43). Note that this approach gives all N kernels in triangular form via (22) from Section 3.1.

Example 3.3 A phase-locked loop for the demodulation of FM signals is diagramed in Figure 3.2. The input is an FM signal

$$r(t) = \sin[\omega t + \phi_1(t)]$$

where

$$\phi_1(t) = \int_0^t u(\sigma)\,d\sigma$$

and $u(t)$ is the message (modulating) signal.

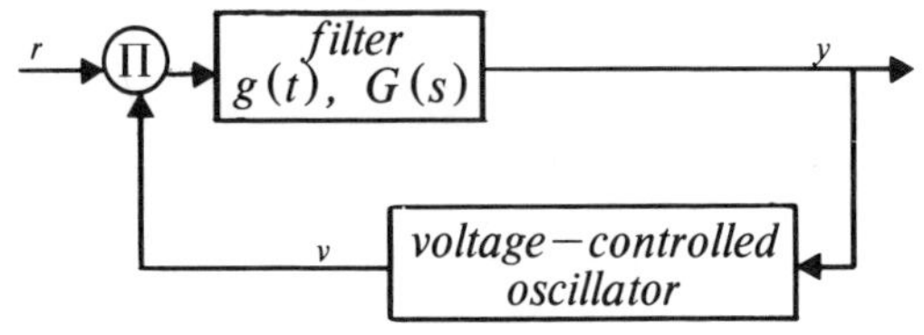

Figure 3.2. A phase-locked loop.

The loop filter is described by the transfer function $G(s)$, and the voltage-controlled oscillator produces the signal

$$v(t) = 2K\cos[\omega t + \phi_2(t)]$$

where

$$\phi_2(t) = \int_0^t y(\sigma)\,d\sigma$$

and $y(t)$ is the output of the phase-locked loop. The output of the multiplier then consists of two terms: a high-frequency term

$$K\sin[2\omega t + \phi_1(t) + \phi_2(t)]$$

and a low-frequency term

$$K\sin[\phi_1(t) - \phi_2(t)]$$

Assuming that the loop filter removes the high-frequency term, the signal $e(t)$ can be considered to contain only the low-frequency term. That is,

$$e(t) = K\sin[\phi_1(t) - \phi_2(t)] = K\sin[x(t)]$$

where the phase error signal $x(t)$ is given by

$$x(t) = \phi_1(t) - \phi_2(t) = \int_0^t u(\sigma)\,d\sigma - \int_0^t y(\sigma)\,d\sigma$$

Then the output of the loop filter, which also is the output of the phase-locked loop, is

$$y(t) = \int_0^t g(t-\tau)e(\tau)\,d\tau$$

From these relationships a differential-integral equation that describes the phase error is

$$\begin{aligned}\dot{x}(t) &= \dot{\phi}_1(t) - \dot{\phi}_2(t) = u(t) - y(t)\\ &= u(t) - \int_0^t g(t-\tau)K\sin[x(\tau)]\,d\tau\end{aligned}$$

This equation suggests the model shown in Figure 3.3. When $x(t)$ is zero the loop is said to be locked, and in this situation $\phi_1(t) = \phi_2(t)$, or $y(t) = u(t)$.

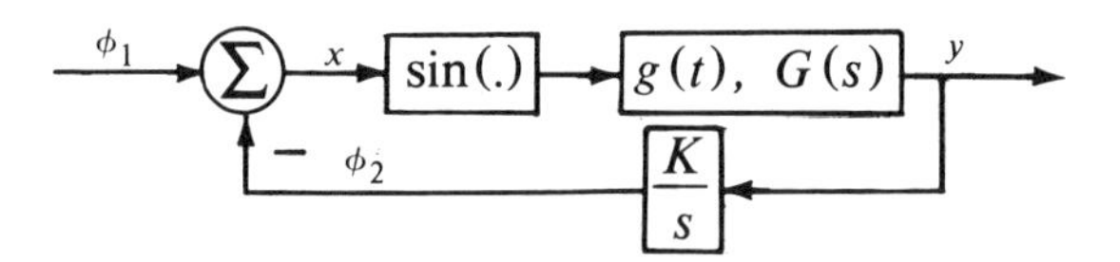

Figure 3.3. A nonlinear model for the phase-locked loop.

The difficulty in analyzing the model depends chiefly on the nature of the loop-filter transfer function $G(s)$. For simplicity, I will consider only the so-called first-order phase-locked loop, wherein $G(s) = 1$ (or $g(t) = \delta_0(t)$). Then the differential equation description for the phase error simplifies to

$$\dot{x}(t) = -K\sin[x(t)] + u(t)$$

and if the loop is locked, $x(0) = 0$. To compute the kernels in this simple case, there is no need to use the general notation. The differential equation can be replaced by the equation

$$\dot{x}(t) = -Kx(t) + \frac{K}{6}x^3(t) + u(t) + \cdots$$

where only those terms that contribute to the first three kernels have been retained explicitly. Since $x^{(j)}(t) = x^j(t)$ for scalar $x(t)$, let

$$x^{\otimes}(t) = \begin{bmatrix} x(t) \\ x^2(t) \\ x^3(t) \end{bmatrix}$$

Then (50) becomes

$$\frac{d}{dt}[x^{\otimes}(t)] = \begin{bmatrix} -K & 0 & K/6 \\ 0 & -2K & 0 \\ 0 & 0 & -3K \end{bmatrix} x^{\otimes}(t) + \begin{bmatrix} 0 & 0 & 0 \\ 2 & 0 & 0 \\ 0 & 3 & 0 \end{bmatrix} x^{\otimes}(t)u(t) + \begin{bmatrix} 1 \\ 0 \\ 0 \end{bmatrix} u(t) + \cdots$$

$$x(t) = [\,1 \;\; 0 \;\; 0\,]\, x^{\otimes}(t)$$

where the phase-error signal is taken to be the output of the state equation. A short calculation yields

$$e^{At} = \begin{bmatrix} e^{-Kt} & 0 & (e^{-Kt} - e^{-3Kt})/12 \\ 0 & e^{-2Kt} & 0 \\ 0 & 0 & e^{-3Kt} \end{bmatrix}$$

and from (23) the first three triangular kernels are

$$h(t,\sigma_1) = e^{-K(t-\sigma_1)}\delta_{-1}(t-\sigma_1)$$

$$h(t,\sigma_1,\sigma_2) = 0$$

$$h(t,\sigma_1,\sigma_2,\sigma_3) = \frac{1}{2}[e^{-Kt}e^{-K\sigma_1}e^{K\sigma_2}e^{K\sigma_3} - e^{-3Kt}e^{K\sigma_1}e^{K\sigma_2}e^{K\sigma_3}]\,\delta_{-1}(t-\sigma_1)\delta_{-1}(\sigma_1-\sigma_2)\delta_{-1}(\sigma_2-\sigma_3)$$

(The unit step functions are there just to emphasize that these are triangular kernels.)

Extending the approach of this section to state equations of the form

$$\dot{x}(t) = f(x(t),u(t),t), \quad x(0) = 0$$
$$y(t) = h(x(t),t), \quad t \geqslant 0 \tag{51}$$

where $u(t)$ and $y(t)$ are scalar signals is not hard. It is messy, to be sure, especially when worked out in detail, but the mechanics are familiar. A power series form of the equation is obtained, and then a set of vector kernels describing $x(t)$ is calculated in much the same way as was done for bilinear state equations. But now the nonlinear dependence of $f(x,u,t)$ upon u means that the kernels must contain impulses. A transparent case will show why.

Example 3.4 For the scalar state equation

$$\dot{x}(t) = u^2(t), \quad x(0) = 0$$
$$y(t) = x^3(t)$$

integration directly yields

$$x(t) = \int_0^t u^2(\sigma)\, d\sigma$$

Writing this as a degree-2 homogeneous term

$$x(t) = \int_0^t \int_0^t h(t,\sigma_1,\sigma_2)u(\sigma_1)u(\sigma_2)\, d\sigma_2 d\sigma_1$$

requires the impulsive kernel

$$h(t,\sigma_1,\sigma_2) = \delta_0(\sigma_1-\sigma_2)$$

Thus, the output is given by

$$y(t) = \int_0^t \delta_0(\sigma_1-\sigma_2)\delta_0(\sigma_3-\sigma_4)\delta_0(\sigma_5-\sigma_6)u(\sigma_1) \cdots u(\sigma_6)\, d\sigma_6 \cdots d\sigma_1$$

which clearly shows that the system is homogeneous of degree 6.

Returning to the differential equation in (51), I assume that $f(0,0,t) = 0$, and, as usual, that $f(x,u,t)$ has properties that suffice to remove worries about existence and uniqueness of solutions for $t \geqslant 0$. Differentiability sufficient to carry out the following development will be assumed, and the argument t will be dropped in part because the calculations are essentially the same for the stationary and nonstationary cases. Using the Kronecker product notation, the expansion of $f(x,u)$ about $x = 0$, $u = 0$ through degree N can be written in the form

$$f(x,u) = F_{01}u + F_{02}u^2 + F_{03}u^3 + F_{10}x^{(1)} + F_{11}x^{(1)}u + F_{12}x^{(1)}u^2$$
$$+ F_{20}x^{(2)} + F_{21}x^{(2)}u + F_{22}x^{(2)}u^2 + \cdots$$

This provides a differential equation for x of the form

$$\dot{x} = \sum_{i=0}^{N} \sum_{j=0}^{N} F_{ij} x^{(i)} u^j + \cdots \tag{52}$$

where $F_{00} = 0$. The procedure for developing differential equations for $x^{(2)}, \ldots, x^{(N)}$ is just as before. Now, however, the equation for

$$x^{\otimes} = \begin{bmatrix} x \\ x^{(2)} \\ \vdots \\ x^{(N)} \end{bmatrix}$$

will have a number of additional terms:

$$\begin{aligned} \frac{d}{dt} x^{\otimes} &= F x^{\otimes} + G_1 x^{\otimes} u + G_2 x^{\otimes} u^2 + \cdots + G_N x^{\otimes} u^{N-1} \\ &\quad + g_1 u + \cdots + g_N u^N + \cdots \end{aligned} \tag{53}$$

From this point, the idea is to mimic the development in the bilinear case. Using a change of variables involving the transition matrix for F, and then integrating both sides of the resulting state equation sets up the iterative resubstitution procedure. Of course, there are many more terms here, but, at this level of notation, applying the procedure and inserting impulses to write the homogeneous terms in the right form is straightforward in principle. Once this has been done, expanding the output equation to degree N,

$$\begin{aligned} h(x) &= y_0(t) + h_1 x + h_2 x^{(2)} + \cdots + h_N x^{(N)} + \cdots \\ &= y_0(t) + h x^{\otimes} + \cdots \end{aligned} \tag{54}$$

leads to a polynomial input/output representation upon deletion of the ellipses. Notice that this last step requires nothing more complicated than using properties of the Kronecker product. In particular no additional impulses need be inserted.

Example 3.5 The nonlinear feedback system shown in Figure 3.4 is described by

$$\begin{aligned} \dot{x}(t) &= Ax(t) + b\psi[u(t) - y(t)] \\ y(t) &= cx(t), \quad t \geqslant 0, \; x(0) = 0 \end{aligned}$$

where the scalar nonlinearity is a polynomial (or power series)

$$\psi(\alpha) = \alpha + \psi_2\alpha^2 + \psi_3\alpha^3 + \cdots$$

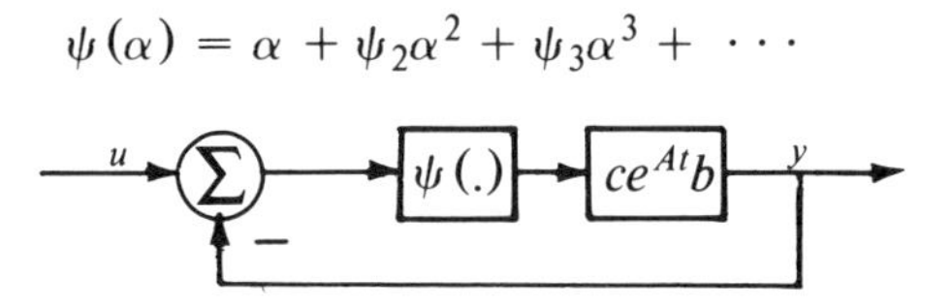

Figure 3.4 Nonlinear system for Example 3.4

Using the approach just outlined, I will compute the first- and second-degree kernels for the closed-loop system. Corresponding to (53), the terms needed for the first two kernels involve setting

$$x^{\otimes} = \begin{bmatrix} x \\ x^{(2)} \end{bmatrix}$$

and replacing the given state equation by a state equation of the form

$$\frac{d}{dt}x^{\otimes} = Fx^{\otimes} + G_1x^{\otimes}u + g_1u + g_2u^2 + \cdots$$

$$y = hx^{\otimes}$$

It should be clear that higher-degree terms in $x^{\otimes}$ and u will not be needed. Moreover, it will turn out that the general notation being employed in the differential equation for $x^{\otimes}$ carries along terms that are superfluous as far as the first two kernels are concerned. In particular, the equation for $x^{\otimes}$ contains terms involving x, $x^{(2)}$, xu, $x^{(2)}u$, u, and u^2. The $x^{(2)}u$ terms are not needed and arbitrarily setting the coefficients to 0 can simplify matters.

The differential equation for x can be written out in the form

$$\begin{aligned}\dot{x} &= Ax + b[(u-cx) + \psi_2(u-cx)^2 + \psi_3(u-cx)^3 + \cdots] \\ &= (A-bc)x + bu + \psi_2bu^2 - 2\psi_2bcxu + \psi_2b(cx)^2 + 3\psi_3b(cx)^2u + \cdots\end{aligned}$$

Using the Kronecker product notation to write

$$(cx)^2 = (cx) \otimes (cx) = c \otimes cx^{(2)}$$

and dropping into the dots terms that do not enter into the $x^{\otimes}$ equation, gives

$$\dot{x} = (A-bc)x + bu + \psi_2bu^2 - 2\psi_2bcxu + \psi_2bc \otimes cx^{(2)} + 3\psi_3bc \otimes cx^{(2)}u + \cdots$$

To develop a differential equation for $x^{(2)}$, the product rule gives

$$\begin{aligned}\frac{d}{dt}[x^{(2)}] &= \dot{x} \otimes x + x \otimes \dot{x} \\ &= [(A-bc) \otimes I + I \otimes (A-bc)]x^{(2)} + [b \otimes I + I \otimes b]xu \\ &\quad + \psi_2[b \otimes I + I \otimes b]xu^2 - 2\psi_2[bc \otimes I + I \otimes bc]x^{(2)}u + \cdots\end{aligned}$$

Again terms that will not contribute to the final result have been dropped. Thus the state equation (53) in terms of $x^{\otimes}$ is

$$\frac{d}{dt}[x^{\otimes}] = \begin{bmatrix} A-bc & \psi_2 bc \otimes c \\ 0 & [(A-bc) \otimes I + I \otimes (A-bc)] \end{bmatrix} x^{\otimes}$$

$$+ \begin{bmatrix} -2\psi_2 bc & 3\psi_3 bc \otimes c \\ [b \otimes I + I \otimes b] & -2\psi_2[(bc) \otimes I + I \otimes (bc)] \end{bmatrix} x^{\otimes} u + \begin{bmatrix} b \\ 0 \end{bmatrix} u + \begin{bmatrix} \psi_2 b \\ 0 \end{bmatrix} u^2$$

$$y = [c \quad 0]x^{\otimes}, \quad x^{\otimes}(0) = 0$$

Now the resubstitution procedure can be applied just as was done for bilinear state equations, with the exception that impulses must be inserted to obtain terms of the correct form. It is easy to show that in the general notation the first two triangular kernels are given by

$$h(t,\sigma) = he^{F(t-\sigma)}g_1$$

$$h(t,\sigma_1,\sigma_2) = he^{F(t-\sigma_1)}G_1e^{F(\sigma_1-\sigma_2)}g_1 + he^{F(t-\sigma_1)}g_2\delta_0(\sigma_1-\sigma_2)$$

To complete the calculation, these kernels can be expressed in terms of the given state equation by showing that

$$e^{Ft} = \begin{bmatrix} e^{(A-bc)t} & \int_0^t e^{(A-bc)(t-\sigma)}\psi_2 bc \otimes ce^{[(A-bc) \otimes I+I \otimes (A-bc)]\sigma} d\sigma \\ 0 & e^{[(A-bc) \otimes I+I \otimes (A-bc)]t} \end{bmatrix}$$

Then the first two triangular kernels are

$$h(t,\sigma) = ce^{(A-bc)(t-\sigma)}b\delta_{-1}(t-\sigma)$$

$$h(t,\sigma_1,\sigma_2) = [-2\psi_2 e^{(A-bc)(t-\sigma_1)}bce^{(A-bc)(\sigma_1-\sigma_2)}b$$

$$+ \psi_2 \int_0^{t-\sigma_1} ce^{(A-bc)(t-\sigma_1-\gamma)}bc \otimes ce^{[(A-bc) \otimes I+I \otimes (A-bc)]\gamma} d\gamma (b \otimes I$$

$$+ I \otimes b)e^{(A-bc)(\sigma_1-\sigma_2)}b$$

$$+ \psi_2 ce^{(A-bc)(t-\sigma_1)}b\delta_0(\sigma_1-\sigma_2)]\delta_{-1}(t-\sigma_1)\delta_{-1}(\sigma_1-\sigma_2)$$

As mentioned at the outset, the terms involving $x^{(2)}u$, in other words, the terms in the second block column of G_1, do not enter the result, and could have been set to zero for simplicity.

3.4 The Variational Equation Approach

In the variational equation approach, a state-equation description is obtained for each degree-k homogeneous subsystem in the input/output representation. It turns out that, although the equation for the degree-k subsystem is coupled nonlinearly to the equations for the lower-degree subsystems, each of the equations has identical first-degree (linear) terms. Thus the various kernels can be computed using the linear-state-equation solution reviewed in Section 3.1.

As in the previous section, I begin by considering the linear-analytic state equation

$$\dot{x}(t) = a(x(t),t) + b(x(t),t)u(t), \quad t \geqslant 0$$
$$y(t) = c(t)x(t) + y_0(t), \quad x(0) = 0 \tag{55}$$

where $a(0,t) = 0$ so that the response to $u(t) = 0$ is $x(t) = 0$, $y(t) = y_0(t)$. The analyticity assumption can be weakened since only a finite number of kernels will be computed, but it is retained here for simplicity. More general state equations without the special assumptions on the unforced response are discussed later in the section.

The homogeneous-subsystem state equations are derived by considering the response of the differential equation in (55) to inputs of the form $\alpha u(t)$, where α is an arbitrary scalar. The response can be written as an expansion in the parameter α of the form (In the present context, subscripts do not indicate components of a vector.)

$$x(t) = \alpha x_1(t) + \alpha^2 x_2(t) + \cdots + \alpha^N x_N(t) + \cdots \tag{56}$$

where the dots contain terms of degree greater than N in α. Viewing the analytic functions $a(x,t)$ and $b(x,t)$ in terms of power series, substituting (56) into (55), and equating coefficients of like powers of α leads to a differential equation for each $x_k(t)$, the degree-k component of $x(t)$.

Just as in the Carleman linearization approach, the first step is to replace the terms in (55) by power series representations. For ease of exposition, only the calculation of the first three kernels will be treated. Thus the state equation

$$\dot{x}(t) = A_1(t)x^{(1)}(t) + A_2(t)x^{(2)}(t) + A_3(t)x^{(3)}(t)$$
$$+ B_0(t)u(t) + B_1(t)x^{(1)}(t)u(t) + B_2(t)x^{(2)}(t)u(t) + \cdots$$
$$y(t) = c(t)x^{(1)}(t) + y_0(t), \quad x^{(1)}(0) = 0, \quad t \geqslant 0 \tag{57}$$

will be considered. Both the assumed input $\alpha u(t)$ and the assumed response in (56) are substituted into (57). Note that from the rules of calculation for Kronecker products,

$$x^{(2)}(t) = [\alpha x_1(t) + \alpha^2 x_2(t) + \cdots] \otimes [\alpha x_1(t) + \alpha^2 x_2(t) + \cdots]$$
$$= \alpha^2 x_1^{(2)}(t) + \alpha^3 [x_1(t) \otimes x_2(t) + x_2(t) \otimes x_1(t)] + \cdots$$
$$x^{(3)}(t) = \alpha^3 x_1^{(3)}(t) + \cdots \tag{58}$$

where, again, only terms of degree 3 or less are explicitly retained. The terms of higher degree in x that have been dropped from (58) would not contribute lower-degree terms in α. That is, substituting (56) into a degree-k function of x yields terms of degree k and higher in α. Now, (57) can be written in the form

$$\begin{aligned}
&\alpha\dot{x}_1(t) + \alpha^2\dot{x}_2(t) + \alpha^3\dot{x}_3(t) + \cdots \\
&\quad = \alpha A_1(t)x_1(t) + \alpha^2[A_1(t)x_2(t)+A_2(t)x_1^{(2)}(t)] \\
&\quad + \alpha^3[A_1(t)x_3(t)+A_2(t)(x_1(t)\otimes x_2(t)+x_2(t)\otimes x_1(t))+A_3(t)x_1^{(3)}(t)] \\
&\quad + \alpha B_0(t)u(t) + \alpha^2 B_1(t)x_1(t)u(t) + \alpha^3[B_1(t)x_2(t)+B_2(t)x_1^{(2)}(t)]u(t) \\
&\quad + \cdots,\ \alpha x_1(0) + \alpha^2 x_2(0) + \alpha^3 x_3(0) + \cdots = 0
\end{aligned} \tag{59}$$

Since this differential equation and the equation for the initial state must hold for all α, coefficients of like powers of α can be equated. This gives the first three variational equations:

$$\begin{aligned}
&\dot{x}_1(t) = A_1(t)x_1(t) + B_0(t)u(t),\ \ x_1(0) = 0 \\
&\dot{x}_2(t) = A_1(t)x_2(t) + A_2(t)x_1^{(2)}(t) + B_1(t)x_1(t)u(t),\ x_2(0) = 0 \\
&\dot{x}_3(t) = A_1(t)x_3(t) + A_2(t)[x_1(t)\otimes x_2(t) + x_2(t)\otimes x_1(t)] \\
&\qquad + A_3(t)x_1^{(3)}(t) + B_1(t)x_2(t)u(t) + B_2(t)x_1^{(2)}(t)u(t),\ \ x_3(0) = 0
\end{aligned} \tag{60}$$

The first equation in (60) is the linearized version of the differential equation in (55). Defining the vector kernel

$$w(t,\sigma) = \Phi(t,\sigma)B_0(\sigma)\delta_{-1}(t-\sigma) \tag{61}$$

where $\Phi(t,\tau)$ is the transition matrix for $A_1(t)$, yields the representation

$$x_1(t) = \int_0^t w(t,\sigma)\,u(\sigma)\,d\sigma \tag{62}$$

Proceeding to the second equation in (60), the term $x_1^{(2)}(t)$ can be written in the form

$$\begin{aligned}
x_1^{(2)}(t) &= [\int_0^t w(t,\sigma)u(\sigma)\,d\sigma] \otimes [\int_0^t w(t,\sigma)u(\sigma)\,d\sigma] \\
&= \int_0^t\int_0^t w(t,\sigma_1)\otimes w(t,\sigma_2)u(\sigma_1)u(\sigma_2)\,d\sigma_2 d\sigma_1
\end{aligned} \tag{63}$$

Substituting (62) and (63) into the second equation in (60), it is found that this is a linear differential equation in $x_2(t)$. (It should be clear that this linearity feature is the key to the method.) Thus

$$\begin{aligned}
x_2(t) = \int_0^t \Phi(t,\sigma)[A_2(\sigma)&\int_0^\sigma\int_0^\sigma w(\sigma,\sigma_1)\otimes w(\sigma,\sigma_2)u(\sigma_1)u(\sigma_2)\,d\sigma_2 d\sigma_1 \\
&+ B_1(\sigma)\int_0^\sigma w(\sigma,\sigma_1)u(\sigma_1)d\sigma_1 u(\sigma)]\,d\sigma
\end{aligned} \tag{64}$$

Using the fact that $w(t,\sigma) = 0$ if $\sigma > t$, (64) can be written in the form

$$x_2(t) = \int_0^t \int_0^t \int_0^t \Phi(t,\sigma)A_2(\sigma)w(\sigma,\sigma_1) \otimes w(\sigma,\sigma_2)u(\sigma_1)u(\sigma_2)\, d\sigma d\sigma_2 d\sigma_1$$
$$+ \int_0^t \int_0^t \Phi(t,\sigma)B_1(\sigma)w(\sigma,\sigma_1)u(\sigma_1)u(\sigma)\, d\sigma d\sigma_1 \tag{65}$$

Thus the degree-2 component of $x(t)$ is given by

$$x_2(t) = \int_0^t \int_0^t w(t,\sigma_1,\sigma_2)u(\sigma_1)u(\sigma_2)\, d\sigma_2 d\sigma_1 \tag{66}$$

where

$$w(t,\sigma_1,\sigma_2) = \int_0^t \Phi(t,\sigma)A_2(\sigma)w(\sigma,\sigma_1) \otimes w(\sigma,\sigma_2)\, d\sigma$$
$$+ \Phi(t,\sigma_2)B_1(\sigma_2)w(\sigma_2,\sigma_1)$$
$$= \int_{max[\sigma_1,\sigma_2]}^t \Phi(t,\sigma)A_2(\sigma)\Phi(\sigma,\sigma_1) \otimes \Phi(\sigma,\sigma_2)\, d\sigma B_0(\sigma_1) \otimes B_0(\sigma_2)$$
$$+ \Phi(t,\sigma_2)B_1(\sigma_2)\Phi(\sigma_2,\sigma_1)B_0(\sigma_1)\delta_{-1}(\sigma_2-\sigma_1),\ 0 \leqslant \sigma_1,\sigma_2 \leqslant t \tag{67}$$

Of course, the same procedure is used to derive a degree-3 vector kernel that describes $x_3(t)$. This straightforward but messy calculation is left to the reader. To determine the degree-3 polynomial representation for the input/output behavior, it is clear that each of the vector kernels should be multiplied by $c(t)$.

Example 3.6 Revisiting the first-order phase-locked loop in Example 3.3 using the variational equation approach will indicate the mechanics in a more detailed fashion, as well as contrast the two methods discussed so far. To compute the first three kernels, the starting point is the state equation for the phase error in power series form:

$$\dot{x}(t) = -Kx(t) + \frac{K}{6}x^3(t) + u(t) + \cdots,\quad x(0) = 0$$

Substituting the expansion

$$x(t) = \alpha x_1(t) + \alpha^2 x_2(t) + \alpha^3 x_3(t) + \cdots$$

into the state equation with the assumed input $\alpha u(t)$ gives the first three variational equations:

$$\dot{x}_1(t) = -Kx_1(t) + u(t),\quad x_1(0) = 0$$
$$\dot{x}_2(t) = -Kx_2(t),\quad x_2(0) = 0$$
$$\dot{x}_3(t) = -Kx_3(t) + \frac{K}{6}x_1^3(t),\quad x_3(0) = 0$$

Solving the first variational equation is a simple matter:

$$x_1(t) = \int_0^t e^{-K(t-\sigma)} u(\sigma)\, d\sigma$$

Thus, the degree-1 kernel for the system is

$$h(t,\sigma) = e^{-K(t-\sigma)} \delta_{-1}(t-\sigma)$$

The second variational equation is even simpler, giving $x_2(t) = 0$ for all $t \geq 0$. Thus the degree-2 kernel is identically 0. The third variational equation gives

$$x_3(t) = \int_0^t e^{-K(t-\sigma)} \frac{K}{6} x_1^3(\sigma)\, d\sigma$$

Writing this in the standard degree-3 homogeneous form is going to take a little more work. The first step is to write

$$x_1^3(\sigma) = [\int_0^\sigma e^{-K(\sigma-\sigma_1)} u(\sigma_1)\, d\sigma_1\,]^3$$

$$= \int_0^\sigma e^{-K(\sigma-\sigma_1)} e^{-K(\sigma-\sigma_2)} e^{-K(\sigma-\sigma_3)} u(\sigma_1)u(\sigma_2)u(\sigma_3)\, d\sigma_1 d\sigma_2 d\sigma_3$$

$$= \int_0^t e^{-K(\sigma-\sigma_1)} e^{-K(\sigma-\sigma_2)} e^{-K(\sigma-\sigma_3)} \delta_{-1}(\sigma-\sigma_1)\delta_{-1}(\sigma-\sigma_2)\delta_{-1}(\sigma-\sigma_3)$$

$$u(\sigma_1)u(\sigma_2)u(\sigma_3)\, d\sigma_1 d\sigma_2 d\sigma_3$$

Substituting this expression into the expression for $x_3(t)$, and rearranging the order of integration, gives

$$x_3(t) = \int_0^t \frac{K}{6} e^{-Kt} [\int_0^t e^{-2K\sigma} \delta_{-1}(\sigma-\sigma_1)\delta_{-1}(\sigma-\sigma_2)\delta_{-1}(\sigma-\sigma_3)\, d\sigma] e^{K(\sigma_1+\sigma_2+\sigma_3)}$$

$$u(\sigma_1)u(\sigma_2)u(\sigma_3)\, d\sigma_1 d\sigma_2 d\sigma_3$$

$$= \int_0^t \frac{-1}{12} [e^{-3Kt} e^{K(\sigma_1+\sigma_2+\sigma_3)} - e^{-Kt} e^{-2K max(\sigma_1,\sigma_2,\sigma_3)} e^{K(\sigma_1+\sigma_2+\sigma_3)}]$$

$$u(\sigma_1)u(\sigma_2)u(\sigma_3)\, d\sigma_1 d\sigma_2 d\sigma_3$$

Now a degree-3 kernel for the system is apparent. Notice, however, that it is not immediately apparent that this result agrees with that in Example 3.3.

The mechanics of the variational equation approach change little when the most general state equations are considered. In fact, the special assumptions on the linearity of the output and on the zero-input response can be relaxed without causing distress. To illustrate this in some detail, consider the general state equation

$$\dot{x}(t) = f(x(t),u(t),t), \quad x(0) = x_0$$
$$y(t) = h(x(t),u(t),t), \quad t \geqslant 0 \tag{68}$$

where $u(t)$ and $y(t)$ are scalars. Suppose that with the fixed initial state and the input $\hat{u}(t)$, the response is $\hat{x}(t)$, $\hat{y}(t)$. In this setting it is of interest to find a polynomial input/output representation that describes the deviation of the output from $\hat{y}(t)$, $y_\delta(t) = y(t) - \hat{y}(t)$, in terms of the deviation of the input from $\hat{u}(t)$, $u_\delta(t) = u(t) - \hat{u}(t)$. This means I am abandoning all the changes of variables previously used to clean up notations. Through degree N, the right side of the differential equation in (68) can be replaced by (dropping most t's)

$$f(x,u,t) = f(\hat{x}+x_\delta,\hat{u}+u_\delta,t)$$
$$= f(\hat{x},\hat{u},t) + \sum_{i=0}^{N}\sum_{j=0}^{N} F_{ij}(t)x_\delta^{(i)}u_\delta^j + \cdots, \quad F_{00} = 0$$

via a Taylor series about $\hat{x}$, $\hat{u}$. Now consider deviation inputs of the form $\alpha u_\delta(t)$, where α is an arbitrary scalar, and assume that the resulting deviation response is expanded in terms of α:

$$x_\delta = \alpha x_{1\delta} + \alpha^2 x_{2\delta} + \cdots$$

(Note that the α^0 term is missing since $\alpha = 0$ implies the input is $\hat{u}$, which implies the response is $\hat{x}$.)

Substituting into the differential equation gives, through degree 3 (again)

$$\alpha\dot{x}_{1\delta} + \alpha^2\dot{x}_{2\delta} + \alpha^3\dot{x}_{3\delta} + \cdots = F_{10}(t)[\alpha x_{1\delta} + \alpha^2 x_{2\delta} + \alpha^3 x_{3\delta}]$$
$$+ F_{20}(t)[\alpha x_{1\delta} + \alpha^2 x_{2\delta}]^{(2)} + F_{30}(t)[\alpha x_{1\delta}]^{(3)} + \alpha F_{01}(t)u_\delta$$
$$+ \alpha^2 F_{02}(t)u_\delta^2 + \alpha^3 F_{03}(t)u_\delta^3 + F_{11}(t)[\alpha x_{1\delta} + \alpha^2 x_{2\delta}]\alpha u_\delta$$
$$+ F_{12}(t)[\alpha x_{1\delta}]\alpha^2 u_\delta^2 + F_{21}(t)[\alpha x_{1\delta}]^{(2)}\alpha u_\delta + \cdots$$

Equating coefficients of like powers of α gives the first three variational equations listed below.

$$\dot{x}_{1\delta} = F_{10}(t)x_{1\delta} + F_{01}(t)u_\delta, \quad x_{1\delta}(0) = 0$$
$$\dot{x}_{2\delta} = F_{10}(t)x_{2\delta} + F_{20}(t)x_{1\delta}^{(2)} + F_{02}(t)u_\delta^2 + F_{11}(t)x_{1\delta}u_\delta, \quad x_{2\delta}(0) = 0$$
$$\dot{x}_{3\delta} = F_{10}(t)x_{3\delta} + F_{20}(t)[x_{1\delta} \otimes x_{2\delta} + x_{2\delta} \otimes x_{1\delta}]$$
$$+ F_{30}(t)x_{1\delta}^{(3)} + F_{03}(t)u_\delta^3 + F_{11}(t)x_{2\delta}u_\delta + F_{12}(t)x_{1\delta}u_\delta^2$$
$$+ F_{21}(t)x_{1\delta}^{(2)}u_\delta, \quad x_{3\delta}(0) = 0$$

But now the computation of vector kernels for each variation proceeds just as before, except that an occasional impulse must be inserted to obtain the standard form of a homogeneous term. Then the process is completed by expanding the output equation, substituting into that expansion, and regathering terms of like degree with perhaps some insertion of more impulses since the output map in (68) is permitted to depend on the input.

Example 3.7 The variational equation approach will be applied to the nonlinear feedback system of Example 3.5:

$$\dot{x}(t) = Ax(t) + b\psi[u(t)-y(t)],\quad x(0) = 0$$

$$y(t) = cx(t),\quad t \geqslant 0$$

where

$$\psi(\alpha) = \alpha + \psi_2\alpha^2 + \psi_3\alpha^3 + \cdots$$

and where $\hat{u}(t) = 0$, $\hat{x}(t) = 0$, and $\hat{y}(t) = 0$ for all $t \geqslant 0$. To compute the kernels through degree 2, the system is written as

$$\dot{x}(t) = Ax(t) + b[u(t)-y(t)] + \psi_2 b[u(t)-y(t)]^2 + \cdots$$

It is appropriate to drop the general notation for this example since it offers no particular advantage. Assuming the input $\alpha u(t)$ and response

$$x(t) = \alpha x_1(t) + \alpha^2 x_2(t) + \cdots$$

substituting, and equating coefficients of like powers of α gives the variational equations

$$\dot{x}_1(t) = [A - bc]x_1(t) + bu(t),\quad x_1(0) = 0$$

$$\dot{x}_2(t) = [A - bc]x_2(t) + \psi_2 b[u(t)-cx_1(t)]^2,\quad x_2(0) = 0$$

From the first equation,

$$x_1(t) = \int_0^t e^{[A-bc](t-\sigma_1)}bu(\sigma_1)\,d\sigma_1$$

and since the output equation is linear, the degree-1 kernel in the polynomial input/output map is

$$h(t-\sigma_1) = ce^{[A-bc](t-\sigma_1)}b\delta_{-1}(t-\sigma_1)$$

The second equation is solved in a similar fashion, although the terms involved are more complicated.

$$x_2(t) = \int_0^t e^{[A-bc](t-\sigma_1)}\psi_2 b[u(\sigma_1) - \int_0^{\sigma_1} h(\sigma_1-\sigma_2)u(\sigma_2)\,d\sigma_2]^2\,d\sigma_1$$

$$= \int_0^t\int_0^{\sigma_1} \psi_2 e^{[A-bc](t-\sigma_1)}b\delta_0(\sigma_1-\sigma_2)u(\sigma_1)u(\sigma_2)\,d\sigma_2 d\sigma_1$$

$$+ \int_0^t\int_0^{\sigma_1} -2\psi_2 e^{[A-bc](t-\sigma_1)}bh(\sigma_1-\sigma_2)u(\sigma_1)u(\sigma_2)\,d\sigma_2 d\sigma_1$$

$$+ \int_0^t\int_0^{\sigma_1}\int_0^{\sigma_1} \psi_2 e^{[A-bc](t-\sigma_1)}bh(\sigma_1-\sigma_2)h(\sigma_1-\sigma_3)u(\sigma_2)u(\sigma_3)\,d\sigma_2 d\sigma_3 d\sigma_1$$

Thus the degree-2 term of the input/output map is

$$\int_0^t \int_0^t h(t,\sigma_1,\sigma_2)u(\sigma_1)u(\sigma_2)\, d\sigma_2 d\sigma_1$$

where, with a little relabeling of variables in the last term,

$$h(t,\sigma_1,\sigma_2) = \psi_2 h(t-\sigma_1)\delta_0(\sigma_1-\sigma_2)\delta_{-1}(\sigma_1-\sigma_2) - 2\psi_2 h(t-\sigma_1)h(\sigma_1-\sigma_2)$$
$$+ \int_0^t \psi_2 h(t-\sigma_3)h(\sigma_3-\sigma_2)h(\sigma_3-\sigma_1)\, d\sigma_3$$

This degree-2 kernel can be written in terms of the given system parameters in the obvious way.

3.5 The Growing Exponential Approach

The properties of growing exponentials discussed in Chapter 2 can be adapted readily to the problem of finding transfer function descriptions from constant-parameter (stationary) state equations. Consider the general form

$$\dot{x}(t) = a(x(t),u(t)),\quad x(0) = 0$$
$$y(t) = cx(t) + y_0(t),\quad t \geqslant 0 \tag{69}$$

where $a(0,0) = 0$ and $a(x,u)$ is analytic in x and u. Briefly stated, the first N symmetric transfer functions corresponding to (69) are computed as follows. First replace $a(x,u)$ by a power series in x and u. Then assume an input of the form

$$u(t) = e^{\lambda_1 t} + \cdots + e^{\lambda_N t} \tag{70}$$

and assume a solution of the form

$$x(t) = \sum_m G_{m_1,\ldots,m_N}(\lambda_1,\ldots,\lambda_N)e^{(m_1\lambda_1 + \cdots + m_N\lambda_N)t} \tag{71}$$

where the notation is precisely that of Chapter 2, and the vector coefficients are undetermined. Substituting into the differential equation, solve for

$$G_{1,0,\ldots,0}(\lambda_1),\ G_{1,1,0,\ldots,0}(\lambda_1,\lambda_2),\ \ldots,\ G_{1,\ldots,1}(\lambda_1,\ldots,\lambda_N)$$

by equating coefficients of like exponentials. Then since the output is a linear function of x,

$$H_1(s) = cG_{1,0,\ldots,0}(s)$$
$$H_{2sym}(s_1,s_2) = \frac{1}{2!}\, cG_{1,1,0,\ldots,0}(s_1,s_2)$$
$$\vdots$$
$$H_{Nsym}(s_1,\ldots,s_N) = \frac{1}{N!}\, cG_{1,\ldots,1}(s_1,\ldots,s_N) \tag{72}$$

I should note that considerable savings in labor is realized if exponentials that obviously will not contribute to the terms of interest are dropped at each stage of the calculation. For example, no term in (71) with $m_j > 1$ for at least one j need be carried.

Example 3.8 To find the first three symmetric transfer functions corresponding to the now familiar state equation description

$$\dot{x}(t) = -Kx(t) + \frac{K}{6}x^3(t) + u(t) + \cdots$$

assume an input of the form

$$u(t) = e^{\lambda_1 t} + e^{\lambda_2 t} + e^{\lambda_3 t}$$

and, dropping arguments for simplicity, a solution of the form

$$\begin{aligned} x(t) = G_{100}\, e^{\lambda_1 t} &+ G_{010}e^{\lambda_2 t} + G_{001}e^{\lambda_3 t} + G_{200}e^{2\lambda_1 t} \\ &+ G_{020}e^{2\lambda_2 t} + G_{002}e^{2\lambda_3 t} + G_{110}e^{(\lambda_1+\lambda_2)t} \\ &+ G_{101}e^{(\lambda_1+\lambda_3)t} + G_{011}e^{(\lambda_2+\lambda_3)t} + G_{111}e^{(\lambda_1+\lambda_2+\lambda_3)t} + \cdots \end{aligned}$$

Of course, in this scalar case with the output identical to the state, the G notation could be replaced by symmetric transfer function notation. Also note that the G_{200}, G_{020}, G_{002} terms are included just to show in the context of an example that they are superfluous. At any rate, an easy calculation gives

$$x^3(t) = 6G_{100}G_{010}G_{001}e^{(\lambda_1+\lambda_2+\lambda_3)t} + \cdots$$

Substituting into the differential equation, and equating the coefficients of

$$e^{\lambda_1 t},\ e^{(\lambda_1+\lambda_2)t},\ e^{(\lambda_1+\lambda_2+\lambda_3)t}$$

respectively, yields the equations

$$\lambda_1\, G_{100} + KG_{100} = 1$$

$$(\lambda_1+\lambda_2)\, G_{110} + KG_{110} = 0$$

$$(\lambda_1+\lambda_2+\lambda_3)\, G_{111} + KG_{111} = KG_{100}\, G_{010}\, G_{001}$$

Solving these in turn gives

$$G_{100}(\lambda_1) = \frac{1}{\lambda_1+K}$$

$$G_{110}(\lambda_1,\lambda_2) = 0$$

$$G_{111}(\lambda_1,\lambda_2,\lambda_3) = \frac{KG_{100}G_{010}G_{001}}{\lambda_1+\lambda_2+\lambda_3+K}$$

Using the obvious facts:

$$G_{010}(\lambda_2) = \frac{1}{\lambda_2+K}\,,\quad G_{001}(\lambda_3) = \frac{1}{\lambda_3+K}$$

the first three symmetric transfer functions are

$$H_1(s) = \frac{1}{s+K}$$

$$H_{2sym}(s_1,s_2) = 0$$

$$H_{3sym}(s_1,s_2,s_3) = \frac{K/6}{(s_1+s_2+s_3+K)(s_1+K)(s_2+K)(s_3+K)}$$

Example 3.9 Consider again the simplest general nonlinear equation of the form (69); the bilinear state equation

$$\dot{x}(t) = Ax(t) + Dx(t)u(t) + bu(t)$$

$$y(t) = cx(t), \quad x(0) = 0$$

To find the first two symmetric transfer functions, let

$$u(t) = e^{\lambda_1 t} + e^{\lambda_2 t}, \quad \lambda_1, \lambda_2 > 0$$

and assume that

$$x(t) = G_{1,0}e^{\lambda_1 t} + G_{0,1}e^{\lambda_2 t} + G_{1,1}e^{(\lambda_1+\lambda_2)t} + \cdots$$

Substituting into the differential equation and equating the coefficients of $e^{\lambda_1 t}$ gives

$$\lambda_1 G_{1,0} = A\ G_{1,0} + b$$

Solving this linear equation yields

$$G_{1,0} = (\lambda_1 I - A)^{-1}b$$

so that the degree-1 transfer function is

$$H(s) = c(sI-A)^{-1}b$$

The coefficients of $e^{(\lambda_1+\lambda_2)t}$ are equated in a similar fashion to yield the equation

$$(\lambda_1+\lambda_2)\ G_{1,1} = A\ G_{1,1} + D\ G_{1,0} + D\ G_{0,1}$$

Substituting $G_{1,0} = (\lambda_1 I - A)^{-1}b$ and $G_{0,1} = (\lambda_2 I - A)^{-1}b$ and solving gives

$$G_{1,1} = [(\lambda_1+\lambda_2)I-A]^{-1}\ D[(\lambda_1 I-A)^{-1}b + (\lambda_2 I-A)^{-1}b]$$

Thus the degree-2 symmetric transfer function is

$$H_{2sym}(s_1,s_2) = \frac{1}{2}\ c[(s_1+s_2)I-A]^{-1}\ D[(s_1 I-A)^{-1}b + (s_2 I-A)^{-1}b]$$

Note that a simpler asymmetric version can be written by inspection, namely

$$H_2(s_1,s_2) = c[(s_1+s_2)I-A]^{-1}\ D(s_1 I-A)^{-1}b$$

I leave it to the reader to show that a degree-3 asymmetric transfer function can be written as

$$H_3(s_1,s_2,s_3) = c[(s_1+s_2+s_3)I-A]^{-1}\ D[(s_1+s_2)I-A]^{-1}\ D(s_1 I-A)^{-1}b$$

From this, a pattern for the higher-degree transfer functions should be clear.

3.6 Systems Described by Nth-Order Differential Equations

Various versions of the methods that have been discussed have appeared in the literature from time to time. Mostly these have been set up for N^{th}-order, nonlinear, differential equations in the older literature. Since some problems are described quite naturally in these terms, I will review the variational expansion method for the equation

$$y^{(N)}(t) + a_{N-1}(t)y^{(N-1)}(t) + \cdots + a_0(t)y(t) + \sum_{k=2}^{K} b_k(t)y^k(t) = u(t) \tag{73}$$

where

$$y(0) = y^{(1)}(0) = \cdots = y^{(N-1)}(0) = 0 \tag{74}$$

so that the solution for $u(t) = 0$ is $y(t) = 0$. Of course, this is a special case, but the ideas generalize in a transparent fashion.

Consider the response to the input $\alpha u(t)$, where α is a scalar, and write

$$y(t) = \sum_{m=1}^{M} \alpha^m y_m(t) + \cdots \tag{75}$$

where only the terms of degree M or less have been explicitly retained. Substituting into the differential equation gives

$$\sum_{m=1}^{M} \alpha^m \sum_{n=0}^{N} a_n(t)y_m^{(n)}(t) + \sum_{k=2}^{K} b_k(t)[\sum_{m=1}^{M} \alpha^m y_m(t)]^k + \cdots = \alpha u(t) \tag{76}$$

where $a_N(t) = 1$, and the initial conditions are

$$y_m^{(n)}(0) = 0, \quad n = 0,1,\ldots,N-1, \quad m = 1,2,\ldots,M$$

Equating coefficients of α on both sides gives

$$\sum_{n=0}^{N} a_n(t)y_1^{(n)}(t) = u(t), \quad y_1^{(n)}(0) = 0, \; n = 0,\ldots,N-1 \tag{77}$$

and the solution of this linear differential equation can be written in the form

$$y_1(t) = \int_0^t h_1(t,\sigma_1)u(\sigma_1)\, d\sigma_1 \tag{78}$$

Equating coefficients of α^2 gives

$$\sum_{n=0}^{N} a_n(t)y_2^{(n)}(t) + b_2(t)y_1^2(t) = 0, \quad y_1^{(n)}(0) = 0, \quad n = 0,\ldots,N-1 \tag{79}$$

The solution of this differential equation can be written in the form

$$y_2(t) = \int_0^t h_1(t,\sigma_1)b_2(\sigma_1)y_1^2(\sigma_1)\, d\sigma_1 \tag{80}$$

To write this in the usual degree-2 homogeneous form requires a substitution for $y_1^2(\sigma_1)$:

$$y_2(t) = \int_0^t h_1(t,\sigma_1)b_2(\sigma_1)\int_0^{\sigma_1} h_1(\sigma_1,\sigma_2)u(\sigma_2)\, d\sigma_2 \int_0^{\sigma_1} h_1(\sigma_1,\sigma_3)u(\sigma_3)\, d\sigma_3 d\sigma_1$$

$$= \int_0^t \int_0^{\sigma_1} \int_0^{\sigma_1} h_1(t,\sigma_1)b_2(\sigma_2)h_1(\sigma_1,\sigma_2)h_1(\sigma_1,\sigma_3)u(\sigma_2)u(\sigma_3)\, d\sigma_2 d\sigma_3 d\sigma_1$$

Inserting unit step functions so the limits of integration can be raised to t, and relabeling variables, gives

$$y_2(t) = \int_0^t \int_0^t h_2(t,\sigma_1,\sigma_2)u(\sigma_1)u(\sigma_2)\, d\sigma_1 d\sigma_2 \tag{81}$$

where

$$h_2(t,\sigma_1,\sigma_2) = \int_0^t h_1(t,\sigma)b_2(\sigma)h_1(\sigma,\sigma_1)h_1(\sigma,\sigma_2)\delta_{-1}(\sigma-\sigma_1)\delta_{-1}(\sigma-\sigma_2)\, d\sigma \tag{82}$$

I can proceed in a similar way to compute the higher-degree kernels in the (truncated) polynomial input/output representation. The only obstacle to a general formulation lies in the nonlinear term in (76). This can be handled by writing

$$[\sum_{m=1}^{M} \alpha^m y_m(t)]^k = \sum_{j=k}^{M_k} \alpha^j y_{j,k}(t) \tag{83}$$

and deriving a recursion for the terms $y_{j,k}(t)$, $j \geqslant k$.

Let

$$f(\alpha) = \sum_{m=1}^{M} \alpha^m y_m(t)$$
$$g(\alpha) = [\sum_{m=1}^{M} \alpha^m y_m(t)]^{k-1} \tag{84}$$

Then, $g(\alpha)$ can be written as

$$g(\alpha) = \sum_{j=k-1}^{M_{k-1}} \alpha^j y_{j,k-1}(t) \tag{85}$$

and

$$f(\alpha)g(\alpha) = \sum_{j=k}^{M_k} \alpha^j y_{j,k}(t) \tag{86}$$

To isolate the α^j, $j \geqslant k$, terms on both sides of this equation, differentiate both sides j times with respect to α and set $\alpha = 0$. Using the product rule

$$\frac{d^j}{d\alpha^j}[f(\alpha)g(\alpha)] = \sum_{i=0}^{j} \binom{j}{i} [\frac{d^i}{d\alpha^i}f(\alpha)][\frac{d^{j-i}}{d\alpha^{j-1}}g(\alpha)] \tag{87}$$

gives

$$\sum_{i=0}^{j} y_i(t) y_{j-i,k-1}(t) = y_{j,k}(t) \tag{88}$$

But the lower limit on the sum can be raised to 1 since $y_0(t) = 0$. Since $j \geqslant k$, and since nonzero summands correspond to $j-i \geqslant k-1$, the upper limit on the sum can be replaced by j-k+1. Thus,

$$y_{j,k}(t) = \sum_{i=1}^{j-k+1} y_i(t) y_{j-i,k-1}(t) \tag{89}$$

where, for $k = 1$, $y_{j,1}(t) = y_j(t)$.

Returning now to the problem at hand, equate the coefficients of α^3 on both sides of the equation

$$\sum_{m=1}^{M} \alpha^m \sum_{n=0}^{N} a_n(t) y_m^{(n)}(t) + \sum_{k=2}^{K} b_k(t) \sum_{j=k}^{M_k} \alpha^j y_{j,k}(t) + \cdots = \alpha u(t),\ y_m^{(n)}(0) = 0 \tag{90}$$

This gives

$$\sum_{n=0}^{N} a_n(t) y_3^{(n)}(t) + b_2(t) y_{3,2}(t) + b_3(t) y_{3,3}(t) = 0 \tag{91}$$

where all initial conditions are zero. The solution can be written in the form

$$y_3(t) = \int_0^t h_1(t,\sigma_1)[b_2(\sigma_1) y_{3,2}(\sigma_1) + b_3(\sigma_1) y_{3,3}(\sigma_1)]\, d\sigma_1 \tag{92}$$

The recursions just developed yield

$$y_{3,2}(\sigma_1) = y_1(\sigma_1) y_{2,1}(\sigma_1) + y_2(\sigma_1) y_{1,1}(\sigma_1) = 2y_1(\sigma_1) y_2(\sigma_1)$$
$$y_{3,3}(\sigma_1) = y_1(\sigma_1) y_{2,2}(\sigma_1) = y_1^3(\sigma_1) \tag{93}$$

so that

$$y_3(t) = \int_0^t h_1(t,\sigma_1)[2b_2(\sigma_1) y_1(\sigma_1) y_2(\sigma_1) + b_3(\sigma_1) y_1^3(\sigma_1)]\, d\sigma_1 \tag{94}$$

Now it is just a matter of substitution for $y_1(\sigma_1)$, $y_2(\sigma_2)$ from (78), (80), and some manipulation of integrals to put this into the form of a degree-3 homogeneous subsystem.

I should also point out that the growing exponential method can be adapted quite easily to N^{th}-order differential equations, as the following example shows.

Example 3.10 The simple pendulum consists of a mass m suspended on a massless rod of length L. The input torque at the pivot is $u(t)$, the damping coefficient at the pivot is a, and the output $y(t)$ is the angle from the vertical. The well known differential equation describing this system is

$$\ddot{y}(t) + \frac{a}{mL^2}\dot{y}(t) + \frac{g}{L} \sin[y(t)] = \frac{1}{mL^2} u(t)$$

and it is assumed that the initial conditions are zero. To compute the first three symmetric transfer functions by the growing exponential method, the first step is to replace $sin[y(t)]$ by its power series expansion. Of course, only terms through order three need be retained explicitly, so the differential equation of interest is

$$\ddot{y}(t) + \frac{a}{mL^2}\dot{y}(t) + \frac{g}{L}y(t) - \frac{g}{3!L}y^3(t) + \cdots = \frac{1}{mL^2}u(t)$$

The growing exponential method can be simplified in this case by arguing, either from the physics of the situation or from the differential equation, that $y(t)$ will contain no homogeneous terms of even degree. That is, if the input signal $u(t)$ produces the output signal $y(t)$, then the input signal $-u(t)$ produces $-y(t)$, and it follows that only odd-degree terms can be present.

To calculate the symmetric transfer functions through degree three, assume an input signal of the form

$$u(t) = e^{\lambda_1 t} + e^{\lambda_2 t} + e^{\lambda_3 t},\quad \lambda_1,\lambda_2,\lambda_3 > 0,\quad t \in (-\infty,\infty)$$

Since all degree-2 terms are known to be zero, assume the response

$$y(t) = H_1(\lambda_1)e^{\lambda_1 t} + H_1(\lambda_2)e^{\lambda_2 t} + H_1(\lambda_3)e^{\lambda_3 t}$$
$$+ H_{3sym}(\lambda_1,\lambda_2,\lambda_3)e^{(\lambda_1+\lambda_2+\lambda_3)t} + \cdots$$

where, as usual, only terms contributing to the final result have been retained. (Notice that the symmetric transfer function notation, rather than the G-notation, has been used, since the calculations involve the output directly.) Substituting into the differential equation gives, again with many terms dropped,

$$\lambda_1^2 H_1(\lambda_1)e^{\lambda_1 t} + (\lambda_1+\lambda_2+\lambda_3)^2 H_{3sym}(\lambda_1,\lambda_2,\lambda_3)e^{(\lambda_1+\lambda_2+\lambda_3)t} + \frac{a}{mL^2}\lambda_1 H_1(\lambda_1)e^{\lambda_1 t}$$
$$+ \frac{a}{mL^2}(\lambda_1+\lambda_2+\lambda_3)H_{3sym}(\lambda_1,\lambda_2,\lambda_3)e^{(\lambda_1+\lambda_2+\lambda_3)t} + \frac{g}{L}H_1(\lambda_1)e^{\lambda_1 t}$$
$$+ \frac{g}{L}H_{3sym}(\lambda_1,\lambda_2,\lambda_3)e^{(\lambda_1+\lambda_2+\lambda_3)t}$$
$$- \frac{g}{L}H_1(\lambda_1)H_1(\lambda_2)H_1(\lambda_3)e^{(\lambda_1+\lambda_2+\lambda_3)t} + \cdots$$
$$= \frac{1}{mL^2}e^{\lambda_1 t} + \frac{1}{mL^2}e^{\lambda_2 t} + \frac{1}{mL^2}e^{\lambda_3 t}$$

Equating coefficients of $e^{\lambda_1 t}$ gives

$$H_1(\lambda_1) = \frac{1/(mL^2)}{\lambda_1^2 + a/(mL^2)\lambda_1 + g/L}$$

Thus, the degree-1 transfer function is

$$H_1(s) = \frac{1/(mL^2)}{s^2 + a/(mL^2)s + g/L}$$

Equating coefficients of $e^{(\lambda_1+\lambda_2+\lambda_3)t}$ yields

$$H_{3sym}(\lambda_1,\lambda_2,\lambda_3) = \frac{g/L}{(\lambda_1+\lambda_2+\lambda_3)^2+a/(mL^2)(\lambda_1+\lambda_2+\lambda_3)+g/L} H_1(\lambda_1)H_1(\lambda_2)H_1(\lambda_3)$$

or, in more compact form,

$$H_{3sym}(s_1,s_2,s_3) = \frac{mgL}{3!} H_1(s_1+s_2+s_3)H_1(s_1)H_1(s_2)H_1(s_3)$$

3.7 Remarks and References

Remark 3.1 The idea of using resubstitution (sometimes called Peano-Baker) techniques or successive approximations (often called Picard iterations) is well known in the theory of differential equations. Perhaps the first suggestion that successive approximations be used to compute kernels was made by J. Barrett in a published discussion appended to the paper

R. Flake, "Volterra Series Representations of Time-Varying Nonlinear Systems," *Proceedings of the Second International Congress of IFAC,* Butterworths, London, pp. 91-97, 1963.

The possibilities for the successive approximation approach in a more general setting were first realized much later in

C. Bruni, G. DiPillo, G. Koch, "On the Mathematical Models of Bilinear Systems," *Ricerche di Automatica,* Vol. 2, pp. 11-26, 1971.

where the general form for the kernels corresponding to a bilinear state vector equation was first derived. The linear-analytic system case was first considered in

R. Brockett, "Volterra Series and Geometric Control Theory," *Automatica* Vol. 12, pp. 167-176, 1976 (addendum with E. Gilbert, p. 635).

The conditions for existence of a uniformly convergent Volterra system representation are derived by combining the successive approximation technique with power series expansion of the analytic functions in the state equation. A crucial step in computing the form of the kernels is the use of the Carleman linearization idea to obtain a bilinear state vector equation that approximates the linear-analytic differential equation. This technique is discussed in

A. Krener, "Linearization and Bilinearization of Control Systems," *Proceedings of the 1974 Allerton Conference,* Electrical Engineering Department, University of Illinois, Urbana-Champaign, Illinois, pp. 834-843, 1974.

This paper contains arguments that justify the bound indicated in (47). For differential equations that have nonlinear terms in the input, the calculation of the kernels is discussed in

R. Brockett, "Functional Expansions and Higher Order Necessary Conditions in Optimal Control," in *Mathematical System Theory,* G. Marchesini, S. Mitter eds., Lecture Notes in Economics and Mathematical Systems, Vol. 131, Springer-Verlag, New York, pp. 111-121, 1976.

In these latter three references, the reduced Kronecker product representation is used. The relationship between successive approximation and the computation of Volterra series is discussed further in

B. Leon, D. Schaefer, "Volterra Series and Picard Iteration for Nonlinear Circuits and Systems," *IEEE Transactions on Circuits and Systems,* Vol. CAS-25, pp. 789-793, 1978.

Remark 3.2 An extensive compilation of properties of the Kronecker product along with some applications in system theory can be found in

J. Brewer, "Kronecker Products and Matrix Calculus in System Theory," *IEEE Transactions on Circuits and Systems,* Vol. CAS-25, pp. 772-781, 1978.

Remark 3.3 The variational expansion method has a long and gloried history in the mathematics of differential equations that are analytic in a parameter. The method was used by Euler, clarified by Cauchy, and made completely rigorous by the convergence proofs of Poincare. A detailed treatment of this theory with convergence proofs for results much like Theorem 3.1 is given in

F. Moulton, *Differential Equations,* Macmillan, New York, 1930.

Another approach to the variational expansion method for nonlinear systems is discussed in

E. Gilbert, "Functional Expansions for the Response of Nonlinear Differential Systems," *IEEE Transactions on Automatic Control,* Vol. AC-22, pp. 909-921, 1977.

Therein functional expansions involving homogeneous terms are discussed in general and compared with functional expansions involving the Frechet differential. Then the homogeneous functional expansion is used to obtain the variational equations corresponding to a given state equation, and these variational equations are solved to obtain the kernels much as I have done. My use of the Kronecker product notation is an attempt to make the mechanics of the approach more explicit. But for rigor and completeness, consult these references.

Remark 3.4 A number of methods for computing kernels or transfer functions corresponding to given differential equations have been omitted from this chapter. One example is the method for computing kernels given in the paper by R. Flake mentioned in Remark 3.1. Another example is the procedure for computing triangular kernels given in

C. Lesiak, A. Krener, "The Existence and Uniqueness of Volterra Series for Nonlinear Systems," *IEEE Transactions on Automatic Control,* Vol. AC-23, pp. 1090-1095, 1978.

A method for calculating transfer functions which is much different from the growing exponential approach is discussed in

R. Parente, "Nonlinear Differential Equations and Analytic System Theory," *SIAM Journal on Applied Mathematics,* Vol. 18, pp. 41-66, 1970.

Remark 3.5 While the methods that have been discussed all solve more or less the same problem, there are important differences. A notable feature of the Carleman linearization approach for linear-analytic systems is that the kernels are prescribed in terms of a simple general form. That this general form involves quantities of very high dimension is clear, and it seems fair to say that the Carleman linearization method trades dimensionality for simplicity. Another feature is that the kernels are found in triangular form. This means, for example, that the Carleman linearization method can be used to find the regular kernels via a simple variable change.

An appealing feature of the variational equation method is that the various degree subsystems are displayed in terms of interlocking differential equations. Also the dimensions of the quantities involved are much lower than in the Carleman linearization method. The biggest difficulty is the lack of a general form for the kernels. Not only can the kernels be difficult to compute, they are not triangular, symmetric, or regular.

The growing exponential approach is somewhat different from the others in that the symmetric transfer functions are obtained. Whether this is an advantage or disadvantage depends largely on the purpose of computing the input/output representation. The main advantage of the method seems to be that it is subtlety-free. The computations become lengthy, perhaps unwieldy, but they are of an extremely simple nature.

Remark 3.6 Elementary discussions of frequency-modulation techniques and the phase-locked loop demodulation method can be found in many books on communications. See for example

S. Haykin, *Communication Systems,* John Wiley, New York, 1978.

Volterra series analysis of the phase-locked loop using, incidentally, the variational equation method, is discussed in

H. Van Trees, "Functional Techniques for the Analysis of the Nonlinear Behavior of Phase-Locked Loops," *Proceedings of the IEEE,* Vol. 52, pp. 894-911, 1964.

The pendulum example is discussed in

R. Parente, "Functional Analysis of Systems Characterized by Nonlinear Differential Equations," MIT RLE Technical Report No. 444, 1966.

and in Chapter 8 of

M. Schetzen, *The Volterra and Wiener Theories of Nonlinear Systems,* John Wiley, New York, 1980.

The basic theory of ideal DC machines is developed in many texts. See for example

A. Fitzgerald, C. Kingsly, *Electric Machinery,* 2nd ed., McGraw-Hill, New York, 1961.

The report by Parente mentioned above also contains Volterra series analyses of series- and shunt-wound DC motors.

3.8 Problems

3.1. If A is a 3 x 3 matrix, show how to compute $A^{[2]}$ from $A^{(2)}$.

3.2. Find the first three kernels corresponding to the scalar state equation

$$\dot{x}(t) = \cos[x(t)] + u(t)$$

$$y(t) = x(t), \quad x(0) = \frac{\pi}{2}, \ t \geqslant 0$$

by both the Carleman linearization method, and the variational equation method. Find the first three symmetric transfer functions using the growing exponential method.

3.3. Suppose $x(t)$ satisfies the differential equation

$$\dot{x}(t) = \begin{bmatrix} 0 & 1 \\ -a_0 & -a_1 \end{bmatrix} x(t), \quad x(0) = x_0$$

Find linear differential equations that are satisfied by $x^{(2)}(t)$ and $x^{[2]}(t)$.

3.4. Write the system described by

$$\ddot{x}(t) = u(t), \quad y(t) = x^2(t)$$

as a bilinear state equation of the form (16).

3.5. Use the growing exponential method to find the first three symmetric transfer functions corresponding to the Nth-order differential equation:

$$y^{(N)}(t) + a_{N-1}y^{(N-1)}(t) + \cdots + a_0 y(t) + \sum_{k=2}^{\infty} b_k y^k(t) = u(t)$$

3.6. Show that for a stationary state equation, the variational equation method implies that the system can be represented as an interconnection structured system.

3.7. Derive an expression for the kernels of the bilinear state equation using the variational equation method.

3.8. Consider the bilinear state equation with polynomial output,

$$\dot{x}(t) = Ax(t) + Dx(t)u(t) + bu(t)$$
$$y(t) = c^1x(t) + c_2x^{(2)}(t) + \cdots + c_Nx^{(N)}(t)$$

Show how to rewrite this as a bilinear state equation in the form (16).

3.9. Locate a proof of the existence of solutions for linear state equations using the method of successive approximations. Using the successive approximations defined for the bilinear state equation,

$$\dot{x}_0(t) = A(t)x_0(t) + b(t)u(t), \quad x_0(0) = 0$$
$$\dot{x}_j(t) = A(t)x_j(t) + D(t)x_{j-1}(t)u(t) + b(t)u(t), \quad x_j(0) = 0,\ j > 0$$

rewrite the proof to show existence of solutions for the bilinear case.

3.10. Verify the Volterra system representation for bilinear state equations by differentiating the expression for $z(t)$ in (22) and substituting into the differential equation in (17).

3.11. Show uniform convergence of the resubstitution procedure in the linear case (see (5)) by filling in the following outline. Assume $A(t)$ is continuous on $[0,T]$, and assume the existence of a unique, continuous solution $x(t)$ on $[0,T]$. Conclude that $\|A(t)\| \leqslant K_1$, $\|x(t)\| \leqslant K_2$ for $t \in [0,T]$. Show that

$$\|\int_0^t A(\sigma_1)\int_0^{\sigma_1} A(\sigma_2)\cdots\int_0^{\sigma_{n-1}} A(\sigma_n)x(\sigma_n)\,d\sigma_n\cdots d\sigma_1\| \leqslant \frac{K_1^nK_2T^n}{n!}$$

Conclude uniform convergence of the Peano-Baker series for $\Phi(t,0)$ on $[0,T]$.

3.12. Show that the bilinear state equation (16) can be written in the form

$$\dot{x}_1(t) = A_1(t)x_1(t) + D_1(t)x_1(t)u(t)$$
$$y(t) = c_1(t)x_1(t), \quad x_1(0) = x_{10}$$

by defining the state vector according to

$$x_1(t) = \begin{bmatrix} 1 \\ x(t) \end{bmatrix}$$

Then show that (16) also can be written in the form

$$\dot{z}(t) = D_2(t)z(t)u(t)$$
$$y(t) = c_2(t)z(t), \quad z(0) = z_0$$

3.13. Use Problem 3.12 to establish the form of the Volterra system representation, and the relevant convergence conditions, for a bilinear state equation using the following device. Let $\Phi(t,\tau)$ be the transition matrix for $u(t)D_2(t)$ and then write the solution of the system in terms of the Peano-Baker series for $\Phi(t,\tau)$.

3.14. Compute the degree-5 symmetric transfer function for the pendulum system in Example 3.8 using the growing exponential method.

3.15. Show that a degree-5 transfer function for the first-order phase-locked loop is

$$H(s_1,\ldots,s_5) = H_1(s_1+\cdots+s_5)[\frac{K}{2}H_1(s_1)H_1(s_2)H_{3sym}(s_3,s_4,s_5)$$
$$-\frac{K}{5!}H_1(s_1)\cdots H_1(s_5)]$$

3.16. From the general expression for the Volterra system representation of a bilinear state equation with constant coefficient matrices, derive a closed-form expression for the solution when A and D commute.

APPENDIX 3.1 Convergence of the Volterra Series Representation for Linear-Analytic State Equations

A sufficient condition for the existence of a convergent Volterra system representation for a system described by a linear-analytic state equation is stated in Theorem 3.1 at the beginning of Section 3.3. The purpose of this appendix is to give a detailed sketch of a proof of that theorem, and to point out that the proof also yields an interesting alternative statement of the convergence conditions. The proof uses the variational equation approach given in Section 3.4, although this is known in mathematics as the Poincare expansion.[1]

I will begin by considering an analytic differential equation containing a real parameter α

$$\dot{x}(t) = f(x(t),t) + \alpha g(x(t),t), \quad t \geq 0, \; x(0) = x_0 \tag{1}$$

The assumptions are that $f(x,t)$ and $g(x,t)$ are $n \times 1$ analytic functions of x and continuous functions of t on $R^n \, X \, [0,\infty)$. (Local versions of these assumptions can be used with no essential complications, but with some loss in simplicity of exposition. Specifically, it can be assumed that $f(x,t)$ and $g(x,t)$ are analytic for x in some neighborhood of the solution of (1) when $\alpha = 0$.) Assume that for $\alpha = 0$ the differential equation has a solution defined for $t \in [0,T]$. Then the variable-change argument at the beginning of Section 3.3 can be applied, and it suffices to consider the case where $x(0) = 0$, and where the solution for $\alpha = 0$ is $x(t) = 0$ for all $t \in [0,T]$. Thus it can be assumed that

$$f(0,t) = 0, \quad t \in [0,T] \tag{2}$$

Following the method in Section 3.4, an expansion in terms of the parameter α

---

1 A very detailed, rigorous, and complete treatment of the Poincare expansion is given in Chapter 3 of the book: F. Moulton, *Differential Equations,* Macmillan, New York, 1930. A somewhat less detailed exposition is given in Chapter 5 of the book: T. Davies, E. James, *Nonlinear Differential Equations,* Addison-Wesley, New York, 1966.

(the parenthetical subscripts are used here to distinguish terms in the expansion from components of a vector),

$$x(t) = \alpha x_{(1)}(t) + \alpha^2 x_{(2)}(t) + \cdots \tag{3}$$

can be computed for the solution of the differential equation. The major part of this appendix is devoted to establishing the convergence of such an expansion. From this convergence result, the existence of a convergent Volterra series representation for $x(t)$ will follow in a straightforward fashion.

Suppose that the n component functions $f_j(x,t)$ of $f(x,t)$ and $g_j(x,t)$ of $g(x,t)$ are represented by their power series expansions in x about $x = 0$. Then each scalar component of the right side of the vector differential equation (1) can be viewed as a power series in x and α, although only terms in α^0 and α occur, that converges for $|\alpha| \leqslant K$, $|x_j| \leqslant r$, $j = 1, 2, \ldots, n$. Here K is any positive number, and r is some sufficiently small positive number that is taken to be independent of j without loss of generality. Furthermore, a number M can be found such that whenever $|x_j| \leqslant r$, $j = 1, \ldots, n$, and $t \in [0,T]$,

$$|f_j(x,t)| \leqslant M, \quad |g_j(x,t)| \leqslant \frac{M}{K}, \quad j = 1, \ldots, n \tag{4}$$

This implies, by the Cauchy bounds for analytic functions, that the various partial derivatives of $f_j(x,t)$ and $g_j(x,t)$ at $x = 0$ are bounded as follows. Using the superscript $(i_1, \ldots, i_n)$ to denote

$$\frac{\partial^{i_1+\cdots+i_n}}{\partial x_1^{i_1} \cdots \partial x_n^{i_n}} \tag{5}$$

then, for $t \in [0,T]$,

$$|f_j^{(i_1,\ldots,i_n)}(0,t)| \leqslant \frac{i_1! \cdots i_n! M}{r^{i_1+\cdots+i_n}}, \quad i_1+\cdots+i_n = 0, 1, 2, \ldots$$

$$|g_j^{(i_1,\ldots,i_n)}(0,t)| \leqslant \frac{i_1! \cdots i_n! M/K}{r^{i_1+\cdots+i_n}}, \quad i_1+\cdots+i_n = 0, 1, 2, \ldots \tag{6}$$

where each i_j is a nonnegative integer.

The bounds (6) can be used to select a dominating function as follows. Consider the real-valued function

$$\phi(x,\alpha) = \frac{M}{K_1 r_1} \frac{(K_1 x_1 + \cdots + K_1 x_n + r_1 \alpha)\,(r_1 K_1 + K_1 x_1 + \cdots + K_1 x_n + r_1 \alpha)}{(r_1 K_1 - K_1 x_1 - \cdots - K_1 x_n - r_1 \alpha)} \tag{7}$$

where $r_1 < r$ and $K_1 < K$ are positive numbers. This is an analytic function of x and α that can be represented by its Taylor series expansion about $x = 0$, $\alpha = 0$. The series will converge for, say, $|x_j| < r_1/2n$, $j = 1, \ldots, n$, and $|\alpha| < K_1$. Moreover, a simple calculation using (6) shows

$$\phi^{(i_1,\ldots,i_n)}(0,0) \geqslant |f_j^{(i_1,\ldots,i_n)}(0,t)|, \quad t \in [0,T]$$

$$[\frac{\partial}{\partial\alpha}\phi]^{(i_1,\ldots,i_n)}(0,0) \geqslant |g_j^{(i_1,\ldots,i_n)}(0,t)|, \quad t \in [0,T] \tag{8}$$

Thus every coefficient in the power series expansion of $\phi(x,\alpha)$ about $x = 0$, $\alpha = 0$ is no less than the absolute value of the corresponding coefficient in the power series expansion of $f_j(x,t) + \alpha g_j(x,t)$, for $t \in [0,T]$.

Now consider the $n \times 1$ vector differential equation

$$\dot{z}(t) = \begin{bmatrix} \phi(z(t),\alpha) \\ \vdots \\ \phi(z(t),\alpha) \end{bmatrix}, \quad z(0) = 0 \tag{9}$$

and suppose the solution can be expressed as a convergent Taylor series in α. Then, going through the procedure in Section 3.4 gives a method for calculating the terms in the expansion

$$z(t) = \alpha z_{(1)}(t) + \alpha^2 z_{(2)}(t) + \cdots \tag{10}$$

The coefficients in the differential equations for each component of $z_{(j)}(t)$ in (10) are given by the coefficients in the power series expansion of $\phi(z,\alpha)$, while the coefficients in the differential equations for each component of $x_{(j)}(t)$ in (3) are given by the coefficients in the power series expansions of $f_j(x,t)$ and $g_j(x,t)$. It is straightforward to show from a comparison of these differential equations using (8) that each component of $x_{(j)}(t)$ is bounded in absolute value by the corresponding component of $z_{(j)}(t)$ for $t \in [0,T]$.[2] Thus, convergence of (10) for some range of α implies convergence of (3) for the same range of α.

To find the radius of convergence of (10), I will carry through the details concerning the solution of (9) and then consider the expansion of this solution in a power series in α. The solution $z(t)$ will have all components identical so that the change of variables

$$z_1(t) = \cdots = z_n(t) = \frac{r_1}{n}[w(t) - \frac{\alpha}{K_1}] \tag{11}$$

can be performed. That is,

$$w(t) = \frac{n}{r_1} z_j(t) + \frac{\alpha}{K_1}, \quad j = 1,\ldots,n \tag{12}$$

Then the new variable $w(t)$ satisfies the scalar differential equation

$$\dot{w}(t) = \frac{nM}{r_1} w(t)[1 + w(t)][1 - w(t)]^{-1}, \quad w(0) = \frac{\alpha}{K_1} \tag{13}$$

2 This is shown in detail for the case $n = 2$ in Davies and James's *Nonlinear Differential Equations*.

The scalar equation can be solved by separation of variables, and the solution is given by

$$w(t) = \frac{1-\alpha/K_1}{2\alpha/K_1} e^{\frac{-nM}{r_1}t}[1+\frac{\alpha}{K_1} - [(1+\frac{\alpha}{K_1})^2 - \frac{4\alpha}{K_1} e^{\frac{nM}{r_1}t}]^{1/2}] \tag{14}$$

This solution can be expanded in a power series in α about $\alpha = 0$, and the radius of convergence is determined by the requirement that

$$(1+\frac{\alpha}{K_1})^2 - \frac{4\alpha}{K_1} e^{\frac{nM}{r_1}t} \geqslant 0 \tag{15}$$

Adding

$$4e^{\frac{2nM}{r_1}t} - 4e^{\frac{nM}{r_1}t}$$

to both sides gives

$$[1+\frac{\alpha}{K_1} - 2e^{\frac{nM}{r_1}t}]^2 \geqslant 4e^{\frac{2nM}{r_1}t} - 4e^{\frac{nM}{r_1}t}$$

from which

$$\frac{\alpha}{K_1} \leqslant -1 + 2e^{\frac{nM}{r_1}t} - 2(e^{\frac{2nM}{r_1}t} - e^{\frac{nM}{r_1}t})^{1/2}$$

Further manipulation yields

$$\frac{\alpha}{K_1} \leqslant \frac{1-(1-e^{\frac{-nM}{r_1}t})^{1/2}}{1+(1-e^{\frac{-nM}{r_1}t})^{1/2}} \tag{16}$$

This condition gives that the solution of the scalar differential equation (13), and thus the solution of the vector differential equation (9), can be expanded in a power series in α that converges for

$$|\alpha| < K_1 \frac{1-[1-e^{\frac{-nM}{r_1}t}]}{1+[1-e^{\frac{-nM}{r_1}t}]} \tag{17}$$

Of course, since the series expansion of the solution is unique, (17) gives a condition for convergence of (10).

The condition (17) for convergence of the variational expansion can be viewed in two ways. For the specified time interval $[0,T]$, (17) gives that the expansion (3) will converge if α is sufficiently small. On the other hand, for a specified value of α, (17) gives that the expansion will converge for T sufficiently small. These two interpretations can be carried through the following application to linear-analytic state equations, although I shall explicitly deal only with the first.

Now consider a linear-analytic state equation

$$\dot{x}(t) = a(x(t),t) + b(x(t),t)u(t), \quad x(0) = x_0 \tag{18}$$

Suppose that the class of input signals is composed of continuous functions $u(t)$ for $t \in [0,T]$, that satisfy $|u(t)| \leqslant \alpha$. For any such input, setting

$$f(x,t) = a(x,t), \quad \alpha g(x,t) = b(x,t)u(t) \tag{19}$$

permits application of the convergence result to conclude that the variational expansion for (18) converges for $t \in [0,T]$ so long as α is sufficiently small. Following the process given in Section 3.4 for replacing the kth term in the variational expansion by a degree-k homogeneous integral representation completes the argument needed to prove:

Theorem A3.1 Suppose a solution to the unforced linear-analytic state equation (18) exists for $t \in [0,T]$. Then there exists an $\alpha > 0$ such that for all continuous input signals satisfying $|u(t)| < \alpha$, $t \in [0,T]$, there is a Volterra series representation for the solution of the state equation that converges for $t \in [0,T]$.

CHAPTER 4

Realization Theory

The realization problem for a given input/output representation can be viewed as the reverse of the problem considered in Chapter 3. That is, realization theory deals with computing and characterizing the properties of state-equation representations that correspond to a specified homogeneous, polynomial, or Volterra system. Of course, the specified system is assumed to be described in terms of a set of kernels or transfer functions. In particular, most of the discussion here will be for stationary systems described by the regular kernel or regular transfer function representation.

After a review of linear realization theory, realizability conditions and procedures for computing bilinear state equation realizations will be discussed for stationary homogeneous systems. Then stationary polynomial and Volterra systems will be addressed. Following a discussion of structural properties of bilinear state equations, realizability conditions for nonstationary systems in terms of (nonstationary) bilinear state equations are considered. Throughout the development, only finite-dimensional realizations are of interest—infinite-dimensional realizations are ruled out of bounds. Furthermore, emphasis is placed on the construction and properties of minimal-dimension bilinear realizations.

4.1 Linear Realization Theory

The basic realization problem in linear system theory can be stated as follows. Given a linear-system transfer function $H(s)$, find a finite-dimensional linear state equation, called a *linear realization* in this context, that has $H(s)$ as its transfer function. The linear state equations of interest take the form

$$\dot{x}(t) = Ax(t) + bu(t),\ t \geq 0$$
$$y(t) = cx(t),\ x(0) = 0 \qquad (1)$$

where $x(t)$ is the $m \times 1$ state vector, for each t an element of the state space R^m, and $u(t)$ and $y(t)$ are scalars. A direct transmission term, $du(t)$, can be added to the output equation without changing the basic development, but that will not be done here. For economy, the linear state equation (1) will be denoted by (A,b,c,R^m).

It is natural to consider the realization problem in two parts. First, find necessary and sufficient conditions on $H(s)$ for linear realizability. That is, find conditions such that a linear realization (finite dimension) exists for the given system. Second, for a linear realizable system find a method for computing A, b, and c. Usually it is of interest to find a *minimal linear realization;* a realization with dimension m as small as possible.

The linear-realizability question is very simple, as the reader is no doubt aware. It is clear that strictly-proper rationality of the transfer function $H(s)$ is a necessary condition for linear realizability of the system, since the transfer function for (1) is the strictly proper rational function $c(sI - A)^{-1}b$. This condition also is sufficient, as can be shown by using well known forms of the state equation (1), which can be written by inspection from the coefficients of $H(s)$. While this familiar development could be pursued to the construction of minimal linear realizations, I will review a different approach, one that extends more easily to the nonlinear case. In fact, because of the similarity of ideas in the linear and nonlinear realization theories, the review of the linear case will be more detailed than usual.

Using the well known series expansion

$$(sI - A)^{-1} = Is^{-1} + As^{-2} + A^2s^{-3} + \cdots \tag{2}$$

the transfer function of the linear state equation (1) can be written as a negative power series

$$c(sI - A)^{-1}b = cbs^{-1} + cAbs^{-2} + cA^2bs^{-3} + \cdots \tag{3}$$

(For simplicity of notation, I leave the dimension of identity matrices to be fixed by conformability requirements.) This makes clear the fact that for linear realizability it suffices to consider only those transfer functions $H(s)$ that can be represented by a negative power series of the form

$$H(s) = h_0s^{-1} + h_1s^{-2} + h_2s^{-3} + \cdots \tag{4}$$

In other words, only transfer functions that are analytic at infinity and that have a zero at infinity need be considered. Comparison of (3) and (4) shows that, from the series viewpoint, the basic mathematical problem in linear realization theory involves finding matrices A, b, and c, of dimensions $m \times m$, $m \times 1$, and $1 \times m$ such that

$$cA^jb = h_j,\ j = 0, 1, 2, \ldots \tag{5}$$

The first step in solving this basic problem will be to construct a particularly simple *abstract realization.* That is, a realization wherein A, b, and c are specified as linear operators involving a specially chosen linear space as the state space. Then matrix representations can be computed for these linear operators when the state space is replaced by R^m.

Suppose $V(s)$ is any negative power series of the form

$$V(s) = v_0 s^{-1} + v_1 s^{-2} + v_2 s^{-3} + \cdots \tag{6}$$

A *shift operator* S is defined according to

$$SV(s) = v_1 s^{-1} + v_2 s^{-2} + v_3 s^{-3} + \cdots \tag{7}$$

In words, the action of the shift operator is to slide the coefficients of the series one position to the left while dropping the original left-most coefficient. Clearly $SV(s)$ is a negative power series so that the shift operator can be applied repeatedly. The usual notation $S^j V(s)$ is used to denote j applications of S.

Using the shift operator and a given transfer function $H(s)$, a linear space of negative power series over the real field, with the usual definitions of addition and scalar multiplication, can be specified as follows. Let

$$U = span\ \{\ H(s),\ SH(s),\ S^2H(s),\ \ldots\ \} \tag{8}$$

Clearly the shift operator is a linear operator on U, $S:U \to U$. Now define the *initialization operator* $L:R \to U$ to be the linear operator specified by the given transfer function (viewed as a series), so that for any real number r,

$$Lr = H(s)r \tag{9}$$

Finally, define the *evaluation operator* $E:U \to R$ by

$$EV(s) = E(v_0 s^{-1} + v_1 s^{-2} + \cdots) = v_0 \tag{10}$$

where $V(s)$ is any element of U.

I should point out that when $H(s)$ is viewed as the function given by the sum of the negative power series, the shift operator and the space U can be reinterpreted. Indeed, when $V(s)$ is a function corresponding to a negative power series in s,

$$SV(s) = sV(s) - [sV(s)]|_{s=\infty}$$
$$EV(s) = [sV(s)]|_{s=\infty} \tag{11}$$

and U becomes a linear space of functions of s. Although the negative power series representation is often most convenient to demonstrate properties and prove results, the interpretation in (11) is usually better for the examples and problems in the sequel.

It is very simple to demonstrate that the linear operators S, L, and E form an abstract realization on the linear space U. This is called the *shift realization*, and it is written as (S,L,E,U). The verification of the realization involves nothing more than the calculations:

$$\begin{aligned} ES^0L &= EH(s) = E(h_0 s^{-1} + h_1 s^{-2} + \cdots) = h_0 \\ ESL &= ESH(s) = E(h_1 s^{-1} + h_2 s^{-2} + \cdots) = h_1 \\ ES^2L &= ES(h_1 s^{-1} + h_2 s^{-2} + \cdots) \\ &= E(h_2 s^{-1} + h_3 s^{-2} + \cdots) = h_2 \end{aligned} \tag{12}$$

and so on. The appropriate interpretation here is that each constant h_j represents a linear operator, $h_j:R \rightarrow R$, which is given by the composition of linear operators E, S^j, and L.

To find a concrete (unabstract) realization of the form (1), it remains to replace U by a linear space R^m, and to find matrix representations for the operators S, L, and E with respect to this replacement. There are many ways to do this, each of which gives a different matrix structure for the realization. But central to the replacement is the following result, the proof of which exhibits one particular construction of a realization.

Theorem 4.1 A linear system described by the transfer function $H(s)$ is linear realizable if and only if U is finite dimensional. Furthermore, if the system is linear realizable, then (S,L,E,U) is a minimal linear realization.

Proof Suppose $H(s)$ is linear realizable, and that (A,b,c,R^m) is any linear realization of $H(s)$. Letting W be the linear space of all negative power series, define a linear operator $\Phi:R^m \rightarrow W$ according to

$$\Phi(x) = cxs^{-1} + cAxs^{-2} + cA^2xs^{-3} + \cdots$$

Clearly $H(s) \in R[\Phi]$ since

$$\Phi(b) = cbs^{-1} + cAbs^{-2} + cA^2bs^{-3} + \cdots$$

Also, from

$$\begin{aligned}\Phi(A^jb) &= cA^jbs^{-1} + cA^{j+1}bs^{-2} + cA^{j+2}bs^{-3} + \cdots \\ &= S^jH(s)\end{aligned}$$

it is clear that $S^jH(s) \in R[\Phi]$. Thus $U \subseteq R[\Phi]$, and it follows that dimension $U \leqslant m$ since Φ is a linear operator on an m-dimensional space. Moreover, this argument shows that the dimension of U is less than or equal to the state-space dimension of any linear realization of $H(s)$. Thus (S,L,E,U) is a minimal-dimension linear realization.

Now suppose that U has finite dimension m. Then Problem 4.3 shows that

$$H(s),\ SH(s),\ S^2H(s),\ \ldots,\ S^{m-1}H(s)$$

is a basis for U, and thus U can be replaced by R^m by choosing the standard ordered basis vectors $e_j \in R^m$ according to

$$e_1 = H(s),\quad e_2 = SH(s),\ \ldots,\ e_m = S^{m-1}H(s)$$

Writing $S^mH(s)$ as a linear combination of $H(s), SH(s), \ldots, S^{m-1}H(s)$, say

$$S^mH(s) = \sum_{j=0}^{m-1} r_{m-1-j}S^jH(s)$$

the shift operator can be represented on R^m by the $m \times m$ matrix

$$A = \begin{bmatrix} 0 & 0 & \cdots & 0 & r_{m-1} \\ 1 & 0 & \cdots & 0 & r_{m-2} \\ 0 & 1 & \cdots & 0 & r_{m-3} \\ \vdots & \vdots & & \vdots & \vdots \\ 0 & 0 & \cdots & 1 & r_0 \end{bmatrix}$$

Viewing the initialization operator as $L{:}R \rightarrow R^m$, it is clear that L can be represented by the $m \times 1$ vector corresponding to $H(s)$:

$$b = e_1 = \begin{bmatrix} 1 \\ 0 \\ \vdots \\ 0 \end{bmatrix}$$

Finally, the evaluation operator can be viewed as $E{:}R^m \rightarrow R$, and a matrix representation can be computed as follows. By the correspondence between U and R^m, it is clear that $Ee_{j+1} = h_j$ for $j = 0, 1, 2, \ldots, m-1$, and thus E is represented by

$$c = [\, h_0 \; h_1 \; \cdots \; h_{n-1} \,]$$

This construction of a realization completes the proof.

Before presenting an example, it is appropriate to use the shift realization formulation to recover the well known result on rationality mentioned at the beginning of this section. There are more direct proofs, but the one given here should clarify the nature of the linear space U.

Theorem 4.2 A linear system described by the transfer function $H(s)$ is linear realizable if and only if $H(s)$ is a strictly proper rational function.

Proof If $H(s)$ is the strictly proper rational function

$$H(s) = \frac{b_{m-1}s^{m-1} + b_{m-2}s^{m-2} + \cdots + b_0}{s^m + a_{m-1}s^{m-1} + \cdots + a_0}$$

then, from (11),

$$\begin{aligned} SH(s) &= \frac{b_{m-1}s^m + b_{m-2}s^{m-1} + \cdots + b_0 s}{s^m + a_{m-1}s^{m-1} + \cdots + a_0} - b_{m-1} \\ &= \frac{(b_{m-2} - b_{m-1}a_{m-1})s^{m-1} + (b_{m-3} - b_{m-1}a_{m-2})s^{m-2} + \cdots + (-b_{m-1}a_0)}{s^m + a_{m-1}s^{m-1} + \cdots + a_0} \end{aligned}$$

Similarly,

$$S^2H(s) = \frac{(b_{m-3}-b_{m-1}a_{m-2}-b_{m-2}a_{m-1}+b_{m-1}a_{m-1}^2)s^{m-1} + \cdots + (-b_{m-2}a_0+b_{m-1}a_{m-1}a_0)}{s^m + a_{m-1}s^{m-1} + \cdots + a_0}$$

It should be clear from just these two calculations that for any $j \geqslant 0$, $S^jH(s)$ is a strictly proper rational function with the same denominator as $H(s)$. Only the numerator polynomial changes with application of the shift operator. Thus every element in U can be viewed as a rational function which is strictly proper with the same denominator. Only the numerator polynomial can differ from element to element. Since polynomials of degree at most $m-1$ form a linear space of dimension at most m, it follows that dimension $U \leqslant m$, and thus that $H(s)$ is linear realizable. The proof of the converse, as mentioned at the beginning of the section, follows by calculation of the transfer function for a linear state equation.

Example 4.1 For the strictly proper rational transfer function

$$H(s) = \frac{4s^2+7s+3}{s^3+4s^2+5s+2}$$

simple calculations give

$$SH(s) = \frac{4s^3+7s^2+3s}{s^3+4s^2+5s+2} - 4 = \frac{-9s^2-17s-8}{s^3+4s^2+5s+2}$$

and

$$S^2H(s) = \frac{-9s^3-17s^2-8s}{s^3+4s^2+5s+2} + 9 = \frac{19s^2+37s+18}{s^3+4s^2+5s+2}$$

It is clear that $H(s)$ and $SH(s)$ are linearly independent in U, but one more calculation shows that

$$S^2H(s) = -3SH(s) - 2H(s)$$

Thus U can be replaced by R^2 by choosing the standard ordered basis elements according to

$$\begin{bmatrix}1\\0\end{bmatrix} = H(s), \quad \begin{bmatrix}0\\1\end{bmatrix} = SH(s)$$

A matrix representation for the initialization operator on this basis clearly is

$$b = \begin{bmatrix}1\\0\end{bmatrix}$$

Also, since the matrix representation for the shift operator must satisfy

$$A\begin{bmatrix}1\\0\end{bmatrix} = \begin{bmatrix}0\\1\end{bmatrix}, \quad A\begin{bmatrix}0\\1\end{bmatrix} = \begin{bmatrix}-2\\-3\end{bmatrix}$$

it follows that

$$A = \begin{bmatrix} 0 & -2 \\ 1 & -3 \end{bmatrix}$$

Finally, since $EH(s) = 4$ and $ESH(s) = -9$, a matrix representation for the evaluation operator is

$$c = [\,4 \quad -9\,]$$

That a dimension-2 realization has been obtained for a degree-3 transfer function can be explained by factoring the numerator and denominator of H(s) to write

$$H(s) = \frac{(4s+3)(s+1)}{(s+2)(s+1)^2}$$

The linear independence calculations involved in constructing the shift realization "automatically" canceled the common factor in the numerator and denominator.

The realization theory just presented can be rephrased to yield a well known rank condition test for realizability. Viewing U as a linear space of negative power series, each element

$$S^jH(s) = h_js^{-1} + h_{j+1}s^{-2} + h_{j+2}s^{-3} + \cdots \ , \ j = 0,1,2,\ldots \tag{13}$$

can be replaced by the corresponding sequence of coefficients

$$(\, h_j, \ h_{j+1}, \ h_{j+2}, \ \ldots), \ j = 0,1,2,\ldots \tag{14}$$

Then it is clear that U is finite dimensional if and only if only a finite number of these sequences are linearly independent. Arranging this idea in an orderly fashion, and including the fact that the dimension of U is the dimension of the minimal linear realizations of $H(s)$, gives a familiar result in linear system theory.

Theorem 4.3 A linear system described by the transfer function $H(s)$ in (4) is linear realizable if and only if the *Behavior matrix*

$$B_H = \begin{bmatrix} h_0 & h_1 & h_2 & \cdots \\ h_1 & h_2 & h_3 & \cdots \\ h_2 & h_3 & h_4 & \cdots \\ \vdots & \vdots & \vdots & \vdots \end{bmatrix} \tag{15}$$

has finite rank. Furthermore, for a linear-realizable system the rank of B_H is the dimension of the minimal linear realizations of $H(s)$.

I could go on from here to outline the construction of minimal realizations directly from the Behavior matrix. But this is intended to be a brief review, so all of that will be left to the references. I will also omit reviewing the equivalence properties of minimal linear realizations of a given $H(s)$, and the connections with the concepts of reachability and observability. These topics will arise in Section 4.4 in conjunction with bilinear realizations, and taking $D = 0$ in that material captures most of the linear theory here being skipped.

However, before leaving the topic of stationary linear realization theory, I should point out that little changes if the starting point is a given kernel, instead of a given transfer function. Since a strictly proper rational transfer function corresponds precisely to a kernel of the exponential form

$$h(t) = \sum_{i=1}^{m} \sum_{j=1}^{\sigma_i} a_{ij} t^{j-1} e^{\lambda_i t}, \quad t \geqslant 0 \tag{16}$$

it is clear that $h(t)$ is realizable by a linear state equation if and only if it has the form (16). To proceed via the shift realization approach, note that a given kernel can be assumed to be analytic for $t \geqslant 0$, for otherwise it is clear that it cannot be realizable by a linear state equation. Expanding $h(t)$ in a power series about $t = 0$, the Laplace transform of $h(t)$ can be written in the form

$$\begin{aligned}
L[h(t)] &= \int_0^\infty h(t) e^{-st}\, dt \\
&= h(0) \int_0^\infty e^{-st}\, dt + h^{(1)}(0) \int_0^\infty \frac{t}{1!} e^{-st}\, dt + h^{(2)}(0) \int_0^\infty \frac{t^2}{2!} e^{-st}\, dt + \cdots \\
&= h(0) s^{-1} + h^{(1)}(0) s^{-2} + h^{(2)}(0) s^{-3} + \cdots
\end{aligned} \tag{17}$$

where

$$h^{(j)}(0) = \frac{d^j}{dt^j} h(t) \Big|_{t=0}$$

Thus, the entries in (4) are specified by the derivatives of the kernel evaluated at 0, $h_j = h^{(j)}(0)$. From this point, construction of the shift realization proceeds just as before.

For multi-input, multi-output linear systems, the theory of realization becomes more subtle. Although only single-input, single-output systems are under consideration in this book, the basic linear realizability result for the multivariable case will arise as a technical consideration in Section 4.2. Therefore, a brief comment is appropriate.

Consider the linear state equation

$$\begin{aligned}
\dot{x}(t) &= Ax(t) + Bu(t), \ t \geqslant 0 \\
y(t) &= Cx(t), \ x(0) = 0
\end{aligned} \tag{18}$$

where $x(t)$ is m x 1, the input $u(t)$ is an r x 1 vector, and $y(t)$ is a q x 1 vector. The corresponding transfer function is the q x r matrix

$$H(s) = C(sI - A)^{-1}B \tag{19}$$

Theorem 4.4 Suppose a linear system is described by the q x r transfer function matrix $H(s)$. Then the system is realizable by a finite-dimensional linear state equation of the form (18) if and only if $H(s)$ is a strictly proper rational matrix. That is, if and only if each element $H_{ij}(s)$ of $H(s)$ is a strictly proper rational function.

The necessity portion of Theorem 4.4 is clear upon writing $(sI - A)^{-1}$ in the classical adjoint over determinant form. Sufficiency is equally easy: each strictly proper, rational $H_{ij}(s)$ can be realized by a state equation of the form (1), and then all these state equations can be combined to give an r-input, q-output state equation of the form (18). It is in the question of minimal-dimension realizations that things get more difficult, but these issues will not arise in the sequel and so they will be ignored here.

The question of realizability is of interest for nonstationary systems also. Recalling the definitions of stationarity and separability in Chapter 1, the results will be stated in terms of the input/output representation

$$y(t) = \int_{-\infty}^{\infty} h(t,\sigma)u(\sigma)\, d\sigma \tag{20}$$

In the nonstationary case, a linear state equation realization with time-variable coefficients will be of interest,

$$\begin{aligned} \dot{x}(t) &= A(t)x(t) + b(t)u(t) \\ y(t) &= c(t)x(t) \end{aligned} \tag{21}$$

It is convenient for technical reasons to require that $A(t)$, $b(t)$, and $c(t)$ be continuous matrix functions. That is, each entry in these coefficient matrices is a continuous function.

Theorem 4.5 The kernel $h(t,\sigma)$ is realizable by a finite-dimensional, time-variable linear state equation if and only if it is separable.

Proof If the kernel is linear realizable, and (21) is a realization of $h(t,\sigma)$, then

$$h(t,\sigma) = c(t)\Phi(t,\sigma)b(\sigma)$$

Writing

$$c(t)\Phi(t,0) = [v_{01}(t) \quad \cdots \quad v_{0n}(t)]\,, \quad \Phi(0,\sigma)b(\sigma) = \begin{bmatrix} v_{11}(\sigma) \\ \vdots \\ v_{1n}(\sigma) \end{bmatrix}$$

shows that $h(t,\sigma)$ is of the separable form

$$h(t,\sigma) = \sum_{i=1}^{n} v_{0i}(t)v_{1i}(\sigma) \tag{22}$$

The continuity required by separability is furnished by the continuity assumptions on the linear state equation.

Now suppose $h(t,\sigma)$ is separable and, in fact, is given by (22). (Since $h(t,\sigma)$ is real-valued, it can be assumed that each $v_{ji}(.)$ is real.) Then setting

$$A(t) = 0\,, \quad b(t) = \begin{bmatrix} v_{11}(t) \\ \vdots \\ v_{1n}(t) \end{bmatrix}, \quad c(t) = [v_{01}(t) \ \cdots \ v_{0n}(t)]$$

in (21) gives a realization for $h(t,\sigma)$. Notice that this realization has continuous coefficient matrices since separability implies that $v_{0i}(t)$ and $v_{1i}(\sigma)$ in (22) are continuous.

An obvious question, and the one of most interest here, deals with when a kernel $h(t,\sigma)$ can be realized by a constant-parameter linear state equation. In other words, when is an input/output representation written in nonstationary form actually realizable by a stationary system?

Theorem 4.6 The kernel $h(t,\sigma)$ is realizable by a finite-dimensional, constant-parameter linear state equation if and only if it is stationary and differentiably separable.

Proof Necessity of the conditions follows directly from the form of $h(t,\sigma)$ given by a constant-parameter realization. The sufficiency proof is more subtle, and so I will begin by considering the special case where the kernel is stationary, differentiably separable, and of the form

$$h(t,\sigma) = v_0(t)v_1(\sigma)$$

where $v_0(t)$ and $v_1(\sigma)$ are (necessarily) real, differentiable functions. The first step is to pick $T > 0$ so that

$$q_1 = \int_{-T}^{T} v_0^2(t)\, dt > 0$$

Of course, it can be assumed that such T exists, for otherwise $h(t,\sigma) = 0$ and the theorem is uninteresting. Now, by stationarity, $h(t,\sigma) = h(0,\sigma - t)$ so that

$$\frac{d}{d\sigma}h(t,\sigma) + \frac{d}{dt}h(t,\sigma) = 0$$

or

$$v_0(t)\dot{v}_1(\sigma) + \dot{v}_0(t)v_1(\sigma) = 0$$

Multiplying this equation by $v_0(t)$ and integrating with respect to t from $-T$ to T gives

$$q_1\dot{v}_1(\sigma) + r_1 v_1(\sigma) = 0$$

where

$$r_1 = \int_{-T}^{T} v_0(t)\dot{v}_0(t)\, dt$$

But $q_1 > 0$ so that the differential equation is nontrivial, and thus $v_1(\sigma)$ is the exponential

$$v_1(\sigma) = v_1(0)e^{-(r_1/q_1)\sigma}$$

Then the stationarity condition gives

$$v_0(t) = v_0(0)e^{(r_1/q_1)t}$$

from which it follows that

$$h(t,\sigma) = v_0(0)v_1(0)e^{(r_1/q_1)(t-\sigma)}$$

In other words, if a kernel is stationary, differentiably separable, and single-term, then it must be a simple exponential. Clearly this kernel is realizable by a linear state equation. To complete the proof, the case where the kernel takes the more general form in (22) must be considered. If each $v_{0_i}(t)$ and $v_{1_i}(t)$ is real-valued, this is easy since an additive parallel connection of linear state equations can be represented by a linear state equation. If some of the functions in (22) are complex-valued, then it is left to the reader to show that, since conjugates must be included, a linear state equation realization with real coefficient matrices can be found.

4.2 Realization of Stationary Homogeneous Systems

For a specified homogeneous nonlinear system, the problem that will be discussed here is the problem of finding realizations in the form of bilinear state equations. That is, state equations of the form

$$\begin{aligned} \dot{x}(t) &= Ax(t) + Dx(t)u(t) + bu(t) \\ y(t) &= cx(t),\quad t \geqslant 0,\quad x(0) = 0 \end{aligned} \tag{23}$$

where $x(t)$ is the $m \times 1$ state vector, for each t an element of the state space R^m, and the input and output are scalars. (Again, much of the theory generalizes neatly to the multi-input, multi-output case.) The choice of zero initial state reflects an interest in the simplest kind of input/output behavior, although the $x(0) \neq 0$ case can be developed in a similar manner if $x(0)$ is an equilibrium state.

Of course, a bilinear state equation in general does not have a homogeneous input/output representation. Thus, the results in this section involve rather specialized bilinear state equations. Also, the bilinear realization problem for homogeneous systems is subsumed by the theory of Section 4.3 for polynomial and Volterra systems. The intent of the discussion here is to provide a leisurely introduction to the ideas, and to establish notation.

The input/output representation to be used in conjunction with (23) is derived in Chapter 3. There it is shown that the bilinear state equation (23) can be described by a Volterra system representation in which the degree-n subsystem can be written in the (nonstationary) triangular form

$$y_n(t) = \int_0^t \int_0^{\sigma_1} \cdots \int_0^{\sigma_{n-1}} h(t,\sigma_1,\ldots,\sigma_n)u(\sigma_1)\cdots u(\sigma_n)\,d\sigma_n\cdots d\sigma_1 \tag{24}$$

where the kernel is given by

$$h(t,\sigma_1,\ldots,\sigma_n) = ce^{A(t-\sigma_1)}De^{A(\sigma_1-\sigma_2)}D\cdots De^{A(\sigma_{n-1}-\sigma_n)}b,$$

$$t \geqslant \sigma_1 \geqslant \cdots \geqslant \sigma_n \geqslant 0 \tag{25}$$

For the purposes of developing a bilinear realization theory, the main emphasis will be on the regular kernel and regular transfer function. To obtain the regular kernel from (25), the first step is to impose stationarity as prescribed in Section 1.2. This gives a stationary triangular kernel:

$$g_{tri}(\sigma_1,\ldots,\sigma_n) = h(0,-\sigma_1,\ldots,-\sigma_n)$$

$$= ce^{A\sigma_1}De^{A(\sigma_2-\sigma_1)}D\cdots De^{A(\sigma_n-\sigma_{n-1})}b,$$

$$\sigma_n \geqslant \sigma_{n-1} \geqslant \cdots \geqslant \sigma_1 \geqslant 0 \tag{26}$$

Rewriting (26) as a triangular kernel over the "first" triangular domain gives

$$h_{tri}(t_1,\ldots,t_n) = ce^{At_n}De^{A(t_{n-1}-t_n)}D\cdots De^{A(t_1-t_2)}b,$$

$$t_1 \geqslant t_2 \geqslant \cdots \geqslant t_n \geqslant 0 \tag{27}$$

Thus, the regular kernel for the degree-n homogeneous subsystem corresponding to the bilinear state equation (23) is of the form

$$h_{reg}(t_1,\ldots,t_n) = ce^{At_n}De^{At_{n-1}}D\cdots De^{At_1}b \tag{28}$$

Of course, when written out in scalar terms the form of $h_{reg}(t_1,\ldots,t_n)$ is much messier. Indeed, taking into account the kinds of terms that can appear in a matrix exponential shows that the regular kernel corresponding to a bilinear state equation can be written in the form

$$h_{reg}(t_1,\ldots,t_n) = \sum_{i_1=1}^{m_1}\sum_{j_1=1}^{\mu_1}\cdots\sum_{i_n=1}^{m_n}\sum_{j_n=1}^{\mu_n} a_{i_1\cdots i_n}^{j_1\cdots j_n} \frac{t_1^{j_1-1}\cdots t_n^{j_n-1}}{(j_1-1)!\cdots(j_n-1)!}e^{-\lambda_{i_1}t_1}\cdots e^{-\lambda_{i_n}t_n} \tag{29}$$

The various coefficients and exponents in this expression can be complex, but since the regular kernel is real, well known conjugacy conditions must be satisfied.

Clearly the regular kernels for a bilinear state equation are of a particularly simple form. And taking the Laplace transform of (28) shows that the regular transfer functions also have a simple form, namely

$$H_{reg}(s_1,\ldots,s_n) = c(s_nI-A)^{-1}D(s_{n-1}I-A)^{-1}D\cdots D(s_1I-A)^{-1}b \tag{30}$$

Writing each $(s_jI-A)^{-1}$ as the classical adjoint over the determinant shows that each regular transfer function for a bilinear state equation is a *strictly proper* rational function in that the numerator polynomial degree in each variable is (strictly) less than the denominator polynomial degree in that variable. Furthermore, $H_{reg}(s_1,\ldots,s_n)$ has the very special property that the denominator polynomial can be expressed as a product of (real-coefficient) single-variable polynomials, so that the regular transfer functions for a bilinear state equation can be written in the form

$$H_{reg}(s_1,\ldots,s_n) = \frac{P(s_1,\ldots,s_n)}{Q_1(s_1)\cdots Q_n(s_n)} \tag{31}$$

A rational function of this form will be called a *recognizable function,* following the terminology of automata theory. (Of course, a recognizable function can be made to appear unrecognizable by the insertion of a common factor, for example $(s_1 + s_2)$, in the numerator and denominator polynomials. However, I will use the terminology in a manner that implicitly assumes such silliness is removed.)

The bilinear realization problem will be discussed here in terms of a given degree-n homogeneous system described by the regular transfer function. It is left understood that all the regular transfer functions of degree $\neq n$ are zero.

What has been shown to this point is that for a stationary, degree-n homogeneous system to be bilinear realizable, it is necessary that the regular transfer function be a strictly proper, recognizable function. That is, it is necessary that the regular kernel have the exponential form (28) or, equivalently, (29). The following argument shows that the condition on the regular transfer function also is sufficient for bilinear realizability of the system. In terms of the time-domain representation, this means that the exponential form of the regular kernel is necessary and sufficient for bilinear realizability of the system.

So, suppose that a degree-n homogeneous system is described by a strictly proper, recognizable, regular transfer function of the form (31), where

$$P(s_1,\ldots,s_n) = \sum_{i_1=0}^{m_1-1} \cdots \sum_{i_n=0}^{m_n-1} p_{i_1\cdots i_n} s_1^{i_1} \cdots s_n^{i_n} \tag{32}$$

and where $Q_j(s_j)$ is a monic polynomial of degree m_j, $j = 1,\ldots,n$. In order to construct a corresponding bilinear realization, it is convenient to write the numerator in a matrix factored form

$$P(s_1,\ldots,s_n) = S_n\, S_{n-1} \cdots S_1\, P \tag{33}$$

where each S_j is a matrix with entries that involve only the variable s_j, and P is a vector of coefficients from $P(s_1,\ldots,s_n)$. This corresponds in the $n = 1$ case to writing a polynomial as a product of a row vector of variables $[1\ s\ s^2 \cdots s^{m_1-1}]$ and a column vector of coefficients $[p_0\ p_1 \cdots p_{m_1-1}]'$. Before giving the somewhat messy prescription for (33) in the general case, an example will show just how simple the construction is.

Example 4.2 For the polynomial

$$P(s_1,s_2,s_3) = s_1s_2s_3 + s_1s_2 + s_1s_3 + s_2s_3 + s_2 + 1$$

it is a simple matter to factor out the dependence on s_3 by writing

$$P(s_1,s_2,s_3) = [1 \quad s_3] \begin{bmatrix} s_1s_2 + s_2 + 1 \\ s_1s_2 + s_1 + s_2 \end{bmatrix}$$

Now, the s_2 dependence in each polynomial on the right side of this expression can be factored out in a similar way:

$$P(s_1,s_2,s_3) = [1 \quad s_3] \begin{bmatrix} 1 & s_2 & 0 & 0 \\ 0 & 0 & 1 & s_2 \end{bmatrix} \begin{bmatrix} 1 \\ s_1+1 \\ s_1 \\ s_1+1 \end{bmatrix}$$

The last step should be obvious, yielding

$$P(s_1,s_2,s_3) = [1 \quad s_3] \begin{bmatrix} 1 & s_2 & 0 & 0 \\ 0 & 0 & 1 & s_2 \end{bmatrix} \begin{bmatrix} 1 & s_1 & 0 & 0 & 0 & 0 & 0 & 0 \\ 0 & 0 & 1 & s_1 & 0 & 0 & 0 & 0 \\ 0 & 0 & 0 & 0 & 1 & s_1 & 0 & 0 \\ 0 & 0 & 0 & 0 & 0 & 0 & 1 & s_1 \end{bmatrix} \begin{bmatrix} 1 \\ 0 \\ 1 \\ 1 \\ 0 \\ 1 \\ 1 \\ 1 \end{bmatrix}$$

The general prescription for (33) in terms of (32) goes as follows. Let

$$S_n = [1 \;\; s_n \cdots s_n^{m_n-1}] \tag{34}$$

and for $j = 1,\ldots,n-1$, define S_j to be the $(m_n \cdots m_{j+1})$ x $(m_n \cdots m_j)$ matrix with i^{th} row

$$[\, 0_{1\,x\,(im_j-m_j)} \;\, 1 \;\, s_j \cdots s_j^{m_n-1} \;\, 0_{1\,x\,(m_n \cdots m_j - im_j)} \,] \tag{35}$$

Then P is the column vector specified by

$$P' = [p_{0\cdots 0}\; p_{10\cdots 0} \cdots p_{m_1-1,0\cdots 0}\; p_{010\cdots 0}\; p_{110\cdots 0}$$
$$\cdots\; p_{m_1-1,10\cdots 0} \cdots p_{0m_2-1,\ldots,m_n-1} \cdots p_{m_1-1,\ldots,m_n-1}] \tag{36}$$

The result of this numerator factorization procedure is that the regular transfer function can be written in the factored form

$$H_{reg}(s_1,\ldots,s_n) = G_n(s_n) \cdots G_1(s_1) \tag{37}$$

where

$$G_1(s_1) = \frac{S_1P}{Q_1(s_1)}\,, \quad G_j(s_j) = \frac{S_j}{Q_j(s_j)}\,, \quad j = 2,\ldots,n \tag{38}$$

are strictly proper, matrix rational functions. Thus each $G_j(s_j)$ has a linear realization, and can be written in the form

$$G_j(s_j) = C_j(s_jI - A_j)^{-1}B_j \tag{39}$$

Now consider the bilinear state equation specified by

$$A = \begin{bmatrix} A_1 & 0 & \cdots & 0 \\ 0 & A_2 & \cdots & 0 \\ \vdots & \vdots & \vdots & \vdots \\ 0 & 0 & \cdots & A_n \end{bmatrix}, \quad b = \begin{bmatrix} B_1 \\ 0 \\ \vdots \\ 0 \end{bmatrix}$$

$$D = \begin{bmatrix} 0 & 0 & \cdots & 0 & 0 \\ B_2C_1 & 0 & \cdots & 0 & 0 \\ 0 & B_3C_2 & \cdots & 0 & 0 \\ \vdots & \vdots & \vdots & \vdots & \vdots \\ 0 & 0 & \cdots & B_nC_{n-1} & 0 \end{bmatrix}, \quad c = [\,0 \ \cdots \ 0 \ C_n\,] \tag{40}$$

The regular transfer functions for this bilinear state equation can be computed via (30). Due to the block-diagonal form of A,

$$(sI - A)^{-1} = \begin{bmatrix} (sI-A_1)^{-1} & & 0 \\ 0 & \ddots & 0 \\ \vdots & & \vdots \\ 0 & & (sI-A_n)^{-1} \end{bmatrix}$$

so a straightforward computation gives

$$H_{reg}(s_1,\ldots,s_k) = 0, \quad k = 1,\ldots,n-1,n+1,n+2,\ldots$$

$$\begin{aligned} H_{reg}(s_1,\ldots,s_n) &= C_n(s_nI-A_n)^{-1}B_nC_{n-1}(s_{n-1}I-A_{n-1})^{-1}B_{n-1}C_{n-2} \\ &\quad \cdots \ B_2C_1(s_1I-A_1)^{-1}B_1 \\ &= G_n(s_n) \ \cdots \ G_1(s_1) \end{aligned} \tag{41}$$

Thus (40) is a degree-n homogeneous bilinear realization for the given regular transfer function. This development can be summarized as follows.

Theorem 4.7 A degree-n homogeneous system described by the regular transfer function $H_{reg}(s_1,\ldots,s_n)$ is bilinear realizable if and only if $H_{reg}(s_1,\ldots,s_n)$ is a strictly proper, recognizable function.

In addition to the realizability condition in Theorem 4.7, the development above indicates that the bilinear realization problem for a degree-n homogeneous system essentially involves a sequence of n linear realization problems. But the simple factorization procedure used to obtain (37) usually leads to a bilinear realization of quite high dimension, even if minimal linear realizations of each $G_j(s_j)$ are used. To construct a minimal-dimension bilinear realization, a more sophisticated factorization procedure can be used, but I will not pursue that approach further. (See Remark 4.5.)

An alternative approach to bilinear realization theory for a given regular transfer function of the form (31) involves the notion of an abstract shift realization similar to that in the linear case. I will present this approach in detail since it directly provides minimal-dimension bilinear realizations, and since it will be the main tool for polynomial and Volterra systems. The shift realization approach is most easily introduced in terms of a negative power series representation of $H_{reg}(s_1,\ldots,s_n)$ of the form

$$H_{reg}(s_1,\ldots,s_n) = \sum_{i_1=0}^{\infty} \cdots \sum_{i_n=0}^{\infty} h_{i_1 \cdots i_n} s_1^{-(i_1+1)} \cdots s_n^{-(i_n+1)} \tag{42}$$

For strictly proper, recognizable transfer functions of the form (30), the validity of this series representation is clear upon repeated use of the expansion (2). The general setting for the shift realization approach can be taken to be the class of regular transfer functions that are analytic at infinity, and have zeros at infinity in each variable; though it is clear from Theorem 4.7 that this generality will not be needed here. At any rate, comparing (42) with the series form for (30) shows that the basic mathematical problem can be stated as follows. Find matrices A, D, b, and c, of dimensions $m \times m$, $m \times m$, $m \times 1$, and $1 \times m$, such that for all nonnegative integers $j_1, j_2, j_3, \ldots,$

$$cA^{j_k}DA^{j_{k-1}}D \cdots DA^{j_1}b = \begin{cases} h_{j_1 \cdots j_n}, & k = n \\ 0, & k \neq n \end{cases} \tag{43}$$

Similar to the linear case, it is convenient to use the notation (A,D,b,c,R^m) to indicate an m-dimensional bilinear realization corresponding to the given regular transfer function.

For any negative power series in k variables,

$$V(s_1,\ldots,s_k) = \sum_{i_1=0}^{\infty} \cdots \sum_{i_k=0}^{\infty} v_{i_1 \cdots i_k} s_1^{-(i_1+1)} \cdots s_k^{-(i_k+1)} \tag{44}$$

define the *shift operator* S by

$$SV(s_1,\ldots,s_k) = \sum_{i_1=0}^{\infty} \cdots \sum_{i_k=0}^{\infty} v_{i_1+1,i_2 \cdots i_k} s_1^{-(i_1+1)} \cdots s_k^{-(i_k+1)} \tag{45}$$

Notice that the shift involves only the s_1 variable, and that for $k = 1$ it reduces to the shift operator defined in Section 4.1. Clearly S is a linear operator, and $SV(s)$ is a negative power series in k variables so that $S^jV(s)$ is well defined.

Also needed is an *index operator* T that is defined on $V(s_1,\ldots,s_k)$ in (44) by

$$TV(s_1,\ldots,s_k) = \sum_{i_1=0}^{\infty} \cdots \sum_{i_{k-1}=0}^{\infty} v_{0i_1 \cdots i_{k-1}} s_1^{-(i_1+1)} \cdots s_{k-1}^{-(i_{k-1}+1)} \tag{46}$$

for the case $k > 1$, and by $TV(s_1) = 0$ for the $k = 1$ case. Note that T is a linear operator and $TV(s_1,\ldots,s_k)$ is a negative power series in $k - 1$ variables. Thus T can be repeatedly applied, though 0 will be obtained after at most k steps. Throughout the following development the symbols S and T will be used regardless of the particular domain of negative power series, in particular, regardless of the number of variables.

Now suppose a given degree-n homogeneous system is described by a regular transfer function $H_{reg}(s_1,\ldots,s_n)$ in the negative power series form. Define a linear space of negative power series according to

$$U_1 = span\ \{\ H_{reg}(s_1,\ldots,s_n),\ SH_{reg}(s_1,\ldots,s_n),\ S^2H_{reg}(s_1,\ldots,s_n),\ldots\ \} \tag{47}$$

Using the notation TU_1 for the image of U_1 under T, let

$$\begin{aligned} U_2 &= span\ \{\ TU_1,\ STU_1,\ S^2TU_1,\ \ldots\ \} \\ &\ \ \vdots \\ U_n &= span\ \{\ TU_{n-1},\ STU_{n-1},\ S^2TU_{n-1},\ \ldots\ \} \end{aligned} \tag{48}$$

Then U_j is a linear space of negative power series in $n+1-j$ variables, $j = 1,\ldots,n$, and $U_i \cap U_j = 0$. Furthermore U_j is invariant with respect to S, that is, $SU_j \subseteq U_j$, and $TU_j \subseteq U_{j+1}$.

Now consider the linear space

$$U = span\ \{\ U_1,\ \ldots,\ U_n\ \} \tag{49}$$

The elements of U are negative power series in n variables or less, and both S and T then can be viewed as linear operators from U into U. Define the *initialization operator* $L{:}R \rightarrow U$ in terms of the given regular transfer function by

$$Lr = H_{reg}(s_1,\ldots,s_n)r \tag{50}$$

and define the *evaluation operator* $E{:}U \rightarrow R$ by

$$EV(s_1,\ldots,s_k) = \begin{cases} 0, & k > 1 \\ EV(s_1), & k = 1 \end{cases} \tag{51}$$

where $EV(s_1)$ is the evaluation operator defined in (10) for the linear case.

The spiritual similarity of this setup to that for linear systems should be apparent. Also, since regular transfer functions that are not necessarily in power series form will be of most interest, it is convenient to interpret the linear operators defined above directly in terms of functions of k variables corresponding to negative power series. Indeed, the easily derived formulas are:

$$\begin{aligned} SV(s_1,\ldots,s_k) &= s_1V(s_1,\ldots,s_k) - [s_1V(s_1,\ldots,s_k)]\big|_{s_1=\infty} \\ TV(s_1,\ldots,s_k) &= [s_1V(s_1,\ldots,s_k)]\Big|_{\substack{s_1=\infty \\ s_2=s_1 \\ \vdots \\ s_k=s_{k-1}}} \\ EV(s_1,\ldots,s_k) &= \begin{cases} 0, & k > 1 \\ [s_1V(s_1)]\big|_{s_1=\infty}, & k = 1 \end{cases} \end{aligned} \tag{52}$$

These interpretations are very important for calculations since the negative power series representations are not at all pleasant to actually manipulate.

To show that (S,T,L,E,U) is an abstract realization for the given regular transfer function is remarkably simple. The identity

$$ES^{j_n}TS^{j_{n-1}}T \cdots TS^{j_1}L = h_{j_1 \cdots j_n} \tag{53}$$

is verified in the following sequence of calculations.

$$S^{j_1}L = S^{j_1}H_{reg}(s_1,\ldots,s_n)$$

$$= \sum_{i_1=0}^{\infty}\sum_{i_2=0}^{\infty}\cdots\sum_{i_n=0}^{\infty} h_{i_1+j_1,i_2\cdots i_n} s_1^{-(i_1+1)} s_2^{-(i_2+1)} \cdots s_n^{-(i_n+1)}$$

$$TS^{j_1}L = \sum_{i_1=0}^{\infty}\cdots\sum_{i_{n-1}=0}^{\infty} h_{j_1 i_1\cdots i_{n-1}} s_1^{-(i_1+1)} \cdots s_{n-1}^{-(i_{n-1}+1)}$$

$$S^{j_2}TS^{j_1}L = \sum_{i_1=0}^{\infty}\sum_{i_2=0}^{\infty}\sum_{i_{n-1}=0}^{\infty} h_{j_1,i_1+j_2,i_2\cdots i_{n-1}} s_1^{-(i_1+1)} s_2^{-(i_2+1)} \cdots s_{n-1}^{-(i_{n-1}+1)}$$

$$TS^{j_2}TS^{j_1}L = \sum_{i_1=0}^{\infty}\cdots\sum_{i_{n-2}=0}^{\infty} h_{j_1 j_2 i_1\cdots i_{n-2}} s_1^{-(i_1+1)} \cdots s_{n-2}^{-(i_{n-2}+1)}$$

$$\vdots$$

$$S^{j_n}TS^{j_{n-1}}T \cdots TS^{j_1}L = \sum_{i_1=0}^{\infty} h_{j_1\cdots j_{n-1}i_1+j_n} s_1^{-(i_1+1)}$$

$$ES^{j_n}TS^{j_{n-1}}T \cdots TS^{j_1}L = h_{j_1\cdots j_n}$$

It is easy to show that the remaining terms in (43) indeed are 0. If $k < n$, then the E operator does the job since its argument will have more than one variable. If $k > n$, then the T's will give 0. Now the realization procedure involves determining if U is finite dimensional, and if so finding matrix representations for the linear operators S, T, E, and L when U is replaced by R^m. The following result and its proof are reminiscent of Theorem 4.1, though the proof is postponed until Section 4.3.

Theorem 4.8 A degree-n homogeneous system described by the regular transfer function $H_{reg}(s_1,\ldots,s_n)$ is bilinear realizable if and only if U is finite dimensional. Furthermore, if the system is bilinear realizable, then (S,T,L,E,U) is a minimal bilinear realization.

I should remark that it is not hard to show that U is finite dimensional if $H_{reg}(s_1,\ldots,s_n)$ is a strictly proper, recognizable function. It might be worthwhile for the reader to work this out following the spirit of the proof of Theorem 4.2, just to gain some familiarity with the linear operators S and T when applied to strictly proper, recognizable functions.

To find a matrix realization when U is finite dimensional, it is convenient to replace U in the following way. If *dimension* $U = m$, choose the standard ordered basis for R^m so that $e_1,\ldots,e_{m_1}$ represents the linearly independent elements of U_1, $e_{m_1+1},\ldots,e_{m_2}$ represents the linearly independent elements of U_2, and so on. Then from the fact that $U_1,\ldots,U_n$, are disjoint, and from the invariance properties mentioned earlier, matrix representations for S and T will have the form

$$A = \begin{bmatrix} A_{11} & 0 & \cdots & 0 \\ 0 & A_{22} & \cdots & 0 \\ \vdots & \vdots & \vdots & \vdots \\ 0 & 0 & \cdots & A_{nn} \end{bmatrix}$$

$$D = \begin{bmatrix} 0 & 0 & \cdots & 0 & 0 \\ D_{21} & 0 & \cdots & 0 & 0 \\ 0 & D_{32} & \cdots & 0 & 0 \\ \vdots & \vdots & \vdots & \vdots & \vdots \\ 0 & 0 & \cdots & D_{n,n-1} & 0 \end{bmatrix} \tag{54}$$

Also, from the special form of the image of L and the null space of E, the respective matrix representations will have the form

$$b = \begin{bmatrix} b_1 \\ 0 \\ \vdots \\ 0 \end{bmatrix}, \quad c = [0\ 0 \cdots c_n] \tag{55}$$

(Actually, the casual basis picker invariably ends up with $b = e_1$.) The dimension of each A_{jj} is $m_j \ x \ m_j$, and D, b, and c are partitioned accordingly. Note that this is precisely the type of block-form realization used to derive Theorem 4.7.

Example 4.3 Given the bilinear-realizable regular transfer function

$$H_{reg}(s_1,s_2) = \frac{1}{(s_1+2)(s_2+3)} = \frac{1}{s_1s_2+3s_1+2s_2+6}$$

the first step in constructing a realization is to compute the spaces U_1 and U_2. Since

$$SH_{reg}(s_1,s_2) = \frac{s_1}{s_1s_2+3s_1+2s_2+6} - \frac{1}{s_2+3}$$

$$= \frac{-2}{s_1s_2+3s_1+2s_2+6} = -2H_{reg}(s_1,s_2)$$

it is clear that

$$U_1 = span\left\{\frac{1}{s_1s_2+3s_1+2s_2+6}\right\}$$

To compute U_2, note that

$$TH_{reg}(s_1,s_2) = \left[\frac{s_1}{s_1s_2+3s_1+2s_2+6}\right]\Bigg|_{\substack{s_1=\infty \\ s_2=s_1}} = \frac{1}{s_1+3}$$

$$STH_{reg}(s_1,s_2) = \frac{s_1}{s_1+3} - 1 = \frac{-3}{s_1+3}$$

Thus,

$$U_2 = span\left\{\frac{1}{s_1+3}\right\}$$

and making the replacement

$$U_1 = span\begin{bmatrix}1\\0\end{bmatrix}, \quad U_2 = span\begin{bmatrix}0\\1\end{bmatrix}$$

the matrix representations for S, T, L, and E can be obtained as follows. If A is the matrix representation for S, then

$$A\begin{bmatrix}1\\0\end{bmatrix} = \begin{bmatrix}-2\\0\end{bmatrix}, \quad A\begin{bmatrix}0\\1\end{bmatrix} = \begin{bmatrix}0\\-3\end{bmatrix}$$

Thus,

$$A = \begin{bmatrix}-2 & 0\\0 & -3\end{bmatrix}$$

If D is the matrix representation of T, then

$$D\begin{bmatrix}1\\0\end{bmatrix} = \begin{bmatrix}0\\1\end{bmatrix}, \quad D\begin{bmatrix}0\\1\end{bmatrix} = \begin{bmatrix}0\\0\end{bmatrix}$$

so that

$$D = \begin{bmatrix}0 & 0\\1 & 0\end{bmatrix}$$

It is clear that the matrix representation of L is

$$b = \begin{bmatrix}1\\0\end{bmatrix}$$

and, finally, since

$$E\,\frac{1}{s_1s_2+3s_1+2s_2+6} = 0, \quad E\,\frac{1}{s_1+3} = 1$$

the matrix representation for E is

$$c = [0 \quad 1]$$

It should be clear that the two approaches to the realization problem for homogeneous systems that have been discussed are analogous to the two main approaches to linear realization theory. The shift realization approach is based mainly on series representations of rational functions, while the other approach is based more on direct manipulations of the polynomials in the rational transfer function. In addition, the shift realization approach for nonlinear systems can be rephrased in terms of a Behavior matrix not too unlike that in the linear case. This formulation will be demonstrated in Section 4.3.

There is a particular form of interconnection structured realization that corresponds to the block-form bilinear state equation specified by (54) and (55), or by (40). Partitioning the state vector $x(t)$ in the form

$$x(t) = \begin{bmatrix} x_1(t) \\ \vdots \\ x_n(t) \end{bmatrix} \tag{56}$$

where $x_j(t)$ is m_j x 1, the block-form realization can be described by the set of state equations:

$$\begin{aligned} \dot{x}_1(t) &= A_{11}x_1(t) + b_1u(t) \\ \dot{x}_2(t) &= A_{22}x_2(t) + D_{21}x_1(t)u(t) \\ &\vdots \\ \dot{x}_n(t) &= A_{nn}x_n(t) + D_{n,n-1}x_{n-1}(t)u(t) \\ y(t) &= c_nx_n(t) \end{aligned} \tag{57}$$

Then the realization corresponds to the cascade connection of multi-input, multi-output linear systems, and vector multipliers shown in Figure 4.1. (A vector quantity, in general, deserves a double line, while scalar quantities get a single line.)

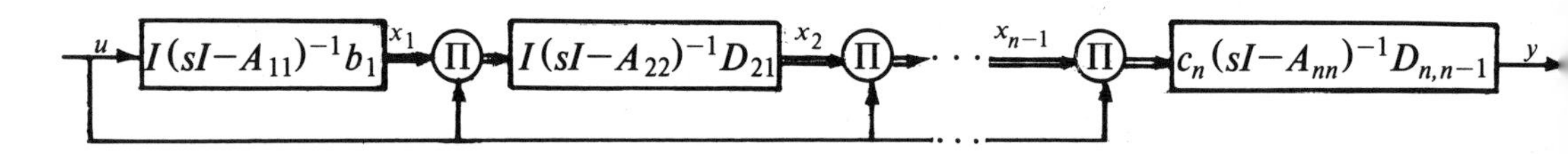

Figure 4.1. An interconnection structured realization.

4.3 Realization of Stationary Polynomial and Volterra Systems

The polynomial system case will be discussed first, and in most detail, though the notation will be chosen in a way that will facilitate consideration of Volterra systems later. Again, finite-dimensional bilinear state equation realizations are of interest, particularly those of minimal dimension. It will be seen that the shift approach extends to this setting in a very simple fashion.

Suppose a degree-N polynomial system of is described by the sequence of regular transfer functions

$$(H(s_1),\ H_{reg}(s_1,s_2),\ \ldots,\ H_{reg}(s_1,\ldots,s_N),\ 0,\ \ldots) \tag{58}$$

where the transfer functions of degree greater than N are all indicated as zeros. The first result shows that the basic realizability condition for polynomial systems follows directly from the homogeneous case.

Theorem 4.9 The polynomial system specified in (58) is bilinear realizable if and only if each regular transfer function is a strictly proper, recognizable function.

Proof Suppose, first, that each transfer function $H_{reg}(s_1,\ldots,s_n)$ is strictly proper and recognizable. Then from Section 4.2 it is clear that each can be realized by a degree-n homogeneous bilinear state equation

$$\dot{x}_j(t) = A_j x_j(t) + D_j x_j(t)u(t) + b_j u(t)$$
$$y_j(t) = c_j x_j(t) \tag{59}$$

where $j = 1,\ldots,N$, and $D_1 = 0$. (The degree-1 realization, of course, is a linear state equation.) Now, consider the additive parallel connection of these state equations. Such a connection can be described by the "block diagonal" bilinear state equation (A,D,b,c,R^m) given by

$$\dot{x}(t) = \begin{bmatrix} A_1 & 0 & \cdots & 0 \\ 0 & A_2 & \cdots & 0 \\ \vdots & \vdots & \vdots & \vdots \\ 0 & 0 & \cdots & A_N \end{bmatrix} x(t) + \begin{bmatrix} D_1 & 0 & \cdots & 0 \\ 0 & D_2 & \cdots & 0 \\ \vdots & \vdots & \vdots & \vdots \\ 0 & 0 & \cdots & D_N \end{bmatrix} x(t)u(t) + \begin{bmatrix} b_1 \\ b_2 \\ \vdots \\ b_N \end{bmatrix} u(t)$$

$$y(t) = [c_1 \ \cdots \ c_N]x(t)$$

Using the block diagonal form for A and D, it is straightforward to compute the degree-k transfer function for this realization:

$$c(s_kI-A)^{-1}D \ \cdots \ D(s_1I-A)^{-1}b = \sum_{j=1}^{N} c_j(s_kI-A_j)^{-1}D_j \cdots D_j(s_1I-A_j)^{-1}b_j$$

But the fact that the jth bilinear state equation is homogeneous of degree j implies that all the summands on the right side are 0 except for when $j = k$. Thus

$$c(s_kI-A)^{-1}D \cdots D(s_1I-A)^{-1}b = c_k(s_kI-A_k)^{-1}D_k \cdots D_k(s_1I-A_k)^{-1}b_k$$
$$= \begin{cases} H_{reg}(s_1,\ldots,s_k), & k = 1,\ldots,N \\ 0, & k > n \end{cases}$$

and (A,D,b,c,R^m) is a bilinear realization for the given polynomial system.

Now suppose the polynomial system is bilinear realizable, and furthermore that (A,D,b,c,R^m) is such a realization. Then by calculation each degree-n regular transfer function is the strictly proper, recognizable function

$$c(s_nI - A)^{-1}D \cdots D(s_1I - A)^{-1}b, \quad n = 1,\ldots,N$$

Thus the proof is complete.

The analog of Theorem 4.9 in the time domain should be obvious. A degree-N polynomial system is bilinear realizable if and only if each regular kernel has the exponential form given in (28) or (29).

The basic bilinear realizability result for polynomial systems also can be developed via a shift realization approach, based on the assumption that each regular transfer function in (58) can be written as a negative power series of the form

$$H_{reg}(s_1,\ldots,s_k) = \sum_{i_1=0}^{\infty} \cdots \sum_{i_k=0}^{\infty} h_{i_1 \cdots i_k} s_1^{-(i_1+1)} \cdots s_k^{-(i_k+1)}, \quad k = 1,\ldots,N \tag{60}$$

Of course, this assumption can be made with no loss of generality as far as bilinear realizations are concerned. It is clear from the formulation in Section 4.2 that constructing a bilinear realization involves finding matrices A, D, b, and c such that for all nonnegative integers $j_1,j_2,\ldots,$

$$cA^{j_k}DA^{j_{k-1}}D \cdots DA^{j_1}b = \begin{cases} h_{j_1 \cdots j_k}, & k = 1,\ldots,N \\ 0, & k > N \end{cases} \tag{61}$$

As has become usual in this chapter, the first step is to construct an abstract shift realization.

Given any finite sequence of negative power series

$$(\, V_1(s_1),\ V_2(s_1,s_2),\ V_3(s_1,s_2,s_3),\ \ldots) \tag{62}$$

define the *shift operator* S by

$$S(\, V_1(s_1),\ V_2(s_1,s_2),\ V_3(s_1,s_2,s_3),\ \ldots) = (\, SV_1(s_1),\ SV_2(s_1,s_2),\ SV_3(s_1,s_2,s_3),\ \ldots) \tag{63}$$

where $SV_k(s_1,\ldots,s_k)$ is the shift operator defined in (44) and (45). Similarly, define the index operator T by

$$T(\, V_1(s_1),\ V_2(s_1,s_2),\ V_3(s_1,s_2,s_3),\ \ldots) = (\, TV_2(s_1,s_2),\ TV_3(s_1,s_2,s_3),\ \ldots) \tag{64}$$

where $TV_k(s_1,\ldots,s_k)$ is the index operator defined in (46). (Of course, if (62) is viewed as a sequence of functions defined by negative power series, then the shift and index operators can be interpreted as per (52).)

In order to proceed further, it is convenient to use the notation

$$\hat{H}(s_1,\ldots,s_N) = (\ H(s_1),\ H_{reg}(s_1,s_2),\ \ldots,\ H_{reg}(s_1,\ldots,s_N),0,\ldots) \tag{65}$$

to indicate the given degree-N polynomial system. Then define the following linear spaces of finite sequences of negative power series.

$$\begin{aligned} U_1 &= span\ \{\hat{H}(s_1,\ldots,s_N),\ S\hat{H}(s_1,\ldots,s_N),\ S^2\hat{H}(s_1,\ldots,s_N),\ \ldots\} \\ U_2 &= span\ \{TU_1,\ STU_1,\ S^2TU_1,\ \ldots\} \\ &\ \vdots \\ U_N &= span\ \{TU_{N-1},\ STU_{N-1},\ S^2TU_{N-1},\ \ldots\} \end{aligned} \tag{66}$$

Letting $U = span\ \{U_1,\ldots,U_N\}$, S and T can be viewed as operators from U into U. Define the *initialization operator* $L:R \rightarrow U$ in terms of the given $\hat{H}(s_1,\ldots,s_N)$ by

$$Lr = \hat{H}(s_1,\ldots,s_N)r \tag{67}$$

and the *evaluation operator* $E:U \rightarrow R$ by

$$E(\ V_1(s_1),\ V_2(s_1,s_2),\ V_3(s_1,s_2,s_3),\ \ldots) = EV_1(s_1) \tag{68}$$

where $EV_1(s_1)$ is defined as in the linear case.

Now the calculations to show that (S,T,L,E,U) is an abstract bilinear realization for the given $\hat{H}(s_1,\ldots,s_N)$ follow directly from the calculations in the homogeneous case. For instance,

$$\begin{aligned} ES^jL &= E(S^jH(s_1),\ S^jH_{reg}(s_1,s_2),\ \ldots,\ S^jH_{reg}(s_1,\ldots,s_N),\ 0,\ \ldots) \\ &= ES^jH(s_1) = h_j,\quad j = 0,1,2,\ldots \\ ES^{j_2}TS^{j_1}L &= E(S^{j_2}TS^{j_1}H(s_1),\ \ldots,\ S^{j_2}TS^{j_1}H_{reg}(s_1,\ldots,s_N),\ 0,\ \ldots) \\ &= ES^{j_2}TS^{j_1}H_{reg}(s_1,s_2) = h_{j_1j_2},\quad j_1,j_2 = 0,1,2,\ldots \end{aligned} \tag{69}$$

Theorem 4.10 A degree-N polynomial system described by the sequence of regular transfer functions $\hat{H}(s_1,\ldots,s_N)$ is bilinear realizable if and only if U is finite dimensional. Furthermore, if the system is bilinear realizable, then (S,T,L,E,U) is a minimal bilinear realization.

Proof Suppose that the polynomial system described by $\hat{H}(s_1,\ldots,s_N)$ is bilinear realizable, that (A,D,b,c,R^m) is any bilinear realization of the system, and that (S,T,L,E,U) is the shift realization of the system. Let W be the linear space of all sequences of negative power series such as (62). Then define a linear operator $\Phi:R^m \rightarrow W$ by

$$\begin{aligned} \Phi(x) = (\ &c(s_1I-A)^{-1}x,\ c(s_2I-A)^{-1}D(s_1I-A)^{-1}x, \\ &c(s_3I-A)^{-1}D(s_2I-A)^{-1}D(s_1I-A)^{-1}x,\ \ldots) \end{aligned}$$

where for brevity I have written the right side as a sequence of strictly proper, recognizable functions instead of the corresponding negative power series. Notice that $\Phi(b) = \hat{H}(s_1,\ldots,s_N)$. Furthermore, using the definition of the shift operator in the homogeneous case,

$$c(s_kI - A)^{-1}D \cdots D(s_1I - A)^{-1}Ab$$
$$= \sum_{i_1=0}^{\infty} \cdots \sum_{i_k=0}^{\infty} cA^{i_k}D \cdots DA^{i_1+1}bs_1^{-(i_1+1)} \cdots s_k^{-(i_k+1)}$$
$$= S\sum_{i_1=0}^{\infty} \cdots \sum_{i_k=0}^{\infty} cA^{i_k}D \cdots DA^{i_1}bs_1^{-(i_1+1)} \cdots s_k^{-(i_k+1)}$$
$$= SH_{reg}(s_1,\ldots,s_k),\quad k = 1,\ldots,N$$

Extending this calculation to the sequence of regular transfer functions $\hat{H}(s_1,\ldots,s_N)$ shows that

$$\Phi(Ab) = S\hat{H}(s_1,\ldots,s_N)$$

Using the definition of T in a similar way,

$$c(s_{k-1}I - A)^{-1}D \cdots D(s_1I - A)^{-1}Db$$
$$= \sum_{i_1=0}^{\infty} \cdots \sum_{i_{k-1}=0}^{\infty} cA^{i_{k-1}}D \cdots DA^{i_1}Dbs_1^{-(i_1+1)} \cdots s_{k-1}^{-(i_{k-1}+1)}$$
$$= T\sum_{i_1=0}^{\infty} \cdots \sum_{i_k=0}^{\infty} cA^{i_k}D \cdots DA^{i_1}bs_1^{-(i_1+1)} \cdots s_k^{-(i_k+1)}$$
$$= TH_{reg}(s_1,\ldots,s_k),\quad k = 2,\ldots,N$$

Again, this calculation, when extended to $\hat{H}(s_1,\ldots,s_N)$, implies that

$$\Phi(Db) = T\hat{H}(s_1,\ldots,s_N)$$

Combining these results gives

$$\Phi(A^{i_{n-1}}D \cdots DA^{i_1}b) = S^{i_{n-1}}T \cdots TS^{i_1}\hat{H}(s_1,\ldots,s_N),\quad n = 1,\ldots,N$$

which shows that $U \subseteq R[\Phi]$. Since Φ is a linear map on an m-dimensional space, it follows that *dimension* $U \leq m$. Thus U is finite dimensional, and furthermore the abstract shift realization is minimal since the dimension of U is no greater than the state space dimension of any other bilinear realization of $\hat{H}(s_1,\ldots,s_N)$.

Assuming now that U has finite dimension m, the following construction yields a minimal bilinear realization (A,D,b,c,R^m) of $\hat{H}(s_1,\ldots,s_N)$. Replacing the space U by R^m with the standard ordered basis choices $e_1,\ldots,e_{m_1}$ for the linearly independent elements of U_1, $e_{m_1+1},\ldots,e_{m_2}$ for the additional linearly independent elements of U_2, and so on, gives a realization as follows. Since $SU_j \subseteq$ *span* $\{U_1,\ldots,U_j\}$, it is clear that the matrix representation for S will have the block-triangular form

$$A = \begin{bmatrix} A_{11} & A_{12} & \cdots & A_{1M} \\ 0 & A_{22} & \cdots & A_{2M} \\ \vdots & \vdots & \vdots & \vdots \\ 0 & 0 & \cdots & A_{MM} \end{bmatrix} \tag{70}$$

where A_{jj} is $m_j \times m_j$. Also $TU_j \subseteq$ *span* $\{U_1,\ldots,U_{j+1}\}$, which implies that the matrix representation for T will have the block (almost triangular) form

$$D = \begin{bmatrix} D_{11} & D_{12} & \cdots & D_{1,M-1} & D_{1M} \\ D_{21} & D_{22} & \cdots & D_{2,M-1} & D_{2M} \\ 0 & D_{32} & \cdots & D_{3,M-1} & D_{3M} \\ & \vdots & \vdots & \vdots & \vdots \\ 0 & 0 & \cdots & D_{M,M-1} & D_{MM} \end{bmatrix} \tag{71}$$

where the blocks are partitioned according to those in A. (Notice that M ($\leqslant N$) blocks are indicated, rather than N. The reason is that a particular U_j may be contained in *span* $\{U_1,\ldots,U_{j-1}\}$.) The matrix representation for L clearly will have the block form

$$b = \begin{bmatrix} b_1 \\ 0 \\ \vdots \\ 0 \end{bmatrix} \tag{72}$$

and a matrix representation for E is found by computing the action of E on each U_j to obtain

$$c = [\, c_{11} \ \cdots \ c_{1M} \,] \tag{73}$$

where each c_{1j} is $1 \times m_j$.

Example 4.6 Consider the degree-2 polynomial system described by the regular transfer functions

$$\hat{H}(s_1,s_2) = \left(\frac{1}{s_1+1} , \frac{1}{(s_1+2)(s_2+3)} , 0, \ldots \right)$$

To find U_1, compute

$$S\hat{H}(s_1,s_2) = \left(\frac{-1}{s_1+1} , \frac{-2}{(s_1+2)(s_2+3)} , 0, \ldots \right)$$

$$S^2\hat{H}(s_1,s_2) = \left(\frac{1}{s_1+1} , \frac{4}{(s_1+2)(s_2+3)} , 0, \ldots \right)$$

$$= -2\hat{H}(s_1,s_2) - 3S\hat{H}(s_1,s_2)$$

Thus,

$$U_1 = span\left\{ \left(\frac{1}{s_1+1}, \frac{1}{(s_1+2)(s_2+3)}, 0, \ldots\right), \left(\frac{-1}{s_1+1}, \frac{-2}{(s_1+2)(s_2+3)}, 0, \ldots\right) \right\}$$

To find U_2 the image of these basis elements for U_1 under T must be computed, and then the subsequent image under repeated shifts must be computed.

$$T\hat{H}(s_1,s_2) = T\left(\frac{1}{s_1+1}, \frac{1}{(s_1+2)(s_2+3)}, 0, \ldots\right) = \left(\frac{1}{s_1+3}, 0, \ldots\right)$$

$$TS\hat{H}(s_1,s_2) = T\left(\frac{-1}{s_1+1}, \frac{-2}{(s_1+2)(s_2+3)} 0, \ldots\right) = \left(\frac{-2}{s_1+3}, 0, \ldots\right)$$

$$= -2T\hat{H}(s_1,s_2)$$

$$ST\hat{H}(s_1,s_2) = S\left(\frac{1}{s_1+3}, 0, \ldots\right) = \left(\frac{-3}{s_1+3}, 0, \ldots\right)$$

$$= -3T\hat{H}(s_1,s_2)$$

Thus,

$$U_2 = span\left\{ \left(\frac{1}{s_1+3}, 0, \ldots\right) \right\}$$

Now replace $U = span\ \{U_1, U_2\}$ by R^3 and choose the standard ordered basis elements according to

$$\begin{bmatrix}1\\0\\0\end{bmatrix} = \left(\frac{1}{s_1+1}, \frac{1}{(s_1+2)(s_2+3)}, 0, \ldots\right)$$

$$\begin{bmatrix}0\\1\\0\end{bmatrix} = \left(\frac{-1}{s_1+1}, \frac{-2}{(s_1+2)(s_2+3)}, 0, \ldots\right)$$

$$\begin{bmatrix}0\\0\\1\end{bmatrix} = \left(\frac{1}{s_1+3}, 0, \ldots\right)$$

This yields the matrix representations

$$A = \begin{bmatrix}0 & -2 & 0\\1 & -3 & 0\\0 & 0 & -3\end{bmatrix}, \quad D = \begin{bmatrix}0 & 0 & 0\\0 & 0 & 0\\1 & -2 & 0\end{bmatrix}$$

The calculations

$$E\hat{H}(s_1,s_2) = 1, \quad ES\hat{H}(s_1,s_2) = -1, \quad ET\hat{H}(s_1,s_2) = 1$$

give

$$c = [1 \quad -1 \quad 1]$$

and, finally,

$$b = \begin{bmatrix} 1 \\ 0 \\ 0 \end{bmatrix}$$

Now consider the bilinear realization problem for a given Volterra system. It is assumed that the system is specified in terms of a sequence of regular transfer functions written in the notation

$$\hat{H}(s_1,\ldots,s_\infty) = (\,H(s_1),\ H_{reg}(s_1,s_2),\ H_{reg}(s_1,s_2,s_3),\ \ldots) \tag{74}$$

As usual, $\hat{H}(s_1,\ldots,s_\infty)$ will be viewed as a sequence of negative power series, each of which takes the form in (60). From this perspective, it is clear that the bilinear realization problem for a Volterra system involves finding matrices A, D, b, and c such that for all $k = 1,2,\ldots$, and all nonnegative integers $j_1,\ldots,j_k$,

$$cA^{j_k}D \cdots DA^{j_1}b = h_{j_1 \cdots j_k} \tag{75}$$

The construction of an abstract shift realization for $\hat{H}(s_1,\ldots,s_\infty)$ proceeds along the same lines as in the polynomial system case, so only a brief review of the mechanics is needed. The shift and index operators are defined, as in the polynomial case, according to

$$S\hat{V}(s_1,\ldots,s_\infty) = (SV_1(s_1),\ SV_2(s_1,s_2),\ \ldots)$$
$$T\hat{V}(s_1,\ldots,s_\infty) = (TV_2(s_1,s_2),\ TV_3(s_1,s_2,s_3),\ \ldots) \tag{76}$$

In terms of these operators and the given Volterra system, a set of linear spaces is defined by

$$U_1 = span\ \{\hat{H}(s_1,\ldots,s_\infty),\ S\hat{H}(s_1,\ldots,s_\infty),\ S^2\hat{H}(s_1,\ldots,s_\infty),\ \ldots\}$$
$$U_2 = span\ \{TU_1,\ STU_1,\ S^2TU_1,\ \ldots\}$$
$$U_3 = span\ \{TU_2,\ STU_2,\ S^2TU_2,\ \ldots\} \tag{77}$$
$$\vdots$$

and, finally,

$$U = span\ \{U_1,\ U_2,\ U_3,\ \ldots\} \tag{78}$$

It is clear that S and T are linear operators from U into U. Define the initialization operator $L{:}R \to U$ in terms of the given system by

$$Lr = \hat{H}(s_1,\ldots,s_\infty)r \tag{79}$$

and the evaluation operator $E{:}U \to R$ by

$$E(V_1(s_1),\ V_2(s_1,s_2),\ \ldots) = EV_1(s_1) \tag{80}$$

The demonstration that (S,T,L,E,U) is an abstract bilinear realization for the given Volterra system follows from by now standard calculations. Also, one answer to the bilinear realizability and minimality questions is easily obtained. If U has finite dimension, then it is clear by the replacement construction that the given system is bilinear realizable. On the other hand, a simple argument using a Φ operator similar to that in the proof of Theorem 4.10 shows that if the given system is bilinear realizable, then U has finite dimension. (And in this case the shift realization is a minimal bilinear realization for the system.) Thus bilinear realizability of a Volterra system is equivalent to finite dimensionality of the linear space U. The search for a more direct characterization begins in the direction of Theorem 4.9.

Theorem 4.11 If the Volterra system specified by $\hat{H}(s_1,\ldots,s_\infty)$ in (74) is bilinear realizable, then each regular transfer function $H_{reg}(s_1,\ldots,s_k)$ is a strictly proper, recognizable function.

A proof of Theorem 4.11 consists of nothing more than taking a bilinear realization of $\hat{H}(s_1,\ldots,s_\infty)$ and observing that by calculation each $H_{reg}(s_1,\ldots,s_k)$ is a strictly proper, recognizable function. This observation, together with Theorem 4.9, also yields the following interesting fact.

Corollary 4.1 If a Volterra system is bilinear realizable, then any polynomial system formed by truncation of the Volterra system is also bilinear realizable.

Unfortunately, the search for a more direct characterization of bilinear realizability for Volterra systems appears to end with the failure of the converse of Theorem 4.11.

Example 4.5 Consider the Volterra system

$$\hat{H}(s_1,\ldots,s_\infty) = \left(\frac{1}{s_1+1}, \frac{1/2!}{(s_1+1)(s_2+1)}, \ldots, \frac{1/n!}{(s_1+1)\cdots(s_n+1)}, \ldots \right)$$

Applying the index operator repeatedly gives

$$T^j\hat{H}(s_1,\ldots,s_\infty) = \left(\frac{1/(j+1)!}{s_1+1}, \frac{1/(j+2)!}{(s_1+1)(s_2+1)}, \ldots, \frac{1/(n+j+1)!}{(s_1+1)\cdots(s_n+1)}, \ldots \right)$$

for $j = 1,2,\ldots$. The denominators of the subsystem transfer functions all behave in the expected way. But since the collection of sequences (of numerators)

$$\left(\frac{1}{(j+1)!}, \frac{1}{(j+2)!}, \ldots, \frac{1}{(j+n+1)!}, \ldots \right),\ \ j = 0,1,\ldots$$

is infinite dimensional, it is clear without even calculating the action of the shift operator that $U = span\ \{U_1,\ U_2,\ldots\}$ will be infinite dimensional. Thus $\hat{H}(s_1,\ldots,s_\infty)$ is not bilinear realizable.

Suppose a Volterra system is given wherein every subsystem regular transfer function is strictly proper and recognizable. To check for bilinear realizability, Example 4.7 indicates that there is no choice but to work through the calculation of the dimension of U. Of course, this raises the issue of having a general form for the sequence of regular transfer functions. When such a general form is available, the calculation of a bilinear realization can be easy if the dimensions are small.

Example 4.6 For the Volterra system

$$\hat{H}(s_1,\ldots,s_\infty) = \left(\frac{1}{s_1+1}, \frac{1}{(s_1+1)(s_2+1)}, \ldots, \frac{1}{(s_1+1)\cdots(s_n+1)}, \ldots\right)$$

a quick calculation shows that

$$S^j\hat{H}(s_1,\ldots,s_\infty) = (-1)^j\hat{H}(s_1,\ldots,s_\infty),\ \ j = 1,2,\ldots$$

and

$$T^j\hat{H}(s_1,\ldots,s_\infty) = \hat{H}(s_1,\ldots,s_\infty),\ \ j = 1,2,\ldots$$

Therefore *dimension* $U = 1$, and another easy calculation shows that a minimal bilinear realization is

$$\dot{x}(t) = -x(t) + x(t)u(t) + u(t)$$
$$y(t) = x(t)$$

In addition to illustrating a sort of easiest-possible Volterra system realization problem, this example in conjunction with Example 4.5 shows that the interrelationships of the subsystem "gains" plays a crucial role in finite-dimensional realizability. That is, bilinear realizability can be created or destroyed simply by changing constants in the numerators of the regular transfer functions in a Volterra system. Another interesting observation can be made by considering the degree-2 polynomial truncation of the system in Example 4.6, namely,

$$\hat{H}(s_1,s_2) = \left(\frac{1}{s_1+1}, \frac{1}{(s_1+1)(s_2+1)}, 0, \ldots\right)$$

The minimal-dimension bilinear realization for this polynomial system has dimension 2. Thus truncation can increase the dimension of the minimal bilinear realization. See Problem 4.7.

A perhaps cleaner statement of the condition for bilinear realizability of a given Volterra system (or, for that matter, polynomial or homogeneous system) can be developed from the shift-realization viewpoint. The approach involves replacing negative power series with sequences so that U is viewed as a linear space of sequences of sequences, arranging these sequences into a matrix, and then noting that rank finiteness of the matrix is equivalent to finite dimensionality of U. Again, this is all very similar in style to the linear case.

Viewing a sequence of regular transfer functions $\hat{H}(s_1,\ldots,s_\infty)$ as a sequence of negative power series

$$\hat{H}(s_1,\ldots,s_\infty) = \left(\sum_{i_1=0}^{\infty} h_{i_1}s_1^{-(i_1+1)}, \sum_{i_1=0}^{\infty}\sum_{i_2=0}^{\infty} h_{i_1i_2}s_1^{-(i_1+1)}s_2^{-(i_2+1)}, \ldots\right) \tag{81}$$

any expression of the form

$$S^{j_k}TS^{j_{k-1}}T\cdots TS^{j_1}\hat{H}(s_1,\ldots,s_\infty) \tag{82}$$

can be viewed in the same way. For example,

$$S\hat{H}(s_1,\ldots,s_\infty) = (\sum_{i_1=0}^{\infty} h_{i_1+1}s_1^{-(i_1+1)}, \sum_{i_1=0}^{\infty}\sum_{i_2=0}^{\infty} h_{i_1+1,i_2}s_1^{-(i_1+1)}s_2^{-(i_2+1)}, \ldots)$$

$$T\hat{H}(s_1,\ldots,s_\infty) = (\sum_{i_1=0}^{\infty} h_{0i_1}s_1^{-(i_1+1)}, \sum_{i_1=0}^{\infty}\sum_{i_2=0}^{\infty} h_{0i_1i_2}s_1^{-(i_1+1)}s_2^{-(i_2+1)}, \ldots) \tag{83}$$

Each of these sequences of negative power series can be viewed as a sequence of sequences. For example,

$$\hat{H}(s_1,\ldots,s_\infty) = (\,(h_0,h_1,h_2,\ldots),\ (h_{00},h_{01},h_{02},\ldots,h_{10},h_{11},h_{12},\ldots),\ \ldots)$$

$$S\hat{H}(s_1,\ldots,s_\infty) = (\,(h_1,h_2,h_3,\ldots),\ (h_{10},h_{11},h_{12},\ldots,h_{20},h_{21},h_{22},\ldots),\ \ldots)$$

$$T\hat{H}(s_1,\ldots,s_\infty) = (\,(h_{00},h_{01},h_{02},\ldots),\ (h_{000},h_{001},h_{002},\ldots,h_{010},h_{011},h_{012},\ldots),\ \ldots)$$

Of course, there are many ways to systematically list the multi-index sequences, but the particular arrangement is immaterial as long as all are listed in the same way.

From this viewpoint, each U_j in (77) and U in (78) can be considered as a linear space of sequences of sequences. The shift and index operators are interpreted as above, and the operators L and E are similarly modified for the sequence interpretation. Then a Behavior matrix for the given system is defined in terms of the sequence interpretation by

$$B_{\hat{H}} = \begin{bmatrix} \hat{H}(s_1,\ldots,s_\infty) \\ S\hat{H}(s_1,\ldots,s_\infty) \\ S^2\hat{H}(s_1,\ldots,s_\infty) \\ \vdots \\ T\hat{H}(s_1,\ldots,s_\infty) \\ ST\hat{H}(s_1,\ldots,s_\infty) \\ S^2T\hat{H}(s_1,\ldots,s_\infty) \\ \vdots \\ S^{j_k}T\cdots TS^{j_1}\hat{H}(s_1,\ldots,s_\infty) \\ \vdots \end{bmatrix}$$

$$= \begin{bmatrix} h_0 & h_1 & h_2 & \cdots & h_{00} & h_{01} & h_{02} & \cdots & h_{10} & h_{11} & h_{12} & \cdots \\ h_1 & h_2 & h_3 & \cdots & h_{10} & h_{11} & h_{12} & \cdots & h_{20} & h_{21} & h_{22} & \cdots \\ h_2 & h_3 & h_4 & \cdots & h_{20} & h_{21} & h_{22} & \cdots & h_{30} & h_{31} & h_{32} & \cdots \\ \vdots & \vdots & \vdots & \vdots & \vdots & \vdots & \vdots & \vdots & \vdots & \vdots & \vdots & \vdots \\ h_{00} & h_{01} & h_{02} & \cdots & h_{000} & h_{001} & h_{002} & \cdots & h_{010} & h_{011} & h_{012} & \cdots \\ h_{01} & h_{02} & h_{03} & \cdots & h_{001} & h_{002} & h_{003} & \cdots & h_{011} & h_{012} & h_{013} & \cdots \\ \vdots & \vdots & \vdots & \vdots & \vdots & \vdots & \vdots & \vdots & \vdots & \vdots & \vdots & \vdots \end{bmatrix} \tag{84}$$

And now the following realizability condition should be an obvious restatement of the finite-dimensionality condition on the linear space U.

Theorem 4.12 The Volterra system described by $\hat{H}(s_1, \ldots, s_\infty)$ is bilinear realizable if and only if the corresponding Behavior matrix $B_{\hat{H}}$ has finite rank. Furthermore, for a bilinear-realizable system, the rank of $B_{\hat{H}}$ is the dimension of the minimal bilinear realizations.

4.4 Properties of Bilinear State Equations

Having focused attention on the bilinear realization question, it is appropriate to discuss some of the features of such state equations. As I have mentioned previously, bilinear state equations have many structural features that are strikingly similar to well known features of linear state equations. These features will be demonstrated in the general situation where the bilinear state equation represents a Volterra system. There is no reason to consider separately the special cases of homogeneous or polynomial systems.

A question that often arises is whether a given bilinear state equation is minimal. That is, whether the state equation is a minimal bilinear realization of its input/output description. A convenient way to address this question is through the appropriate concepts of reachability and observability. These concepts will be developed and related to minimality. Also, certain equivalence properties of minimal bilinear realizations will be discussed.

The appropriate definition of reachability for the bilinear state equation

$$\dot{x}(t) = Ax(t) + Dx(t)u(t) + bu(t)$$
$$y(t) = cx(t), \quad t \geq 0, \quad x(0) = 0 \tag{85}$$

begins with the notion of a reachable state. As usual, $x(t) \in R^m$, and $u(t)$ and $y(t)$ are scalars.

Definition 4.1 A state x_1 of the bilinear state equation (85) is called *reachable* (from $x(0) = 0$) if there exists a piecewise continuous input signal such that for some $t_1 < \infty$, $x(t_1) = x_1$.

I should note that the specification of piecewise continuity for the input signal is more or less a matter of convenience. Both more general and more restrictive classes of inputs can be chosen without changing the results. (But not to specify the class of admissible inputs would be in poor taste.)

It would be nice if the set of reachable states for a bilinear state equation formed a linear subspace of the state space R^m. Unfortunately this is not the case; linear combinations of reachable states may not be reachable. Thus, a somewhat weaker notion of reachability is used so that the techniques of linear algebra can be applied.

Definition 4.2 The bilinear state equation (A,D,b,c,R^m) is called *span reachable* if the set of reachable states spans R^m.

The first step in establishing a criterion for span reachability of a given system is to characterize the span of the reachable states. To this end, let $L_{A,D}(b)$ denote the least dimension subspace of R^m containing b and invariant under A and D.

Lemma 4.1 The subspace $X_{sr} \subseteq R^m$ spanned by the reachable states of (A,D,b,c,R^m) is given by $X_{sr} = L_{A,D}(b)$.

Proof Suppose x_1 is a reachable state, so that for some input $u(t)$ and some $t_1 < \infty$, $x(t_1) = x_1$. Then x_1 can be written using the expression derived in Chapter 3 for the solution $x(t)$ of a bilinear state equation. For the case of $x(0) = 0$ and constant coefficient matrices, the first few terms are

$$x_1 = \int_0^{t_1} e^{A(t_1-\sigma)} bu(\sigma)\, d\sigma + \int_0^{t_1}\int_0^{\sigma_1} e^{A(t_1-\sigma_1)} De^{A(\sigma_1-\sigma_2)} bu(\sigma_1)u(\sigma_2)\, d\sigma_2 d\sigma_1 + \cdots \tag{86}$$

Expressing the matrix exponentials as power series, and using the uniform convergence of these series to interchange summation and integration, a messier expression is obtained, the first few terms of which are

$$x_1 = b\int_0^{t_1} u(\sigma)\, d\sigma + Ab\int_0^{t_1} (t-\sigma)u(\sigma)\, d\sigma + Db\int_0^{t_1}\int_0^{\sigma_1} u(\sigma_1)u(\sigma_2)\, d\sigma_2 d\sigma_1 + \cdots \tag{87}$$

This expression shows that x_1 is a linear combination of products of A and D times b, so that $x_1 \in L_{A,D}(b)$. Thus $X_{sr} \subseteq L_{A,D}(b)$ since there is a set of reachable states that forms a basis for X_{sr}.

To obtain the reverse containment, it is not hard to show that if $x(t)$ is contained in a subspace for all $t \geqslant 0$, then $\dot{x}(t)$ is contained in the same subspace for all $t \geqslant 0$. Thus, for any constant input $u(t) = u$, and any reachable state x_1,

$$(A + Du)x_1 + bu \in X_{sr}$$

In particular, $x_1 = 0$ is reachable, and thus $b \in X_{sr}$. Therefore, if u is any real number, and x_1 is any reachable state,

$$(A + Du)x_1 \in X_{sr}$$

Since there is a set of reachable states that spans X_{sr}, for any u the image of X_{sr} under $(A + Du)$ satisfies

$$(A + Du)X_{sr} \subseteq X_{sr}$$

It is left to Problem 4.14 to show that this implies that X_{sr} is invariant under both A and D. Since X_{sr} contains b and is invariant under A and D, $L_{A,D}(b) \subseteq X_{sr}$. This completes the proof.

A characterization of $L_{A,D}(b)$ can be obtained by recursively defining

$$p_1 = b, \quad p_i = [Ap_{i-1} \quad Dp_{i-1}], \; i = 2, 3, \ldots \tag{88}$$

and letting

$$P_i = [p_1 \, p_2 \; \cdots \; p_i] \tag{89}$$

Lemma 4.2 The linear subspaces $L_{A,D}(b)$ and $R[P_m]$ are identical.

Proof The linear subspace $R[P_k]$ is the subspace spanned by the columns of P_k. The columns of P_k contain those of P_{k-1}, and the additional columns are generated by multiplication by A and D. Therefore,

$$R[P_1] \subseteq R[P_2] \subseteq \cdots \subseteq R^m$$

In particular, there exists a $k < m$ such that

$$R[P_1] \subset R[P_2] \subset \cdots \subset R[P_{k-1}] = R[P_k] = \cdots = R[P_m] \subseteq R^m$$

and therefore $R[P_{k-1}] = R[P_m]$ is invariant under A and D and contains b. It remains to show that $R[P_m]$ is the least-dimension such subspace. So suppose $X \subseteq R^m$ is any subspace that contains b and that is invariant under A and D. But X must contain b, Ab, Db, . . ., that is $X \supseteq R[P_m]$. Consequently $R[P_m]$ is of least dimension.

This result leads directly to a criterion for span reachability because *rank* P_m is precisely the dimension of $R[P_m]$.

Theorem 4.13 The m-dimensional bilinear state equation (85) is span reachable if and only if *rank* $P_m = m$.

I now turn to the problem of developing a suitable observability property for bilinear state equations. Again, the concept to be used will be defined in a somewhat weaker fashion than observability for linear state equations.

Definition 4.3 A state x_0 of the bilinear state equation (85) is called *indistinguishable* (from 0) if the response $y(t)$ with $x(0) = x_0$ is identical to the response with $x(0) = 0$ for every piecewise continuous input signal.

Here, as before, piecewise continuity is specified just for definiteness. Notice that the definition implies nothing about the ability to compute a distinguishable initial state from knowledge of the response $y(t)$. This issue will be regarded as extraneous to the structure theory under discussion.

Definition 4.4 The bilinear state equation (85) is called *observable* if there are no indistinguishable states.

To characterize the concept of observability for the bilinear state equation (A,D,b,c,R^m), it is convenient to let $G_{A,D}(c) \subseteq R^m$ be the largest subspace contained in $N[c]$ that is invariant under A and D.

Lemma 4.3 The subset of all indistinguishable states of (A,D,b,c,R^m) is a linear subspace that is given by $X_i = G_{A,D}(c)$.

Proof (Sketch) Using the representation derived in Chapter 3, the response of the bilinear system to arbitrary initial state x_0 and input $u(t)$ can be written as a series, the first few terms of which are

$$y(t) = ce^{At}x_0 + \int_0^t ce^{A(t-\sigma)}De^{A\sigma}x_0u(\sigma)\,d\sigma + \int_0^t ce^{A(t-\sigma)}bu(\sigma)\,d\sigma + \cdots$$

Expanding the matrix exponentials yields terms of the form

$$y(t) = cx_0 + cAx_0t + cDx_0\int_0^t u(\sigma)\,d\sigma$$

$$+ cb\int_0^t u(\sigma)\,d\sigma + cAb\int_0^t (t-\sigma)u(\sigma)\,d\sigma + \cdots \tag{90}$$

It should be clear from this expression that the set of indistinguishable states forms a linear subspace. Also it is easy to see that if $x_0 \in G_{A,D}(c)$, then x_0 is indistinguishable. In other words, $G_{A,D}(c) \subseteq X_i$. The reverse containment is obtained by showing that for all real numbers u,

$$(A + Du)X_i \subseteq X_i$$

and that

$$X_i \subseteq N[c]$$

The details are not hard to fill in, and thus are omitted.

To characterize X_i now involves characterizing $G_{A,D}(c)$. Let

$$q_1 = c,\quad q_i = \begin{bmatrix} q_{i-1}A \\ q_{i-1}D \end{bmatrix},\quad i = 2, 3, \ldots \tag{91}$$

and

$$Q_i = \begin{bmatrix} q_1 \\ q_2 \\ \vdots \\ q_i \end{bmatrix} \tag{92}$$

Then the following result is proved in a manner similar to Lemma 4.2.

Lemma 4.4 The linear subspaces X_i and $N[Q_m]$ are identical.

Now the obvious application of linear algebra gives an observability criterion.

Theorem 4.14 The m-dimensional bilinear state equation (85) is observable if and only if $\mathit{rank}\ Q_m = m$.

While these concepts are of interest in themselves, the intent here is to use them in conjunction with the theory of minimal bilinear realizations. To this end there is one more fact about the matrices P_i and Q_i that is crucial.

Lemma 4.5 For any $j,k = 1,2...$, the product Q_jP_k is the same for all bilinear realizations of a given system.

Proof Suppose that (A,D,b,c,R^m) and $(\hat{A},\hat{D},\hat{b},\hat{c},R^{\hat{m}})$ both are bilinear realizations of a given system. Then for $n = 1,2,...$ the regular kernels of the two systems give

$$ce^{A\sigma_n}De^{A\sigma_{n-1}}D \cdots De^{A\sigma_1}b = \hat{c}e^{\hat{A}\sigma_n}\hat{D}e^{\hat{A}\sigma_{n-1}}\hat{D} \cdots \hat{D}e^{\hat{A}\sigma_1}\hat{b}$$

for all $\sigma_1,...,\sigma_n \geqslant 0$. Replacing every matrix exponential by its power series expansion and equating coefficients of like arguments shows that

$$cA^{i_n}D \cdots DA^{i_1}b = \hat{c}\hat{A}^{i_n}\hat{D} \cdots \hat{D}\hat{A}^{i_1}\hat{b}$$

for every $n = 1,2,...$, and every $i_j \geqslant 0$. This completes the proof since every element of the product Q_jP_k has precisely this form.

At this point, almost all the tools needed to characterize minimality for bilinear realizations are at hand. The one remaining calculation involves showing that if (A,D,b,c,R^m) is a realization for a given system, then for any invertible, $m \times m$ matrix T, $(TAT^{-1},TDT^{-1},Tb,cT^{-1},R^m)$ also is a realization for the system. This is left as an easy exercise.

Theorem 4.15 A bilinear realization of a specified Volterra system is minimal if and only if it is span reachable and observable.

Proof Suppose (85) is a bilinear realization of dimension m for the given Volterra system, but that it is not span reachable. I will show how to construct another bilinear realization of dimension $< m$. Since (A,D,b,c,R^m) is not span reachable, $R[P_m] \subset R^m$ and I can write $R^m = R[P_m] \oplus V$, where $\oplus$ denotes direct sum, and V is a linear subspace of dimension at least 1. Pick a basis for R^m that is the union of a basis $w_1, ...,w_r$ for $R[P_m]$ and a basis $w_{r+1},...,w_m$ for V. Letting T^{-1} be the $m \times m$ matrix with i^{th} column w_i, then $(TAT^{-1},TDT^{-1},Tb,cT^{-1},R^m)$ also is an m-dimensional bilinear realization of the given Volterra system. Furthermore, since $R[P_m]$ contains b and is invariant under A and D, this new realization is in the partitioned form

$$TAT^{-1} = \begin{bmatrix} A_{11} & A_{12} \\ 0 & A_{22} \end{bmatrix}, \quad TDT^{-1} = \begin{bmatrix} D_{11} & D_{12} \\ 0 & D_{22} \end{bmatrix}$$

$$Tb = \begin{bmatrix} b_1 \\ 0 \end{bmatrix}, \quad c = [c_1 \quad c_2] \tag{93}$$

The 0 blocks in TAT^{-1} and TDT^{-1} are $(m-r) \times r$, the 0 block in Tb is $(m-r) \times 1$, and c_1 is $r \times 1$. Now it is an easy calculation to show that for $n = 1, 2, \ldots$ and $\sigma_1, \ldots, \sigma_n \geqslant 0$,

$$ce^{A\sigma_n}De^{A\sigma_{n-1}}D \cdots De^{A\sigma_1}b = c_1 e^{A_{11}\sigma_n}D_{11}e^{A_{11}\sigma_{n-1}}D_{11} \cdots De^{A_{11}\sigma_1}b_1$$

Thus $(A_{11}, D_{11}, b_1, c_1, R^r)$ is a bilinear realization of dimension $r < m$. In a very similar fashion it can be shown that if a bilinear realization is not observable, then it is not minimal.

Now suppose (A,D,b,c,R^m) and $(\hat{A},\hat{D},\hat{b},\hat{c},R^{\hat{m}})$ are span-reachable and observable bilinear realizations of dimension m and $\hat{m}$, respectively, for the given Volterra system. Letting $M = max[m,\hat{m}]$, Lemma 4.4 gives

$$Q_M P_M = \hat{Q}_M \hat{P}_M \tag{94}$$

But the m rows of P_M and m columns of Q_M are linearly independent, and the $\hat{m}$ rows of $\hat{P}_M$ and $\hat{m}$ columns of $\hat{Q}_M$ are linearly independent. Thus, leaving the details to the reader, (94) implies $m = \hat{m}$. That is, all span-reachable and observable realizations of a given Volterra system have the same dimension. In the first part of the proof it was shown that a minimal bilinear realization is span reachable and observable. Thus all span-reachable and observable realizations of a given Volterra system are minimal.

The last step in the characterization of minimal bilinear realizations will be to show that all such realizations of a given Volterra system are related by a change of variables.

Theorem 4.16 Suppose (A,D,b,c,R^m) is a minimal bilinear realization of a given system. Then $(\hat{A},\hat{D},\hat{b},\hat{c},R^m)$ also is a minimal bilinear realization of the system if and only if there is an invertible matrix T such that $A = T\hat{A}T^{-1}$, $D = T\hat{D}T^{-1}$, $b = T\hat{b}$, $c = \hat{c}T^{-1}$.

Proof If such a T exists, then sufficiency follows from an easy exercise suggested earlier. For necessity, suppose that both state equations are minimal bilinear realizations of the given system. Then by Lemma 4.4,

$$Q_k P_j = \hat{Q}_k \hat{P}_j \,, \quad k,j = 1,2,\ldots \tag{95}$$

and, by Theorem 4.15, Q_m, $\hat{Q}_m$, P_m, and $\hat{P}_m$ all have rank m. In particular, this implies that $(Q_m{}'Q_m)$ is invertible, so if

$$T = (Q_m{}'Q_m)^{-1}Q_m{}'\hat{Q}_m$$

then

$$Q_m{}'Q_m T = Q_m{}'\hat{Q}_m$$

and

$$Q_m{}'Q_m T\hat{P}_m P_m{}' = Q_m{}'\hat{Q}_m\hat{P}_m P_m{}' = Q_m{}'Q_m P_m P_m{}'$$

Since $P_m P_m{}'$ is invertible, this gives that T is invertible and

$$T^{-1} = \hat{P}_m P_m{}'(P_m P_m{}')^{-1}$$

Now (95) with $k = 1,\ j = m$ implies $cP_m = \hat{c}\hat{P}_m$, which implies

$$cP_mP_m' = \hat{c}\hat{P}_mP_m'$$

or, $c = \hat{c}T^{-1}$. Similarly, with $k = m,\ j = 1$, (95) becomes $Q_mb = \hat{Q}_m\hat{b}$, which gives $b = T\hat{b}$. Now note that the columns of AP_m are contained in the columns of P_{m+1}, and the columns of DP_m are contained in the columns of P_{m+1}. Thus (95) implies

$$Q_mAP_m = \hat{Q}_m\hat{A}\hat{P}_m,\quad Q_mDP_m = \hat{Q}_m\hat{D}\hat{P}_m$$

Taking, for example, the first of these equalities,

$$Q_m'Q_mAP_mP_m' = Q_m'\hat{Q}_m\hat{A}\hat{P}_mP_m'$$

or

$$A = (Q_m'Q_m)^{-1}Q_m'\hat{Q}_m\hat{A}\hat{P}_mP_m'(P_mP_m')^{-1} = T\hat{A}T^{-1}$$

The similar calculation for the second equality completes the proof.

4.5 The Nonstationary Case

The transform-domain tools that have been used so extensively in the preceding sections cannot be used fruitfully for nonstationary systems. Also, the regular kernel has been developed only for stationary systems, so this leaves the choice of using either triangular or symmetric kernels in the input/output representation of nonstationary systems. Since bilinear realizations are of interest, the triangular kernel developed in Chapter 3 for such state equations will be used, though the results could be rephrased rather easily in terms of symmetric kernels.

A nonstationary bilinear state equation takes the form

$$\begin{aligned}\dot{x}(t) &= A(t)x(t) + D(t)x(t)u(t) + b(t)u(t)\\ y(t) &= c(t)x(t)\end{aligned}\tag{96}$$

where all the dimensions are as usual, and the coefficient matrices are nominally assumed to be continuous functions of t. In Chapter 3 it was shown that such a state equation with $x(0) = 0$ yields a Volterra system representation

$$y(t) = \sum_{n=1}^{\infty}\int_0^t\int_0^{\sigma_1}\cdots\int_0^{\sigma_{n-1}} h(t,\sigma_1,\ldots,\sigma_n)u(\sigma_1)\cdots u(\sigma_n)\,d\sigma_n\cdots d\sigma_1\tag{97}$$

where the nth triangular kernel is given by

$$\begin{aligned}h(t,\sigma_1,\ldots,\sigma_n) = {}& c(t)\Phi(t,\sigma_1)D(\sigma_1)\Phi(\sigma_1,\sigma_2)D(\sigma_2)\Phi(\sigma_2,\sigma_3)\\ &\cdots D(\sigma_{n-1})\Phi(\sigma_{n-1},\sigma_n)b(\sigma_n)\end{aligned}\tag{98}$$

and where $\Phi(t,\sigma)$ is the transition matrix for $A(t)$.

To consider the bilinear realization problem for a general Volterra system of the form (97) is a difficult task. About all that can be said at present is that the Volterra system is bilinear realizable if and only if there exist appropriately dimensioned, continuous matrix functions $A(t)$, $D(t)$, $b(t)$, and $c(t)$ such that the kernels can be written in the form (98); rather like saying it is bilinear realizable if and only if it is bilinear realizable. The difficulty is similar in nature to the difficulties that arise in the stationary case. Bilinear realizability of a Volterra system depends both on properties of the individual kernels, and on the way the kernels interrelate. However, the outlook is brighter for homogeneous and polynomial systems, and I will concentrate on these cases.

Theorem 4.17 A degree-n homogeneous system described by

$$y(t) = \int_0^t \int_0^{\sigma_1} \cdots \int_0^{\sigma_{n-1}} h(t,\sigma_1,\ldots,\sigma_n)u(\sigma_1) \cdots u(\sigma_n)\, d\sigma_n \cdots d\sigma_1 \tag{99}$$

is bilinear realizable if and only if the kernel $h(t,\sigma_1,\ldots,\sigma_n)$ is separable.

Proof If the system is bilinear realizable, then the kernel can be written in the form (98). From properties of the transition matrix, it follows that the kernel is separable. (Just as in the linear case, the continuity required by separability is furnished by the continuity assumptions on the bilinear state equation.)

Now suppose that the kernel is separable,

$$h(t,\sigma_1,\ldots,\sigma_n) = \sum_{i=1}^{m} v_{0i}(t)v_{1i}(\sigma_1) \cdots v_{ni}(\sigma_n)$$

For the case of $m = 1$, it is easy to show that the bilinear state equation

$$\dot{x}(t) = \begin{bmatrix} 0 & v_{1m}(t) & 0 & \cdots & 0 \\ 0 & 0 & v_{2m}(t) & \cdots & 0 \\ \vdots & \vdots & \vdots & \vdots & \vdots \\ 0 & 0 & 0 & \cdots & v_{n-1,m}(t) \\ 0 & 0 & 0 & \cdots & 0 \end{bmatrix} x(t)u(t) + \begin{bmatrix} 0 \\ 0 \\ \vdots \\ 0 \\ v_{nm}(t) \end{bmatrix} u(t)$$

$$y(t) = [v_{0m}(t)\ 0\ \cdots\ 0]x(t)$$

is a degree-n homogeneous system with kernel

$$h(t,\sigma_1,\ldots,\sigma_n) = v_{0m}(t)v_{1m}(\sigma_1) \cdots v_{nm}(\sigma_n)$$

The proof is now almost complete since an additive parallel connection of these simple bilinear state equations can be used in the general case. The reason the proof is not complete is that when $m = 1$ the $v_{ji}(.)$ must be real functions, but for $m > 1$ they might be complex. Consideration of these details is left to the reader.

It also is interesting to characterize those homogeneous systems that, although represented in terms of a nonstationary triangular kernel, actually are realizable by a constant-parameter bilinear state equation. Once again, the results are similar to the linear-system results.

Theorem 4.18 A degree-n homogeneous system of the form (99) is realizable by a constant-parameter bilinear state equation if and only if the kernel $h(t,\sigma_1,\ldots,\sigma_n)$ is stationary and differentiably separable.

Proof If the degree-n homogeneous system has a constant-parameter bilinear realization, then stationarity and differentiable separability follow easily from the familiar general form of the kernel.

Now suppose the triangular kernel is stationary and differentiably separable. For simplicity I will consider the special case

$$h(t,\sigma_1,\ldots,\sigma_n) = v_0(t)v_1(\sigma_1)\cdots v_n(\sigma_n)$$

where each $v_j(.)$ is a real function. (Just as in the linear case, the generalization of the proof is easy except when complex-valued functions are involved. Then more fussy arguments are required to show that a real-coefficient realization can be obtained.) The main part of the proof will be devoted to showing that the kernel can be written in the form

$$h(t,\sigma_1,\ldots,\sigma_n) = c_1 e^{a_1(t-\sigma_1)} e^{a_2(\sigma_1-\sigma_2)} \cdots e^{a_n(\sigma_{n-1}-\sigma_n)}$$

for real numbers $c_1,a_1,\ldots,a_n$. Once this is established, a bilinear realization is given by

$$\dot{x}(t) = \begin{bmatrix} a_1 & 0 & \cdots & 0 \\ 0 & a_2 & \cdots & 0 \\ \vdots & \vdots & & \vdots \\ 0 & 0 & \cdots & 0 \\ 0 & 0 & \cdots & a_n \end{bmatrix} x(t) + \begin{bmatrix} 0 & 1 & 0 & \cdots & 0 \\ 0 & 0 & 1 & \cdots & 0 \\ \vdots & \vdots & \vdots & & \vdots \\ 0 & 0 & 0 & \cdots & 1 \\ 0 & 0 & 0 & \cdots & 0 \end{bmatrix} x(t)u(t) + \begin{bmatrix} 0 \\ 0 \\ \vdots \\ 0 \\ 1 \end{bmatrix} u(t)$$

$$y(t) = [c_1\ 0\ \cdots\ 0]\, x(t)$$

as is readily verified by the usual calculation. The basic approach involves proving that each $v_j(.)$ satisfies a constant-coefficient linear differential equation of first order. To show this for, say, $v_1(\sigma_1)$, let

$$q_1 = \int_{-T}^{T}\cdots\int_{-T}^{T} v_0^2(t)v_2^2(\sigma_2)\cdots v_n^2(\sigma_n)\, dt\, d\sigma_2 \cdots d\sigma_n$$

where T has been chosen so that $q_1 > 0$. Note that if no such T exists, then the kernel is identically 0, a trivial case. By stationarity

$$h(t,\sigma_1,\ldots,\sigma_n) = h(0,\sigma_1-t,\ldots,\sigma_n-t)$$

so that

$$\sum_{k=1}^{n} \frac{\partial}{\partial\sigma_k} h(t,\sigma_1,\ldots,\sigma_n) + \frac{\partial}{\partial t} h(t,\sigma_1,\ldots,\sigma_n) = 0$$

Computing the derivatives using the separable form gives

$$v_0(t)\dot{v}_1(\sigma_1)v_2(\sigma_2)\cdots v_n(\sigma_n) + v_0(t)v_1(\sigma_1)\dot{v}_2(\sigma_2)v_3(\sigma_3)\cdots v_n(\sigma_n)$$
$$+\cdots+ v_0(t)v_1(\sigma_1)\cdots v_{n-1}(\sigma_1)\dot{v}_n(\sigma_n) + \dot{v}_0(t)v_1(\sigma_1)\cdots v_n(\sigma_n) = 0$$

Multiplying this equation by $v_0(t)v_2(\sigma_2)\cdots v_n(\sigma_n)$ and rearranging gives

$$[v_0^2(t)v_2^2(\sigma_2)\cdots v_n^2(\sigma_n)]\dot{v}_1(\sigma_1) + [v_0^2(t)v_2(\sigma_2)\dot{v}_2(\sigma_2)v_3^2(\sigma_3)\cdots v_n^2(\sigma_n)$$
$$+\cdots+v_0^2(t)v_2^2(\sigma_2)\cdots v_n(\sigma_n)\dot{v}_n(\sigma_n) + v_0(t)\dot{v}_0(t)v_2^2(\sigma_2)\cdots v_n^2(\sigma_n)]v_1(\sigma_1) = 0$$

Both sides of this expression can be integrated with respect to $t, \sigma_2, \ldots, \sigma_n$ to obtain

$$q_1\dot{v}_1(\sigma_1) + r_1 v_1(\sigma_1) = 0$$

with the obvious definition of r_1. Thus $v_1(\sigma_1)$ satisfies a constant-parameter linear differential equation (nontrivial since $q_1 \neq 0$). This means that

$$v_1(\sigma_1) = c_1 e^{a_1\sigma_1}$$

for suitable a_1 and c_1. A similar development can be carried out to show that $v_j(\sigma_j) = c_j e^{a_j\sigma_j}$ for $j = 2, 3, \ldots, n$. Now the stationarity condition can be written as

$$v_0(t)c_1e^{a_1\sigma_1}\cdots c_ne^{a_n\sigma_n} = v_0(0)c_1e^{a_1(\sigma_1-t)}\cdots c_ne^{a_n(\sigma_n-t)}$$
$$= v_0(0)e^{(a_1+\cdots+a_n)(-t)}c_1e^{a_1\sigma_1}\cdots c_ne^{a_n\sigma_n}$$

which gives

$$v_0(t) = v_0(0)e^{-(a_1+\cdots+a_n)t}$$

Thus, with the appropriate redefinition of c_1,

$$h(t,\sigma_1,\ldots,\sigma_n) = c_1e^{-(a_1+\cdots+a_n)t}e^{a_1\sigma_1}\cdots e^{a_n\sigma_n}$$
$$= c_1e^{-(a_1+\cdots+a_n)(t-\sigma_1)}e^{-(a_2+\cdots+a_n)(\sigma_1-\sigma_2)}\cdots e^{-a_n(\sigma_{n-1}-\sigma_n)}$$

and the proof is complete.

These results for homogeneous systems directly provide bilinear realizability results for polynomial systems. That is, bilinear realizability for a polynomial system depends on bilinear realizability of each and every homogeneous subsystem. The easy proof of the following formalization is left to Section 4.7, with the hint that the proof of Theorem 4.9 merits imitation.

Theorem 4.19 A degree-N polynomial system has a (constant parameter) bilinear realization if and only if each of the N triangular kernels is (stationary, and differentiably) separable.

4.6 Remarks and References

Remark 4.1 There is an abundance of material on the linear realization problem, and only a few references will be listed here. An elementary review for stationary systems, including the multi-input, multi-output case, can be found in

C. Chen, *Introduction to Linear System Theory,* Holt, Rinehart, and Winston, New York, 1970.

An elementary treatment that emphasizes Hankel (Behavior) matrices and connections with algebraic properties of rational functions is given in a book that modesty almost prevents me from mentioning:

W. Rugh, *Mathematical Description of Linear Systems,* Marcel Dekker, New York, 1975.

A more research-oriented review of the Hankel matrix approach, along with an interesting discussion of perspectives and open problems is given in

R. Kalman, "Realization Theory of Linear Dynamical Systems," in *Control Theory and Topics in Functional Analysis,* Vol. 2, International Atomic Energy Agency, Vienna, pp. 235-256, 1976.

The abstract shift realization used in Section 4.1 is developed from the approach in

E. Gilbert, "Realization Algorithms for Linear Systems and the Role of the Restricted Backward Shift Realization," *Proceedings of the 1978 Conference on Information Sciences and Systems,* Electrical Engineering Department, The Johns Hopkins University, Baltimore, pp. 145-151, 1978.

Finally, the realization problem for nonstationary linear systems is discussed in

L. Silverman, "Realization of Linear Dynamical Systems," *IEEE Transactions on Automatic Control,* Vol. AC-16, pp. 554-568, 1971.

R. Brockett, *Finite Dimensional Linear Systems,* John Wiley, New York, 1970.

Remark 4.2 An early treatment of the nonlinear realization problem in terms of interconnections of linear systems and multipliers is given in

M. Schetzen, "Synthesis of a Class of Nonlinear Systems," *International Journal of Control,* Vol. 1, pp. 401-414, 1965.

In the degree-2 case, the basic interconnection structure is a cascade connection of a linear system following a multiplicative parallel connection of two linear systems. Additive parallel connections of these basic structures also are used. Realizability tests and realization procedures are developed based on the structural form for the transfer function, $H_1(s_1)H_2(s_2)H_3(s_1+s_2)$, which arises naturally from interconnection results. The issue of realizability in terms of a standard form for the transfer function (say, the symmetric transfer function) is not discussed.

Further development of realization ideas based on structural features of symmetric transfer functions of particular kinds of interconnections of linear systems and multipliers can be found in the following papers.

W. Smith, W. Rugh, "On the Structure of a Class of Nonlinear Systems," *IEEE Transactions on Automatic Control,* Vol. AC-19, pp. 701-706, 1974.

K. Shanmugam, M. Lal, "Analysis and Synthesis of a Class of Nonlinear Systems," *IEEE Transactions on Circuits and Systems,* Vol. CAS-23, pp. 17-25, 1976.

T. Harper, W. Rugh, "Structural Features of Factorable Volterra Systems," *IEEE Transactions on Automatic Control,* Vol. AC-21, pp. 822-832, 1976.

Treatments of the interconnection realization problem that are not based on particular interconnection structures are given for degree-2 homogeneous systems in

G. Mitzel, W. Rugh, "On a Multidimensional S-Transform and the Realization Problem for Homogeneous Nonlinear Systems," *IEEE Transactions on Automatic Control,* Vol. AC-22, pp. 825-830, 1977.

E. Gilbert, "Bilinear and 2-Power Input-Output Maps: Finite Dimensional Realizations and the Role of Functional Series," *IEEE Transactions on Automatic Control,* Vol. AC-23, pp. 418-425, 1978.

In the first of these papers, an algebraic approach to the Laplace transform is developed based on formal series representations. Using the recognizability property, interconnection realizations are developed from a partial fraction expansion of the given transfer function. The second paper uses a specialization of an interconnection structure for so-called bilinear input/output maps (to be discussed in Chapter 6) to arrive at realizations in the homogeneous case.

Remark 4.3 There are many names in the literature for what are called here bilinear state equations, including "regular systems," "internally bilinear systems," and "internally bi-affine systems." There are good reasons for any of these, and the reader is urged to switch rather than fight. On more substantive matters, an early treatment of the bilinear realization problem for a given Volterra system appeared in

A. Isidori, A. Ruberti, "Realization Theory of Bilinear Systems," in *Geometric Methods in System Theory,* D. Mayne, R. Brockett eds., D. Reidel, Dordrecht, Holland, pp. 81-130, 1973.

Two approaches to the problem are presented. The first is a (nonconstructive) factorization approach for the sequence of triangular kernels, while the second involves a so-called generalized Hankel matrix essentially the same as the Behavior matrix in Section 4.3. It is interesting to note the implicit use of the regular kernel in this development. The concepts of span reachability and observability are introduced, and are essential tools in the realization theory. Much of the basic content of this paper also can be found in the papers

P. D'Alessandro, A. Isidori, A. Ruberti, "Realization and Structure Theory of Bilinear Dynamical Systems," *SIAM Journal on Control,* Vol. 12, pp. 517-535, 1974.

A. Isidori, "Direct Construction of Minimal Bilinear Realizations," *IEEE Transactions on Automatic Control,* Vol. AC-18, pp. 626-631, 1973.

Another early paper dealing with bilinear realization is

R. Brockett, "On the Algebraic Structure of Bilinear Systems," in *Theory and Application of Variable Structure Systems,* R. Mohler, A. Ruberti eds., Academic Press, New York, pp. 153-168, 1972.

Equivalences for various forms of bilinear state equations and the notions of span reachability and observability are emphasized in this paper.

Remark 4.4 A much different approach to the bilinear realization problem is given in

M. Fliess, "Sur la Realization des Systemes Dynamiques Bilineaires," *C. R. Academie Science,* Paris, Series A, Vol. 277, pp. 923-926, 1973.

though I recommend the less terse account in

M. Fliess, "Un Outil Algebrique: les Series Formelles Noncommutatives," in *Mathematical System Theory* G. Marchesini, S. Mitter, eds., Lecture Notes in Economics and Mathematical Systems, Vol. 131, Springer-Verlag, New-York, pp. 122-148, 1976.

This approach involves representing input/output behavior in terms of formal series in noncommuting variables. To indicate in simple terms the nature of the formulation, consider a Volterra system representation in triangular form:

$$y(t) = h_0(t) + \int_0^t h_1(t,\sigma_1)u(\sigma_1)\, d\sigma_1$$

$$+ \int_0^t \int_0^{\sigma_1} h_2(t,\sigma_1,\sigma_2)u(\sigma_1)u(\sigma_2)\, d\sigma_2 d\sigma_1 + \cdots$$

Suppose that $h_0(t)$ is analytic for $t \geqslant 0$, and that each of the kernels is analytic on its respective domain $t \geqslant \sigma_n \geqslant \ldots \geqslant \sigma_1 \geqslant 0$. Then power series representations of the form

$$h_0(t) = \sum_{j=0}^{\infty} h_j \frac{t^j}{j!}$$

$$h_1(t,\sigma_1) = \sum_{j_0=0}^{\infty} \sum_{j_1=0}^{\infty} h_{j_0 j_1} \frac{(t-\sigma_1)^{j_1}\sigma_1^{j_0}}{j_0! j_1!}$$

$$h_2(t,\sigma_1,\sigma_2) = \sum_{j_0=0}^{\infty} \sum_{j_1=0}^{\infty} \sum_{j_2=0}^{\infty} h_{j_0 j_1 j_2} \frac{(t-\sigma_2)^{j_2}(\sigma_2-\sigma_1)^{j_1}\sigma_1^{j_0}}{j_0! j_1! j_2!}$$

$$\vdots$$

can be used. These kernel representations provide a means of associating to the system a noncommutative formal series in two variables (or, a formal series in two noncommuting variables), ·

$$W = \sum_{j=0}^{\infty} h_j w_0^j + \sum_{j_0=0}^{\infty} \sum_{j_1=0}^{\infty} h_{j_0 j_1} w_0^{j_1} w_1 w_0^{j_0} + \sum_{j_0=0}^{\infty} \sum_{j_1=0}^{\infty} \sum_{j_2=0}^{\infty} h_{j_0 j_1 j_2} w_0^{j_2} w_1 w_0^{j_1} w_1 w_0^{j_0} + \cdots$$

The correspondence between the Volterra system representation and the noncommutative series representation should be clear from just these first "few" terms. Notice that the noncommutativity is crucial, for if w_0 and w_1 commute, then it is impossible to distinguish between terms. For example, commutativity would imply

$$w_0 w_1 w_0^2 w_1 w_0^3 = w_0^6 w_1^2 = w_0^2 w_1 w_0^2 w_1 w_0^2$$

Now, input/output properties of the system can be interpreted as properties of the series. For example, W represents linear input/output behavior if and only if each nonzero term contains precisely one occurrence of the variable w_1. Also, simple manipulations show that the system represented by W is stationary if and only if each nonzero term in W ends with the variable w_1, except for the constant term. In other words, if and only if W has the form

$$W = h_0 + \sum_{j_1=0}^{\infty} h_{0j_1} w_0^{j_1} w_1 + \sum_{j_1=0}^{\infty} \sum_{j_2=0}^{\infty} h_{0j_1j_2} w_0^{j_2} w_1 w_0^{j_1} w_1 + \cdots$$

The bilinear realization problem for a system described by W is set up most naturally in terms of bilinear state equations of the form

$$\dot{x}(t) = Ax(t) + Dx(t)u(t)$$
$$y(t) = cx(t),\ x(0) = x_0$$

(Recall Problem 3.12.) Applying the resubstitution method to this state equation yields a series expression that can be written in the form

$$y(t) = c[I + A\int_0^t d\sigma + D\int_0^t u(\sigma)\,d\sigma + A^2\int_0^t\int_0^{\sigma_1} d\sigma_2 d\sigma_1 + AD\int_0^t\int_0^{\sigma_1} u(\sigma_2)\,d\sigma_2 d\sigma_1$$
$$+ DA\int_0^t u(\sigma_1)\int_0^{\sigma_1} d\sigma_2 d\sigma_1 + D^2\int_0^t u(\sigma_1)\int_0^{\sigma_1} u(\sigma_2)\,d\sigma_2 d\sigma_1$$
$$+ A^3\int_0^t\int_0^{\sigma_1}\int_0^{\sigma_2} d\sigma_3 d\sigma_2 d\sigma_1 + \cdots]x_0$$

Notice that the coefficient matrix products correspond in a natural way to the order of the iterated integrals of either 1 or $u(t)$. These iterated integrals can be denoted by a monomial in two variables, w_0 and w_1, to yield a noncommutative series representation for the response of the bilinear system,

$$y = cx_0 + cAx_0w_0 + cDx_0w_1 + cA^2x_0w_0^2 + cADx_0w_0w_1 + cDAx_0w_1w_0$$
$$+ cD^2x_0w_1^2 + cA^3x_0w_0^3 + \cdots$$

Of course, this is a noncommutative series because

$$\int_0^t \int_0^{\sigma_1} u(\sigma_2)\, d\sigma_2 d\sigma_1 \neq \int_0^t u(\sigma_1) \int_0^{\sigma_1} d\sigma_2 d\sigma_1$$

that is, $w_0 w_1 \neq w_1 w_0$.

Now a bilinear-realizability result can be stated immediately. A system represented by W is bilinear realizable if and only if there exist two $m \times m$ matrices A and D, an $m \times 1$ vector x_0, and a $1 \times m$ vector c such that the coefficient of $w_0^{j_k} w_1 w_0^{j_{k-1}} \cdots w_1 w_0^{j_0}$ in W is given by $cA^{j_k} DA^{j_{k-1}} \cdots DA^{j_0} x_0$. This condition is equivalent to a rationality condition in the algebraic theory of noncommutative series, and a quick glance at the references will show that this is only the beginning of the story. The concepts of minimality, span reachability, observability, and even a Behavior matrix, all can be formulated from the theory. In fact the regular transfer function representation for stationary Volterra systems can be defined as a commutative series that can be obtained by associating the k-variable commutative monomial $s_1^{-(j_1+1)} \cdots s_k^{-(j_k+1)}$ to the 2-variable noncommutative monomial $w_0^{j_k} w_1 \cdots w_0^{j_1} w_1$. This connection is discussed in

M. Fliess, "A Remark on Transfer Functions and the Realization of Homogeneous Continuous-Time Systems," *IEEE Transactions on Automatic Control,* Vol. AC-24, pp. 507-508, 1979.

Remark 4.5 The shift realization approach to bilinear realization theory that I have used so extensively is based on

A. Frazho, "A Shift Operator Approach to Bilinear System Theory," *SIAM Journal on Control and Optimization,* Vol. 18, pp. 640-658, 1980.

The polynomial factorization approach for homogeneous systems is taken from

G. Mitzel, S. Clancy, W. Rugh, "On Transfer Function Representations for Homogeneous Nonlinear Systems," *IEEE Transactions on Automatic Control,* Vol. AC-24, pp. 242-249, 1979.

Remark 4.6 The importance of bilinear systems can be further substantiated by considering the approximation result established in

H. Sussman, "Semigroup Representations, Bilinear Approximation of Input-Output Maps, and Generalized Inputs," in *Mathematical System Theory,* G. Marchesini, S. Mitter, eds., Lecture Notes in Economics and Mathematical Systems, Vol. 131, Springer-Verlag, New York, 1976.

For single-input, single-output systems, the result can be outlined as follows. The input space U consists of all measurable functions $u(t)$ defined on $[0,T]$ and satisfying $|u(t)| \leq M$ for all $t \in [0,T]$, where T and M are fixed. The output signal is given in operator notation by $y = F[u]$. It is assumed that F is causal, and continuous in the sense that the sequence of output signals $F[u_k]$, $k = 0, 1, \ldots$ converges uniformly to $F[u]$ whenever the sequence of input signals converges weakly to the input u. Then for every $\epsilon > 0$ there is a bilinear realization whose operator representation $y = B[u]$ satisfies $|F[u] - B[u]| < \epsilon$ for all $t \in [0,T]$ and all $u \in U$.

Similar results have been obtained using the noncommutative series representations discussed in Remark 4.4. See

M. Fliess, "Series de Volterra et Series Formelles Non Commutatives," *C. R. Academie Science,* Paris, Series A, Vol. 280, pp. 965-967, 1975.

M. Fliess, "Topologies pour Certaines Functions de Lignes Non Lineaires; Application aux Asservissements," *C. R. Academie Science,* Paris, Series A, Vol. 282, pp. 321-324, 1976.

Remark 4.7 Of course, other kinds of realizations can be discussed in addition to bilinear realizations. Linear-analytic state equations have been studied in this regard, though not nearly to the extent of bilinear state equations. See

R. Brockett, "Volterra Series and Geometric Control Theory," *Automatica,* Vol. 12, pp. 167-176, 1976 (addendum with E. Gilbert, Vol. 12, p. 635).

It is not hard to show that a homogeneous or polynomial system is linear-analytic realizable if and only if it is bilinear realizable. The point is that a minimal linear-analytic realization can be of lower dimension than a minimal bilinear realization. For the homogeneous case, a procedure for computing a minimal linear-analytic realization for a bilinear-realizable system is given in

M. Evans, "Minimal Realizations of k-Powers," *Proceedings of the 1980 Conference on Information Sciences and Systems,* Department of Electrical Engineering and Computer Science, Princeton University, Princeton, New Jersey, pp. 241-245, 1980.

For polynomial systems, the minimal linear-analytic realization problem is discussed in

P. Crouch, "Dynamical Realizations of Finite Volterra Series," *SIAM Journal on Control and Optimization,* Vol. 19, 1981.

In the Volterra system case, much remains to be done. The kinds of things that happen when considering linear-analytic realizations are indicated by a rather simple example. Consider the system with input/output behavior

$$y(t) = \tanh \left[\int_0^t u(\sigma)\,d\sigma\right]$$

Using the power series expansion of the hyperbolic tangent about 0 gives a Volterra system representation of the form

$$y(t) = \sum_{n=1}^{\infty} \int_0^t \cdots \int_0^t \frac{\alpha_n}{n!}\, u(\sigma_1) \cdots u(\sigma_n)\; d\sigma_1 \cdots d\sigma_n$$

where alphas are used because the actual coefficients are rather complicated. In triangular form, the Volterra system can be reresented by

$$y(t) = \sum_{n=1}^{\infty} \int_0^t \int_0^{\sigma_1} \cdots \int_0^{\sigma_{n-1}} \alpha_n \, u(\sigma_1) \cdots u(\sigma_n) \, d\sigma_n \cdots d\sigma_1$$

This Volterra system has a scalar linear-analytic realization, namely,

$$\dot{x}(t) = [1 - x^2(t)]u(t)$$

$$y(t) = x(t), \; x(0) = 0$$

but no finite-dimensional bilinear realization. In addition to showing that linear-analytic realizability and bilinear realizability for Volterra systems are not equivalent, this example shows why infinite-dimensional bilinear realizations might be of interest. For a simple calculation of triangular kernels shows that the Volterra system has a realization of the form

$$\frac{d}{dt}\begin{bmatrix} x_1(t) \\ x_2(t) \\ x_3(t) \\ \vdots \end{bmatrix} = \begin{bmatrix} 0 & 0 & 0 & \cdots \\ 1 & 0 & 0 & \cdots \\ 0 & 1 & 0 & \cdots \\ \vdots & \vdots & \vdots & \vdots \end{bmatrix} \begin{bmatrix} x_1(t) \\ x_2(t) \\ x_3(t) \\ \vdots \end{bmatrix} u(t) + \begin{bmatrix} 1 \\ 0 \\ 0 \\ \vdots \end{bmatrix} u(t)$$

$$y(t) = [\alpha_1 \; \alpha_2 \; \alpha_3 \; \cdots] \begin{bmatrix} x_1(t) \\ x_2(t) \\ x_3(t) \\ \vdots \end{bmatrix}$$

Infinite-dimensional bilinear realizations are discussed in the paper by Frazho mentioned in Remark 4.5, and in

G. Koch, "A Realization Theorem for Infinite Dimensional Bilinear Systems," *Ricerche di Automatica,* Vol. 3, 1972.

R. Brockett, "Finite and Infinite Dimensional Bilinear Realization," *Journal of the Franklin Institute,* Vol. 301, pp. 509-520, 1976.

W. Wong, "Volterra Series, Universal Bilinear Systems, and Fock Representations," *Proceedings of the 1979 Conference on Information Sciences and Systems,* Electrical Engineering Department, The Johns Hopkins University, Baltimore, pp. 207-213, 1979.

Of course, realizations in terms of state equations more general than linear-analytic also can be considered. A transform-domain characterization of realizability and minimality for degree-2 homogeneous systems in terms of very general state equations is given in

E. Gilbert, "Minimal Realizations for Nonlinear I-O Maps:The Continuous-Time 2-Power Case," *Proceedings of the 1978 Conference on Information Sciences and Systems,* Electrical Engineering Department, The Johns Hopkins University, Baltimore, pp. 308-316, 1978.

Further results, including a canonical form for minimal realizations and the fact that the state spaces of minimal realizations are related by a particular type of homeomorphism, are to appear in

E. Gilbert, "Minimal Realizations for Continuous-Time 2-Power Input-Output Maps," *IEEE Transactions on Automatic Control,* Vol. AC-26, 1981.

4.7 Problems

4.1. Suppose the Behavior matrix B_H in (15) for a given linear system has rank n. Show that the first n columns of B_H are linearly independent.

4.2. If the Behavior matrix B_H in (15) for a given linear system has rank n, let

$$A_1 = \begin{bmatrix} h_0 & h_1 & \cdots & h_{n-1} \\ h_1 & h_2 & \cdots & h_n \\ \vdots & \vdots & \vdots & \vdots \\ h_{n-1} & h_n & \cdots & h_{2n-2} \end{bmatrix}, \quad A_2 = \begin{bmatrix} h_1 & h_2 & \cdots & h_n \\ h_2 & h_3 & \cdots & h_{n+1} \\ \vdots & \vdots & \vdots & \vdots \\ h_n & h_{n+1} & \cdots & h_{2n-1} \end{bmatrix}$$

Show that

$$A = A_2 A_1^{-1}, \quad b = \begin{bmatrix} h_0 \\ h_1 \\ \vdots \\ h_{n-1} \end{bmatrix}, \quad c = [1\ 0\ \cdots\ 0]$$

is a minimal realization of the system. (Note that A_1^{-1} exists by Problem 4.1.)

4.3. If $H(s)$ is a strictly proper rational function, and the linear space $U = span\ \{H(s),\ SH(s),\ S^2H(s),\ \ldots\}$ has dimension m, show that $H(s)$, $SH(s)$, . . ., $S^{m-1}H(s)$ is a basis for U.

4.4. Suppose a degree-2 homogeneous system is described by the strictly proper regular transfer function

$$H_{reg}(s_1,s_2) = \frac{1}{s_1 s_2 + 1}$$

Compute the dimension of the linear space U defined in (48) and (49).

4.5. Find a minimal bilinear realization for the square-integral computer discussed in Example 1.4. Give another state-equation realization that has lower dimension than the minimal bilinear realization.

4.6. Compute a minimal bilinear realization for the degree-2 polynomial system

$$\hat{H}(s_1,s_2) = (\ \frac{1}{s_1+1}\ ,\ \frac{1}{(s_1+2)(s_2+1)}\ ,\ 0,\ \ldots\ )$$

4.7. Show that a degree-N truncation of the Volterra system in Example 4.6 has a minimal bilinear realization of dimension N.

4.8. For the Volterra system

$$\hat{H}(s_1,\ldots,s_\infty) = (\ \frac{k_1}{s_1+1}\ ,\ \frac{k_2}{(s_1+1)(s_2+1)}\ ,\ \frac{k_3}{(s_1+1)(s_2+1)(s_3+1)}\ ,\ \ldots\ )$$

suppose the numerator coefficients are such that

$$k_1 s^{-1} + k_2 s^{-2} + k_3 s^{-3} + \cdots$$

corresponds to a strictly proper rational function. Show that the system is bilinear realizable.

4.9. Does the shift realization approach yield a simple, block-partitioned structure for bilinear realizations in the Volterra system case?

4.10. Show that the system described by

$$H_{reg}(s_1,\ldots,s_n) = \frac{1}{Q_1(s_1)Q_2(s_2)\cdots Q_n(s_n)}$$

is realized by the interconnection structured system shown below.

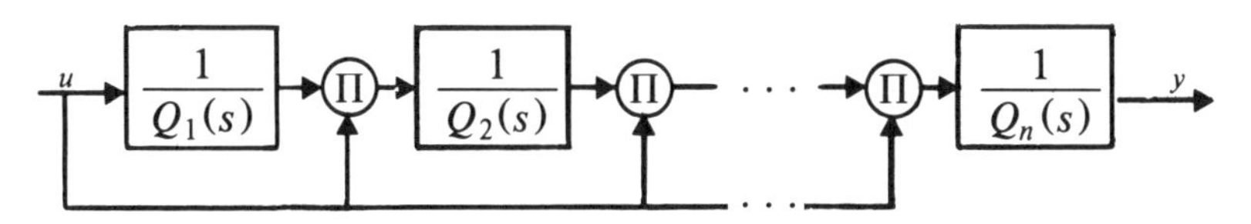

4.11. For the case where the numerator of the regular transfer function in (31) is a constant, show that the bilinear realization in (40) is minimal if and only if every linear realization in (39) is minimal.

4.12. For the bilinear state equation

$$\dot{x}(t) = Ax(t) + Dx(t)u(t) + bu(t)$$

$$y(t) = cx(t)$$

suppose that the state vector is changed according to $z(t) = Tx(t)$, where T is an $n \times n$, invertible matrix. Find the state equation in terms of $z(t)$.

4.13. Show that the bilinear state equation (A,D,b,c,R^m) is span reachable if and only if $(TAT^{-1},TDT^{-1},Tb,cT^{-1},R^m)$ is span reachable.

4.14. Suppose A and D are $m \times m$ matrices, and X is a linear space that is invariant under $(A + Du)$ for all real numbers u. Show that X is invariant under both A and D.

4.15. Prove Theorem 4.19.

4.16. Suppose a degree-n homogeneous system is described by a strictly proper, recognizable regular transfer function wherein all the roots of the denominator polynomials have negative real parts. Show that if (A,D,b,c,R^m) is a minimal bilinear realization of the system, then all the eigenvalues of A will have negative real parts. Show also that the system is bounded-input, bounded-output stable.

4.17. Suppose a bilinear-realizable system is connected in cascade with a linear-realizable system. Show that the overall system is bilinear realizable, regardless of the ordering of the two systems in the cascade. (Do not peek at Appendix 4.1.)

4.18. Polynomial systems of certain types can be represented by the *sum* of the regular transfer functions of the homogeneous subsystems. Show how this works by redoing Example 4.6 beginning with

$$\hat{H}(s_1,s_2) = \frac{1}{s_1+1} + \frac{1}{(s_1+2)(s_2+3)} = \frac{s_1s_2+4s_1+2s_2+7}{(s_1+1)(s_1+2)(s_2+3)}$$

and slightly modifying the realization procedure.

4.19. This problem considers further the representation suggested in Problem 4.18. Show that a degree-N polynomial system is bilinear realizable if and only if the sum of the subsystem regular transfer functions is a recognizable function which is strictly proper in s_1 and proper in $s_2, \ldots, s_N$.

APPENDIX 4.1 Interconnection Rules for the Regular Transfer Function

The derivation of interconnection rules for the regular transfer function representation apparently best proceeds in a manner closely tied to the structure of bilinear state equations. This approach is in contrast to the development of interconnection rules for the other transfer function representations. At any rate, Chapters 3 and 4 have provided the theory needed to generate a table of regular transfer functions, and it is the purpose of this appendix to present such a table, and to show how it is derived.

The types of systems to be considered are interconnections of linear systems and homogeneous bilinear systems (all finite dimensional). The linear systems will be described in terms of a state equation

$$\dot{z}(t) = Fz(t) + gu(t)$$
$$y(t) = hz(t)$$

or a (strictly proper, rational) transfer function

$$H(s) = h(sI - F)^{-1}g$$

The homogeneous bilinear systems will be described in terms of a state equation

$$\dot{x}(t) = Ax(t) + Dx(t)u(t) + bu(t)$$
$$y(t) = cx(t)$$

or a (strictly proper, recognizable) regular transfer function as given in (30) of Section 4.2:

$$H_{reg}(s_1,\ldots,s_n) = c(s_nI-A)^{-1}D(s_{n-1}I-A)^{-1}D \cdots (s_1I-A)^{-1}b$$

(As usual, the dimension of identity matrices will be fixed by conformability requirements, and will not be indicated by notation.)

The fact that a degree-n homogeneous bilinear state equation can be assumed to be in the block partitioned form (40) will be important. This block form is repeated below for convenience.

$$x(t) = \begin{bmatrix} x_1(t) \\ x_2(t) \\ \vdots \\ x_n(t) \end{bmatrix}, \quad A = \begin{bmatrix} A_1 & 0 & \cdots & 0 \\ 0 & A_2 & \cdots & 0 \\ \vdots & \vdots & \vdots & \vdots \\ 0 & 0 & \cdots & A_n \end{bmatrix}, \quad D = \begin{bmatrix} 0 & 0 & \cdots & 0 & 0 \\ D_1 & 0 & \cdots & 0 & 0 \\ 0 & D_2 & \cdots & 0 & 0 \\ \vdots & \vdots & \vdots & \vdots & \vdots \\ 0 & 0 & \cdots & D_{n-1} & 0 \end{bmatrix}$$

$$c = [0 \ \cdots \ 0 \ c_n] \ , \quad b = \begin{bmatrix} b_1 \\ 0 \\ \vdots \\ 0 \end{bmatrix}$$

In terms of this block form, the regular transfer function can be written as

$$H_{reg}(s_1, \ldots, s_n) = c_n(s_nI - A_n)^{-1}D_{n-1}(s_{n-1}I - A_{n-1})^{-1}D_{n-2} \cdots (s_1I - A_1)^{-1}b_1$$

To generate a table of interconnection formulas, the basic idea is similar to Carleman linearization. The first step is to write a composite state equation for the overall system in terms of subsystem state equations. The second step is to derive a differential equation for the new "Kronecker product variables" that appear in the composite state equation. Finally, if all these equations can be written as a big bilinear state equation, then the regular transfer function of the overall system is easy to compute, in principle.

To illustrate this procedure, consider the next-to-last entry in Table 4.1; the multiplicative parallel connection of a linear system and a degree-2 homogeneous bilinear system. Making use of the block form of the homogeneous bilinear state equation, the overall system can be described by:

$$\begin{aligned} \dot{z}(t) &= Fz(t) + gu(t) \\ \dot{x}_1(t) &= A_1x_1(t) + b_1u(t) \\ \dot{x}_2(t) &= A_2x_2(t) + D_1x_1(t)u(t) \\ y(t) &= hz(t)c_2x_2(t) \end{aligned}$$

Using the Kronecker product, the (scalar) output equation can be written in the form

$$y(t) = [hz(t)] \otimes [c_2x_2(t)] = [h \otimes c_2][z(t) \otimes x_2(t)]$$

Now, a bilinear equation for $z(t) \otimes x_2(t)$ is easily computed:

$$\frac{d}{dt}[z(t) \otimes x_2(t)] = \dot{z}(t) \otimes x_2(t) + z(t) \otimes \dot{x}_2(t)$$

$$= [F \otimes I + I \otimes A_2][z(t) \otimes x_2(t)] + [g \otimes I]x_2(t)u(t)$$

$$+ [I \otimes D_1][z(t) \otimes x_1(t)]u(t)$$

For the new term $z(t) \otimes x_1(t)$, a similar calculation gives a bilinear equation

$$\frac{d}{dt}[z(t) \otimes x_1(t)] = [F \otimes I + I \otimes A_1][z(t) \otimes x_1(t)] + [g \otimes I]x_1(t)u(t)$$

$$+ [I \otimes b_1]z(t)u(t)$$

Collecting all the state equations, and letting

$$\hat{x}(t) = \begin{bmatrix} z(t) \\ x_1(t) \\ x_2(t) \\ z(t) \otimes x_1(t) \\ z(t) \otimes x_2(t) \end{bmatrix}$$

gives a bilinear state equation description of the multiplicative connection in block partitioned form:

$$\frac{d}{dt}\hat{x}(t) = \begin{bmatrix} F & 0 & 0 & 0 & 0 \\ 0 & A_1 & 0 & 0 & 0 \\ 0 & 0 & A_2 & 0 & 0 \\ 0 & 0 & 0 & [F \otimes I + I \otimes A_1] & 0 \\ 0 & 0 & 0 & 0 & [F \otimes I + I \otimes A_2] \end{bmatrix} \hat{x}(t)$$

$$+ \begin{bmatrix} 0 & 0 & 0 & 0 & 0 \\ 0 & 0 & 0 & 0 & 0 \\ 0 & D_1 & 0 & 0 & 0 \\ I \otimes b_1 & g \otimes I & 0 & 0 & 0 \\ 0 & 0 & g \otimes I & I \otimes D_1 & 0 \end{bmatrix} \hat{x}(t)u(t) + \begin{bmatrix} g \\ b_1 \\ 0 \\ 0 \\ 0 \end{bmatrix} u(t)$$

Finally, a simple calculation gives the regular transfer function corresponding to this block-partitioned bilinear state equation, as shown in Table 4.1.

The remaining entries in Table 4.1 are derived in the same way. In each case, a bilinear state equation is derived for the overall system, and then the regular transfer function is computed from the state equation. Of course, the reader will recall from Chapter 3 that the reduced Kronecker product can be used to obtain interconnection formulas that are more economical of dimension, but less explicit.

Table 4.1
Interconnection Table for Regular Transfer Functions*

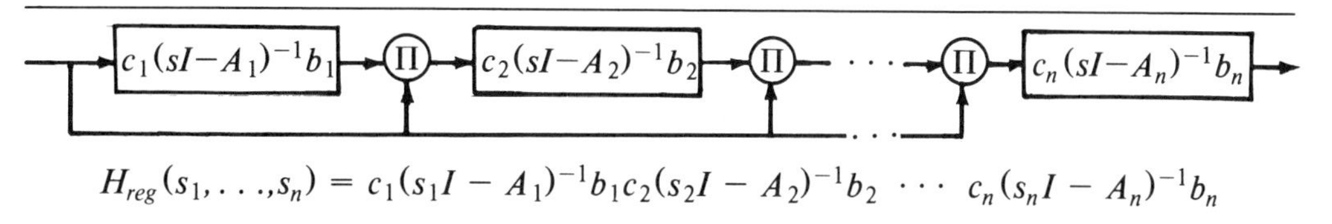

$$H_{reg}(s_1,\ldots,s_n) = c_1(s_1I - A_1)^{-1}b_1c_2(s_2I - A_2)^{-1}b_2 \cdots c_n(s_nI - A_n)^{-1}b_n$$

$c_2(s_2I-A_2)^{-1}D_1(s_1I-A_1)^{-1}b_1$ → Π → $h(sI-F)^{-1}g$

$$H_{reg}(s_1,s_2,s_3) = h(s_3I - F)^{-1}gc_2(s_2I - A_2)^{-1}D_1(s_1I - A_1)^{-1}b_1$$

$h(sI-F)^{-1}g$ → Π → $c_2(s_2I-A_2)^{-1}D_1(s_1I-A_1)^{-1}b_1$

$$H_{reg}(s_1,\ldots,s_4) = c_2(s_4I - A_2)^{-1}(D_1 \otimes h)(s_3I - A_1 \otimes I - I \otimes F)^{-1}$$
$$[(I \otimes g)(s_2I - A_1)^{-1}b_1h + ((b_1h) \otimes I)(s_2I - F \otimes I - I \otimes F)^{-1}$$
$$(g \otimes I + I \otimes g)](s_1I - F)^{-1}g$$

$c_n(s_nI-A_n)^{-1}D_{n-1}(s_{n-1}I-A_{n-1})^{-1}D_{n-2}\cdots(s_1I-A_1)^{-1}b_1$ → $h(sI-F)^{-1}g$

$$H_{reg}(s_1,\ldots,s_n) = h(s_nI - F)^{-1}gc_n(s_nI - A_n)^{-1}D_{n-1}(s_{n-1}I - A_{n-1})^{-1}D_{n-2}$$
$$\cdots D_1(s_1 - A_1)^{-1}b_1$$

$h(sI-F)^{-1}g$ → $c_2(s_2I-A_2)^{-1}D_1(s_1I-A_1)^{-1}b_1$

$$H_{reg}(s_1,s_2) = c_2(s_2I - A_2)^{-1}(D_1 \otimes h)(s_2I - A_1 \otimes I - I \otimes F)^{-1}$$
$$[(I \otimes g)(s_1I - A_1)^{-1}b_1h + ((b_1h) \otimes I)(s_2I - F \otimes I - I \otimes F)^{-1}$$
$$(g \otimes I + I \otimes g)](s_1I - F)^{-1}g$$

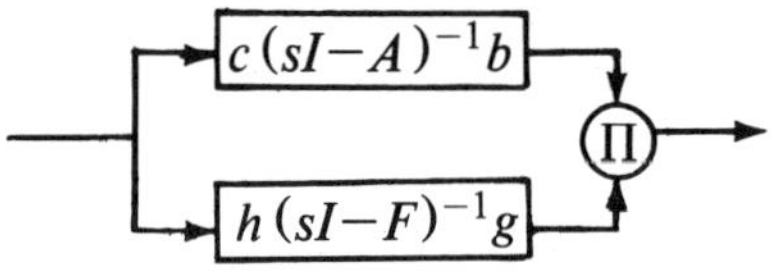

$c(sI-A)^{-1}b$

Π

$h(sI-F)^{-1}g$

$$H_{reg}(s_1,s_2) = (c \otimes h)(s_2 I - A \otimes I - I \otimes F)^{-1}[(I \otimes g)(s_1 I - A)^{-1}b + (b \otimes I)(s_1 I - F)^{-1}g]$$

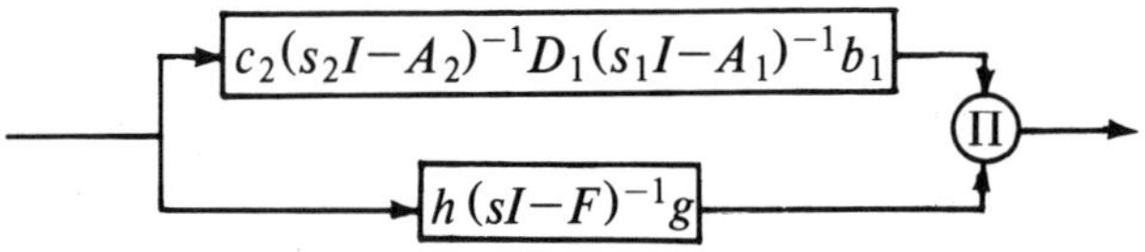

$c_2(s_2I-A_2)^{-1}D_1(s_1I-A_1)^{-1}b_1$

Π

$h(sI-F)^{-1}g$

$$H_{reg}(s_1,s_2,s_3) = (h \otimes c_2)(s_3 I - F \otimes I - I \otimes A_2)^{-1} \{(g \otimes I)(s_2 I - A_2)^{-1} D_1 (s_1 I - A_1)^{-1} b_1 + (I \otimes D)(s_2 I - F \otimes I - I \otimes A_1)^{-1}[(I \otimes b_1)(s_1 I - F)^{-1} g + (g \otimes I)(s_1 I - A_1)^{-1} b_1]\}$$

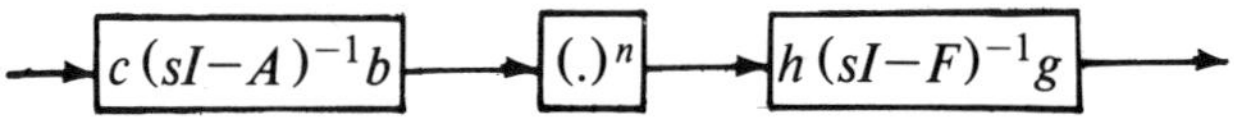

$c(sI-A)^{-1}b$ → $(.)^n$ → $h(sI-F)^{-1}g$

$$H_{reg}(s_1,\ldots,s_n) = h(s_n I - F)^{-1} g \hat{c} (s_n I - \hat{A}_n)^{-1} \hat{b}_{n-1} (s_{n-1} I - \hat{A}_{n-1})^{-1} \hat{b}_{n-2} \cdots (s_1 I - \hat{A}_1)^{-1} \hat{b}_1$$

where $\hat{c} = c \otimes c \otimes \cdots \otimes c$ *(n factors)*

$$\hat{A}_j = I \otimes \cdots \otimes I \otimes A + I \otimes \cdots \otimes I \otimes A \otimes I + \cdots + A \otimes I \otimes \cdots \otimes I \quad (j \text{ terms})$$

$$\hat{b}_j = I \otimes \cdots \otimes I \otimes b + I \otimes \cdots \otimes I \otimes b \otimes I + \cdots + b \otimes I \otimes \cdots \otimes I \quad (j \text{ terms})$$

---

* Π denotes time-domain multiplication, $\otimes$ denotes Kronecker product.

CHAPTER 5

Response Characteristics of Stationary Systems

Methods for computing the response of a homogeneous system to a specified input signal have been discussed in previous chapters. The integrations can be carried out in the time-domain representation, or the association of variables method can be used in the transform domain. In terms of the regular transfer function, a more explicit approach can be used when the input is a sum of exponentials. Response computation for a polynomial system is simply a matter of adding the homogeneous-subsystem responses, though a convenient notation can be hard to find. The same is true of Volterra systems, with the additional complication of convergence issues.

For specific types of input signals, the response of a homogeneous system has special features that generalize well known properties of linear systems. This is especially true in the stationary system case, and thus I will discuss only that situation. The response to impulse inputs, the steady-state response to sinusoidal inputs, and properties of the response to stochastic inputs will be considered. Most of the discussion will be in terms of the symmetric kernel or symmetric transfer function. This is both a matter of tradition, and a result of the fact that the formulas usually appear in a simple form when expressed in terms of symmetric representations. The material in this chapter will be useful in connection with the identification problem to be discussed in Chapter 7.

5.1 Response to Impulse Inputs

In this section the response of homogeneous systems to inputs composed of impulse functions will be computed. For the polynomial or Volterra system cases, not much can be done other than to add up the homogeneous-subsystem responses. The symmetric kernel representation will be used throughout this section. Of course, these kernels are assumed to be impulse free, so that the impulse response is guaranteed to be defined.

Surely I will bore the reader by reminding that for the linear system

$$y(t) = \int_0^t h(t-\sigma)u(\sigma)\, d\sigma \tag{1}$$

the input $u(t) = \delta_0(t)$ yields the response $y(t) = h(t)$, $t \geqslant 0$. That is, the impulse response of a linear system traces out the kernel. For a degree-n (> 1) homogeneous system

$$y(t) = \int_0^t h_{sym}(t-\sigma_1,\ldots,t-\sigma_n)u(\sigma_1)\cdots u(\sigma_n)\, d\sigma_1 \cdots d\sigma_n \tag{2}$$

the input $\delta_0(t)$ yields the response $h_{sym}(t,\ldots,t)$, $t \geqslant 0$.

More interesting calculations arise when inputs composed of sums of impulses are considered. For example, suppose the input to (2) is

$$u(t) = \delta_0(t) + \delta_0(t-T), \quad T > 0 \tag{3}$$

One way to compute the response is to multiply out the expression

$$u(\sigma_1)\cdots u(\sigma_n) = [\delta_0(\sigma_1) + \delta_0(\sigma_1-T)]\cdots[\delta_0(\sigma_n) + \delta_0(\sigma_n-T)] \tag{4}$$

and integrate over each term. This is not difficult because symmetry and some simple combinatorics come to the rescue. The indices can be permuted so that the general term arising in the product in (4) takes the form

$$\delta_0(\sigma_1)\cdots\delta_0(\sigma_m)\delta_0(\sigma_{m+1}-T)\cdots\delta_0(\sigma_n-T)$$

without changing the outcome of the integrations. In fact, there will be $\binom{n}{m}$ terms from (4) that can be written in this particular form. Thus, the response is

$$y(t) = \sum_{m=0}^{n} \binom{n}{m} h_{sym}(\underbrace{t,\ldots,t}_{m},\underbrace{t-T,\ldots,t-T}_{n-m}) \tag{5}$$

Now consider the general case where the input to the degree-n system (2) is

$$u(t) = \delta_0(t) + \delta_0(t-T_1) + \cdots + \delta_0(t-T_{p-1}) \tag{6}$$

where $T_1,\ldots,T_{p-1}$ is a set of distinct positive numbers. (Portions of the following analysis should be reminiscent of Section 2.4.) Again, the procedure is to expand the product

$$u(\sigma_1)\cdots u(\sigma_n) = [\delta_0(\sigma_1)+\cdots+\delta_0(\sigma_1-T_{p-1})]\cdots[\delta_0(\sigma_n)+\cdots+\delta_0(\sigma_n-T_{p-1})]$$

and then perform the integration over each term. But permutation of indices does not affect these integrations, and so the general term in the product can be written in the form

$$\delta_0(\sigma_1)\cdots\delta_0(\sigma_{m_1})\delta_0(\sigma_{m_1+1}-T_1)\cdots\delta_0(\sigma_{m_1+m_2}-T_1)$$
$$\cdots\delta_0(\sigma_{n-m_p+1}-T_{p-1})\cdots\delta_0(\sigma_n-T_{p-1})$$

Counting the number of terms that can be written in this way for a particular $m_1,\ldots,m_p$ yields multinomial coefficients, and the response is given by

$$y(t) = \sum_m \frac{n!}{m_1!\cdots m_p!} h_{sym}(\underbrace{t,\ldots,t}_{m_1};\ldots;\underbrace{t-T_{p-1},\ldots,t-T_{p-1}}_{m_p}) \tag{7}$$

where $\sum_m$ is a p-fold summation over all integer indices $m_1,\ldots,m_p$ such that

$$0 \leqslant m_i \leqslant n \text{ and } m_1 + \cdots + m_p = n.$$

5.2 Steady-State Response to Sinusoidal Inputs

For the remainder of this chapter, steady-state response properties will be the subject of principal interest. Thus consideration of input/output stability properties is needed to insure that the steady-state response is bounded. In the time domain, it is apparent from bounding calculations in Section 1.3 that a sufficient condition for bounded input, bounded output stability of a degree-n homogeneous system is

$$\int_{-\infty}^{\infty} |h_{sym}(t_1, \ldots, t_n)| \, dt_1 \cdots dt_n < \infty$$

But in terms of transform representations, conditions are more difficult to find. A well known condition for linear systems described by reduced rational transfer functions is that a system is bounded input, bounded output stable if and only if all the poles of the transfer function have negative real parts. In the degree-n case $(n > 1)$, a sufficient condition of similar type can be given for systems described by strictly proper, recognizable, regular transfer functions (Problem 4.16). Unfortunately, this result is much less simple to state in terms of the symmetric transfer function. Furthermore, the difficulty in factoring general symmetric polynomials makes conditions on the factors very hard to check. Thus the stability properties needed for a valid steady-state analysis will simply be assumed.

For a stationary linear system described by

$$y(t) = \int_0^t h(\sigma) u(t-\sigma) \, d\sigma \tag{8}$$

consider the response to the one-sided input signal

$$u(t) = 2A\cos(\omega t), \; t \geqslant 0 \tag{9}$$

It is more convenient to write this input in the complex exponential form

$$u(t) = Ae^{i\omega t} + Ae^{-i\omega t} \tag{10}$$

for then

$$y(t) = A \int_0^t h(\sigma) e^{i\omega(t-\sigma)} \, d\sigma + A \int_0^t h(\sigma) e^{-i\omega(t-\sigma)} \, d\sigma \tag{11}$$

or

$$y(t) = A\left[\int_0^t h(\sigma) e^{-i\omega\sigma} \, d\sigma\right] e^{i\omega t} + A\left[\int_0^t h(\sigma) e^{i\omega\sigma} \, d\sigma\right] e^{-i\omega t} \tag{12}$$

Assuming the system is stable, as $t \to \infty$ the integrals converge to $H(i\omega)$ and $H(-i\omega)$, respectively, where

$$H(s) = \int_0^{\infty} h(\sigma) e^{-s\sigma} \, d\sigma \tag{13}$$

is the system transfer function. Thus, by picking T large enough, it can be guaranteed that for all $t \geqslant T$ the system response is within a specified tolerance of the so-called steady-state response

$$y_{ss}(t) = AH(i\omega)e^{i\omega t} + AH(-i\omega)e^{-i\omega t} \tag{14}$$

Of course using standard identities this steady-state response can be rewritten in the forms

$$y_{ss}(t) = 2A \; Re[H(i\omega)]cos(\omega t) - 2A \; Im[H(i\omega)]sin(\omega t) \tag{15}$$

or

$$y_{ss}(t) = 2A|H(i\omega)|cos[\omega t + \angle H(i\omega)] \tag{16}$$

where standard notations have been used for real part, imaginary part, magnitude, and angle. These calculations simply make explicit the well known fact that the steady-state response of a linear system to a sinusoidal input of frequency ω is a sinusoid of the same frequency, with amplitude and phase determined by the magnitude and angle of the transfer function evaluated at $s = i\omega$. (I should point out that there is another way to view the steady state. The input can be considered to begin at $t = -\infty$, and then the response at any finite t is the steady-state response.)

Now consider the generalization of these results to homogeneous systems described by

$$y(t) = \int_0^t h_{sym}(\sigma_1,\ldots,\sigma_n)u(t-\sigma_1)\cdots u(t-\sigma_n)\,d\sigma_1\cdots d\sigma_n \tag{17}$$

But before I begin, it seems wise to point out a common pitfall in discussing the response of a nonlinear system to sinusoidal inputs. When working with linear systems, it is common to consider the input (9) as the real part of the phasor $2Ae^{i\omega t}$. Then the response of the system to this complex input is calculated, and the response of the system to (9) is found simply by taking the real part of the response to the phasor. However, this shortcut depends crucially on the assumption of linearity, as the following example shows.

Example 5.1 To compute the response of the system $y(t) = u^2(t)$ to the input (9), application of the input $u_1(t) = 2Ae^{i\omega t}$ gives the response $y_1(t) = 4A^2e^{i2\omega t}$. Then an erroneous conclusion is that the system response to (9) is $y(t) = 4A^2cos(2\omega t)$. It is erroneous because direct application of (9) gives $y(t) = 4A^2\cos^2(\omega t) = 2A^2 + 2A^2\cos(2\omega t)$.

With the one-sided input signal (10) the response of (17) can be computed from

$$y(t) = \int_0^t h_{sym}(\sigma_1,\ldots,\sigma_n)\prod_{j=1}^n [Ae^{i\omega(t-\sigma_j)} + Ae^{-i\omega(t-\sigma_j)}]\,d\sigma_1\cdots d\sigma_n \tag{18}$$

To put this expression in more useful form, I will mimic the double-exponential-input development of Section 2.4. Letting $\lambda_1 = i\omega$ and $\lambda_2 = -i\omega$ facilitates this development, for then

$$y(t) = A^n \int_0^t h_{sym}(\sigma_1, \ldots, \sigma_n) \sum_{k_1=1}^{2} \cdots \sum_{k_n=1}^{2} \exp[\sum_{j=1}^{n} \lambda_{k_j}(t-\sigma_j)]\, d\sigma_1 \cdots d\sigma_n$$

$$= A^n \sum_{k_1=1}^{2} \cdots \sum_{k_n=1}^{2} [\int_0^t h_{sym}(\sigma_1, \ldots, \sigma_n) \exp(-\sum_{j=1}^{n} \lambda_{k_j}\sigma_j)$$

$$d\sigma_1 \cdots d\sigma_n \exp(\sum_{j=1}^{n} \lambda_{k_j} t) \tag{19}$$

In a manner similar to the linear case, consider the response for large values of t. Assuming stability of the system, the bracketed term in (19) approaches $H_{sym}(\lambda_{k_1}, \ldots, \lambda_{k_n})$ as $t \to \infty$. Thus $y(t)$ in (19) becomes arbitrarily close to the steady-state response defined by

$$y_{ss}(t) = A^n \sum_{k_1=1}^{2} \cdots \sum_{k_n=1}^{2} H_{sym}(\lambda_{k_1}, \ldots, \lambda_{k_n})\exp(\sum_{j=1}^{n} \lambda_{k_j} t) \tag{20}$$

an expression that clearly is analogous to (14). Collecting together those terms with identical exponents $[k\lambda_1 + (n-k)\lambda_2]$, and recalling the definitions of λ_1 and λ_2, (20) can be written as

$$y_{ss}(t) = A^n \sum_{k=0}^{n} G_{k,n-k}(i\omega, -i\omega) e^{i(2k-n)\omega t} \tag{21}$$

where

$$G_{k,n-k}(\lambda_1, \lambda_2) = \sum_{\substack{k_1=1 \\ k_1+\cdots+k_n=2n-k}}^{2} \cdots \sum_{k_n=1}^{2} H_{sym}(\lambda_{k_1}, \ldots, \lambda_{k_n})$$

$$= \binom{n}{k} H_{sym}(\underbrace{\lambda_1, \ldots, \lambda_1}_{k}; \underbrace{\lambda_2, \ldots, \lambda_2}_{n-k}) \tag{22}$$

One useful identity that follows from (22) is

$$G_{k,n-k}(i\omega, -i\omega) = G_{n-k,k}(-i\omega, i\omega)$$

It is convenient to rearrange the terms in (21) as follows. First write

$$y_{ss}(t) = A^n[G_{n,0}(i\omega, -i\omega) e^{in\omega t} + G_{0,n}(i\omega, -i\omega) e^{-in\omega t}]$$

$$+ A^n[G_{n-1,1}(i\omega, -i\omega) e^{i(n-2)\omega t} + G_{1,n-1}(i\omega, -i\omega) e^{-i(n-2)\omega t}]$$

$$+ \cdots + \begin{cases} A^n G_{n/2,n/2}(i\omega, -i\omega), & n \text{ even} \\ A^n[G_{\frac{n+1}{2},\frac{n-1}{2}}(i\omega, -i\omega) e^{i\omega t} + G_{\frac{n-1}{2},\frac{n+1}{2}}(i\omega, -i\omega) e^{-i\omega t}], & n \text{ odd} \end{cases}$$

$$= A^n[G_{n,0}(i\omega, -i\omega) e^{in\omega t} + G_{n,0}(-i\omega, i\omega) e^{-in\omega t}]$$

$$+ A^n[G_{n-1,1}(i\omega,-i\omega)e^{i(n-2)\omega t} + G_{n-1,1}(-i\omega,i\omega)e^{-i(n-2)\omega t}]$$

$$+\cdots+\begin{cases} A^n G_{n/2,n/2}(i\omega,-i\omega), \ n \text{ even} \\ A^n[G_{\frac{n+1}{2},\frac{n-1}{2}}(i\omega,-i\omega)e^{i\omega t}+G_{\frac{n+1}{2},\frac{n-1}{2}}(-i\omega,i\omega)e^{-i\omega t}], \ n \text{ odd} \end{cases} \tag{23}$$

Now, using standard identities,

$$y_{ss}(t) = 2A^n|G_{n,0}(i\omega,-i\omega)|\cos[n\omega t + \angle G_{n,0}(i\omega,-i\omega)]$$

$$+ 2A^n|G_{n-1,1}(i\omega,-i\omega)|\cos[(n-2)\omega t + \angle G_{n-1,1}(i\omega,i\omega)]$$

$$+\cdots+\begin{cases} A^n G_{n/2,n/2}(i\omega,-i\omega), \ n \text{ even} \\ 2A^n|G_{\frac{n+1}{2},\frac{n-1}{2}}(i\omega,-i\omega)|\cos[\omega t + \angle G_{\frac{n+1}{2},\frac{n-1}{2}}(i\omega,-i\omega)], \ n \text{ odd} \end{cases} \tag{24}$$

Thus the steady-state response of a degree-n homogeneous system to a cosinusoidal input of frequency ω is composed of cosinusoidal components at frequencies $n\omega$, $(n-2)\omega, \ldots, 0$ (n even) or ω (n odd).

Now consider a degree-N polynomial system

$$y(t) = \sum_{n=1}^{N} \int_{-\infty}^{\infty} h_{nsym}(\sigma_1,\ldots,\sigma_n)u(t-\sigma_1)\cdots u(t-\sigma_n)\, d\sigma_1 \cdots d\sigma_n \tag{25}$$

with the input signal $u(t) = 2A\cos(\omega t)$. The steady-state response is obtained by adding the contributions of each homogeneous subsystem. Each degree-n subsystem where n is odd contributes terms at frequencies ω, $3\omega, \ldots, n\omega$. Each degree-n subsystem where n is even contributes a constant term and terms at frequencies 2ω, $4\omega, \ldots, n\omega$. From (23) the contribution of the degree-n subsystem to frequency $k\omega$, assuming $k \leqslant n$ and k and n have the same parity, is

$$A^n\, G_{\frac{n+k}{2},\frac{n-k}{2}}(i\omega,-i\omega)e^{ik\omega t} + A^n\, G_{\frac{n+k}{2},\frac{n-k}{2}}(-i\omega,i\omega)\, e^{-ik\omega t} \tag{26}$$

(It is useful to observe that the sum of the subscripts on G indicates the degree of the subsystem, and the difference of the subscripts the harmonic.) Thus letting N_k be the greatest integer $\leqslant N$ with the same parity as k, the steady-state response of (25) can be written as

$$y_{ss}(t) = f_0(A,i\omega) + \sum_{k=1}^{N} [f_k(A,i\omega)\, e^{ik\omega t} + f_k(A,-i\omega)\, e^{-ik\omega t}] \tag{27}$$

where

$$f_0(A,i\omega) = A^2\, G_{1,1}(i\omega,-i\omega) + A^4\, G_{2,2}(i\omega,-i\omega)$$

$$+ \cdots + A^{N_2}\, G_{\frac{N_2}{2},\frac{N_2}{2}}(i\omega,-i\omega) \tag{28}$$

$$f_1(A,i\omega) = A\ G_{1,0}(i\omega,-i\omega) + A^3\ G_{2,1}(i\omega,-i\omega) + \cdots + A^{N_1}\ G_{\frac{N_1+1}{2},\frac{N_1-1}{2}}(i\omega,-i\omega) \tag{29}$$

$$f_2\ (A,i\omega) = A^2\ G_{2,0}(i\omega,-i\omega) + A^4\ G_{3,1}(i\omega,-i\omega) + \cdots + A^{N_2}\ G_{\frac{N_2+2}{2},\frac{N_2-2}{2}}(i\omega,-i\omega) \tag{30}$$

and so on. The general terms can be written in the forms

$$f_0(A,i\omega) = \sum_{j=1}^{N_2/2} A^{2j}G_{jj}(i\omega,-i\omega)$$

$$f_k(A,i\omega) = \sum_{j=0}^{(N_k-k)/2} A^{k+2j}G_{k+j,j}(i\omega,-i\omega)\ ,\ k=1,2,\ldots,N \tag{31}$$

As is by now usual, (27) can be written as

$$y_{ss}(t) = f_0(A,i\omega) + 2\sum_{k=1}^{N} |f_k(A,i\omega)| \cos[k\omega t + \angle f_k(A,i\omega)] \tag{32}$$

which makes explicit in real terms the fact that the steady-state response of a polynomial system to sinusoidal inputs can be expressed as a finite Fourier series. Furthermore, the Fourier coefficients are polynomials in the input amplitude A with coefficients that are functions of the input frequency ω.

Example 5.2 Consider the pendulum system in Example 3.8. Using the symmetric transfer functions through degree 3 calculated there, the steady-state response to $u(t) = 2A\cos(\omega t)$ is given by

$$y_{ss}(t) = 2|AH(i\omega) + 3A^3H_{3sym}(i\omega,i\omega,-i\omega) + \cdots| \cos[\omega t + \phi_1(\omega)] + 2|A^3H_{3sym}(i\omega,i\omega,i\omega) + \cdots| \cos[3\omega t + \phi_3(\omega)] + \cdots$$

It is convenient to let

$$W(s) = \frac{g/L}{s^2 + (a/mL^2)s + g/L}$$

and write the transfer functions computed in Example 3.8 in the form

$$H(s) = \frac{1}{mgL}\ W(s)$$

$$H_{3sym}(s_1,s_2,s_3) = \frac{1}{3!(mgL)^3}\ W(s_1)W(s_2)W(s_3)W(s_1+s_2+s_3)$$

Then

$$y_{ss}(t) = 2|\frac{A}{mgL}\ W(i\omega) + \frac{A^3}{2(mgL)^3}\ W^3(i\omega)W(-i\omega) + \cdots| \cos[\omega t + \phi_1(\omega)] + 2|\frac{A^3}{3(mgL)^3}\ W^3(i\omega)W(i3\omega) + \cdots| \cos[3\omega t + \phi_3(\omega)] + \cdots$$

A simple analysis of this formula can be used to show the possibility of resonance phenomena in the pendulum system at frequencies higher than the input frequency ω. This phenomenon can occur even for very small input amplitudes A, but it is not predicted by the usual linearized model of the pendulum. To be specific, suppose that the damping coefficient a is very small in relation to $(g/L)^{1/2}$. Then the poles of $W(s)$ are very close to the undamped natural frequency $\omega_o = (g/L)^{1/2}$. In this situation, if $\omega = \omega_o/3$, then $|W(i3\omega)|$ can be very large in comparison to $|W(i\omega)|$ so that the dominant term in $y_{ss}(t)$ is the third harmonic. Of course, both the third harmonic term and the fundamental terms in the output depend on higher-degree transfer functions that have been ignored. But it can be shown that these missing terms do not eliminate the possibility of resonance. In fact, the higher-degree terms indicate the possibility of harmonic resonance at many other choices of input frequency.

Although the discussion so far has been in terms of the symmetric transfer function, similar results can be derived for the triangular and regular transfer functions. One way to do this is to use the relationships between the various transfer functions that were derived in Chapter 2. However, it is interesting to take a direct approach in the case of bilinear-realizable regular transfer functions because the required stability property can be explicitly stated.

Suppose

$$H_{reg}(s_1,\ldots,s_n) = \frac{P(s_1,\ldots,s_n)}{Q_1(s_1)\cdots Q_n(s_n)} \tag{33}$$

is a strictly proper, recognizable, regular transfer function. With the input signal $u(t) = 2A\cos(\omega t)$, Theorem 2.10 in Section 2.3 with $\gamma_1 = i\omega$, $\gamma_2 = -i\omega$ gives the response formula

$$Y(s) = A^n \sum_{i_1=1}^{2} \cdots \sum_{i_{n-1}=1}^{2} H_{reg}(s+\gamma_{i_1}+\cdots+\gamma_{i_{n-1}},\ldots,s+\gamma_{i_{n-1}},s)$$
$$\left[\frac{1}{s+\gamma_{i_1}+\cdots+\gamma_{i_{n-1}}+i\omega} + \frac{1}{s+\gamma_{i_1}+\cdots+\gamma_{i_{n-1}}-i\omega}\right] \tag{34}$$

Since each term in (34) is a strictly proper, rational function in s, the steady-state response can be computed via partial fraction expansion. If it is assumed that all poles of $H_{reg}(s_1,\ldots,s_n)$ have negative real parts, that is, all roots of each $Q_j(s_j)$ have negative real parts, then the pole factors contributed by the transfer function can be ignored as far as steady-state response is concerned. Furthermore, since the poles contributed by the input terms in (34) occur at

$$s = \pm in\omega,\ \pm i(n-2)\omega,\ \ldots,\ \begin{cases} \pm i\omega\ , & n\ odd \\ 0\ , & n\ even \end{cases}$$

it is clear that the steady-state response is bounded.

To compute the steady-state response, let $A^n K_k(i\omega)$ be the partial fraction expansion coefficient corresponding to the factor $(s-ik\omega)$ on the right side of (34). Then, discarding all the terms that will yield zero,

$$A^nK_k(i\omega) = (s-ik\omega)Y(s)\big|_{s=ik\omega}$$

$$= A^n \sum_{\substack{i_1=1 \\ \gamma_{i_1}+\cdots+\gamma_{i_{n-1}}=-i(k+1)\omega}}^{2} \cdots \sum_{i_{n-1}=1}^{2} H_{reg}(s+\gamma_{i_1}+\cdots+\gamma_{i_{n-1}},\ldots,s+\gamma_{i_{n-1}},s)\big|_{s=ik\omega}$$

$$+ A^n \sum_{\substack{i_1=1 \\ \gamma_{i_1}+\cdots+\gamma_{i_{n-1}}=-i(k+1)\omega}}^{2} \cdots \sum_{i_{n-1}=1}^{2} H_{reg}(s+\gamma_{i_1}+\cdots+\gamma_{i_{n-1}},\ldots,s+\gamma_{i_{n-1}},s)\big|_{s=ik\omega} \qquad (35)$$

This expression can be simplified by combining the two constrained, multiple summations into one implicit sum, and then replacing s by $ik\omega$. This gives

$$K_k(i\omega) = \sum_{\substack{\gamma_1,\ldots,\gamma_{n-1}=\pm i\omega \\ \gamma_1+\cdots+\gamma_{n-1}=-i(k\pm 1)\omega}} H_{reg}(ik\omega+\gamma_1+\cdots+\gamma_{n-1},\ldots,ik\omega+\gamma_{n-1},ik\omega) \qquad (36)$$

While the general term in (36) is messy, note that for small n it is not hard to write out. And, in general,

$$K_n(i\omega) = H_{reg}(i\omega,i2\omega,\ldots,in\omega)$$

The last step is to take the inverse Laplace transform of each term

$$\frac{K_k(i\omega)}{(s-ik\omega)}$$

in the partial fraction expansion. Using standard trigonometric identities, the steady-state response is given by

$$\begin{aligned} y_{ss}(t) &= 2A^n|K_n(i\omega)|\cos[n\omega t + \angle K_n(i\omega)] \\ &\quad + 2A^n|K_{n-2}(i\omega)|\cos[(n-2)\omega t + \angle K_{n-2}(i\omega)] \\ &\quad + \cdots + \begin{cases} A^nK_0(i\omega), & n \text{ even} \\ 2A^n|K_1(i\omega)|\cos[\omega t + \angle K_1(i\omega)], & n \text{ odd} \end{cases} \end{aligned} \qquad (37)$$

For polynomial systems, the contributions of the various homogeneous subsystems can be added together just as discussed earlier.

5.3 Steady-State Response to Multi-Tone Inputs

When a sum of sinusoids is applied to a homogeneous system of degree greater than 1, the response is complicated by the nonlinear interactions between terms of different frequencies. To introduce this topic, I will begin with the so-called two-tone input:

$$\begin{aligned} u(t) &= 2A_1\cos(\omega_1 t) + 2A_2\cos(\omega_2 t), \quad t \geq 0 \\ &= A_1e^{i\omega_1 t} + A_1e^{-i\omega_1 t} + A_2e^{i\omega_2 t} + A_2e^{-i\omega_2 t} \end{aligned} \qquad (38)$$

Again, the growing exponential development in Chapter 2 can be used, this time for the case of four exponentials:

$$\lambda_1 = i\omega_1,\ \lambda_2 = -i\omega_1,\ \lambda_3 = i\omega_2,\ \lambda_4 = -i\omega_2$$

For a degree-n system with symmetric transfer function $H_{nsym}(s_1,\dots,s_n)$, copying (73) of Section 2.4 with the appropriate changes gives

$$y_{ss}(t) = \sum_m A_1^{m_1+m_2} A_2^{m_3+m_4} G_{m_1m_2m_3m_4}(\lambda_1,\lambda_2,\lambda_3,\lambda_4)e^{(m_1\lambda_1+\cdots+m_4\lambda_4)t} \tag{39}$$

where

$$G_{m_1m_2m_3m_4}(\lambda_1,\lambda_2,\lambda_3,\lambda_4) = \frac{n!}{m_1!m_2!m_3!m_4!} H_{nsym}(\underbrace{\lambda_1,\dots,\lambda_1}_{m_1};\dots;\underbrace{\lambda_4,\dots,\lambda_4}_{m_4}) \tag{40}$$

and $\sum_m$ is a four-fold summation over $m_1,\dots,m_4$ such that $0 \leqslant m_i \leqslant n$ and $m_1+\cdots+m_4 = n$. Substituting for the λ's gives

$$y_{ss}(t) = \sum_m A_1^{m_1+m_2} A_2^{m_3+m_4} G_{m_1m_2m_3m_4}(i\omega_1,-i\omega_1,i\omega_2,-i\omega_2)e^{i[(m_1-m_2)\omega_1+(m_3-m_4)\omega_2]t} \tag{41}$$

Example 5.3 It is perhaps instructive to catalog the terms in (41) for the case $n = 2$. There are ten terms in the summation, and these are shown in Table 5.1. To write the output in terms of real quantities, properties of $G_{m_1m_2m_3m_4}$ with regard to complex conjugation can be used. For example,

$$G_{0110}(i\omega_1,-i\omega_1,i\omega_2,-i\omega_2) = 2H_{2sym}(-i\omega_1,i\omega_2)$$

and

$$G_{1001}(i\omega_1,-i\omega_1,i\omega_2,-i\omega_2) = 2H_{2sym}(i\omega_1,-i\omega_2)$$

so it is clear that (dropping arguments) $G_{0110} = \overline{G}_{1001}$, where the overbar indicates complex conjugate. Similarly,

$$G_{1010} = \overline{G}_{0101},\quad G_{2000} = \overline{G}_{0200},\quad G_{0020} = \overline{G}_{0002}$$

Thus standard trigonometric identities yield the expression

$$\begin{aligned} y_{ss}(t) &= A_1^2G_{1100} + A_2^2G_{0011} + 2A_1A_2|G_{0110}|cos[(\omega_2-\omega_1)t + \angle G_{0110}] \\ &+ 2A_1A_2|G_{1010}|cos[(\omega_1+\omega_2)t + \angle G_{1010}] + 2A_1^2|G_{2000}|cos[2\omega_1t + \angle G_{2000}] \\ &+ 2A_2^2|G_{0020}|cos[2\omega_2t + \angle G_{0020}] \end{aligned}$$

Note that all these frequency components need not occur at distinct frequencies. For example, consider the case $\omega_2 = 3\omega_1$.

Table 5.1
Frequency-Response Terms for Example 5.3

Summation Indices				Summand
1	1	0	0	$A_1^2 G_{1100}$
0	1	1	0	$A_1 A_2 G_{0110} e^{i(\omega_2-\omega_1)t}$
0	0	1	1	$A_2^2 G_{0011}$
1	0	0	1	$A_1 A_2 G_{1001} e^{i(\omega_1-\omega_2)t}$
1	0	1	0	$A_1 A_2 G_{1010} e^{i(\omega_1+\omega_2)t}$
0	1	0	1	$A_1 A_2 G_{0101} e^{-i(\omega_1+\omega_2)t}$
2	0	0	0	$A_1^2 G_{2000} e^{i2\omega_1 t}$
0	2	0	0	$A_1^2 G_{0200} e^{-i2\omega_1 t}$
0	0	2	0	$A_2^2 G_{0020} e^{i2\omega_2 t}$
0	0	0	2	$A_2^2 G_{0002} e^{-i2\omega_2 t}$

When higher-degree homogeneous systems are considered, the number of terms in the steady-state response increases dramatically. Therefore, it seems more useful to derive an expression that gives the coefficient of a particular complex exponential term in the output. As many or as few terms as desired then can be considered, and conjugate exponential terms can be combined easily if the real form is wanted.

The terms in (41) corresponding to the exponential $e^{i[M\omega_1+N\omega_2]t}$, $M \geqslant 0,\ N \geqslant 0$, can be written as follows:

$$\sum_{\substack{m_1=0 \\ m_1+m_2+m_3+m_4=n \\ m_1=m_2+M,\ m_3=m_4+N}}^{n} \sum_{m_2=0}^{n} \sum_{m_3=0}^{n} \sum_{m_4=0}^{n} A_1^{2m_2+M} A_2^{2m_4+N} G_{m_2+M,m_2,m_4+N,m_4}(i\omega_1,-i\omega_1,i\omega_2,-i\omega_2)$$

But now the four-fold summation can be simplified by replacing m_1 and m_3 using the indicated constraints to obtain

$$\sum_{\substack{m_2=0 \\ m_2+m_4=\frac{n-M-N}{2}}}^{n} \sum_{m_4=0}^{n} A_1^{2m_2+M} A_2^{2m_4+N} G_{m_2+M,m_2,m_4+N,m_4}(i\omega_1,-i\omega_1,i\omega_2,-i\omega_2) \qquad (42)$$

With this notation there are several relationships in the subscripts of G that are convenient for checking. The sum of the subscripts is the degree of the system and the difference of the first two (last two) is the associated harmonic of ω_1 (ω_2). Although I have assumed $M,\ N \geqslant 0$, to obtain the coefficient of the term $e^{-i[M\omega_1+N\omega_2]t}$ simply change the sign of every frequency argument in every G. To obtain the coefficient of, say, $e^{i[M\omega_1-N\omega_2]t}$, change the sign of every argument ω_2 in every G. Note that changing the sign of the frequency does not change the input signal, so that (42) remains valid.

Of course, the coefficient of $e^{i[M\omega_1+N\omega_2]t}$ can be expressed directly in terms of the symmetric transfer function $H_{nsym}(s_1,\ldots,s_n)$ using (40). This gives, using a collapsed notation for the arguments of the transfer function,

$$\sum_{\substack{m_2=0 \\ m_2+m_4=\frac{n-M-N}{2}}}^{n} \sum_{m_4=0}^{n} \frac{n!A_1^{2m_2+M}A_2^{2m_4+N}}{(m_2+M)!m_2!(m_4+N)!m_4!} H_{nsym}(\underbrace{i\omega_1}_{m_2+M};\underbrace{-i\omega_1}_{m_2};\underbrace{i\omega_2}_{m_4+N};\underbrace{-i\omega_2}_{m_4}) \quad (43)$$

The same rule is used in (43) as in (42) to obtain the coefficient when M and/or N is negative. I should emphasize that the exponential frequency terms $e^{i[M\omega_1+N\omega_2]t}$ may not be distinct. For example, if $\omega_1 = 2\omega_2$, then $[\omega_1+2\omega_2] = 2\omega_1$ so that the coefficients of these two terms can be combined.

Example 5.4 The contribution of a degree-5 homogeneous system to the (assumed distinct) frequency component $e^{i[\omega_1+2\omega_2]t}$ will be computed. In this case (43) specializes to

$$\sum_{\substack{m_2=0 \\ m_2+m_4=1}}^{5} \sum_{m_4=0}^{5} \frac{5!A_1^{2m_2+1}A_2^{2m_4+2}}{(m_2+1)!m_2!(m_4+2)!m_4!} H_{5sym}(\underbrace{i\omega_1}_{m_2+1};\underbrace{-i\omega_1}_{m_2};\underbrace{i\omega_2}_{m_4+2};\underbrace{-i\omega_2}_{m_4})$$

There are two terms in the summation, corresponding to the index pairs 0, 1 and 1, 0. Thus the summation gives

$$\frac{5!A_1A_2^4}{3!} H_{5sym}(i\omega_1,i\omega_2,i\omega_2,i\omega_2,-i\omega_2)+ \frac{5!A_1^3A_2^2}{2!2!} H_{5sym}(i\omega_1,i\omega_1,-i\omega_1,i\omega_2,i\omega_2)$$

It is instructive also to compute the coefficient of the frequency component $e^{i[\omega_1-2\omega_2]t}$, for there are two ways to proceed. The easiest is that mentioned above: take the coefficient just derived and replace every ω_2 by $-\omega_2$ to obtain

$$\frac{5!A_1A_2^4}{3!} H_{5sym}(i\omega_1,-i\omega_2,-i\omega_2,-i\omega_2,i\omega_2)$$
$$+ \frac{5!A_1^3A_2^2}{2!2!} H_{5sym}(i\omega_1,i\omega_1,-i\omega_1,-i\omega_2,-i\omega_2)$$

A straightforward application of (43) also works, although terms with negative factorials, negative powers, and negative subscripts, which arise because of the implicit nature of the formula, must be deleted. Specifically, (43) becomes, with $M = 1,\ N = -2$,

$$\sum_{\substack{m_2=0 \\ m_2+m_4=3}}^{5} \sum_{m_4=0}^{5} \frac{5!A_1^{2m_2+1}A_2^{2m_4-2}}{(m_2+1)!m_2!(m_4-2)!m_4!} H_{5sym}(\underbrace{i\omega_1}_{m_2+1};\underbrace{-i\omega_1}_{m_2};\underbrace{i\omega_2}_{m_4-2};\underbrace{-i\omega_2}_{m_4})$$

The index pairs contributing to the summation are: 0, 3; 1, 2; 2, 1; and 3, 0. But the last two pairs can be dropped as extraneous so that the coefficient of $e^{i[\omega_1-2\omega_2]t}$ is

$$\frac{5!A_1A_2^4}{3!}H_{5sym}(i\omega_1,i\omega_2,-i\omega_2,-i\omega_2,-i\omega_2)$$
$$+\frac{5!A_1^3A_2^2}{2!2!}H_{5sym}(i\omega_1,i\omega_1,-i\omega_1,-i\omega_2,-i\omega_2)$$

which agrees with the earlier result.

For polynomial or Volterra systems, it should be clear that the analysis just completed can be applied readily. To obtain the coefficient of $e^{i[M\omega_1+N\omega_2]t}$ in the steady-state response, the coefficients in (43) must be added together for $n = 1,2,\ldots$. Thus, the coefficient can be written for a Volterra system in terms of the symmetric transfer functions as

$$\sum_{m_2=0}^{\infty}\sum_{m_4=0}^{\infty}\frac{(2m_2+2m_4+M+N)!A_1^{2m_2+M}A_2^{2m_4+N}}{(m_2+M)!m_2!(m_4+N)!m_4!}$$

$$H_{(2m_2+2m_4+M+N)sym}(\underbrace{i\omega_1}_{m_2+M};\underbrace{-i\omega_1}_{m_2};\underbrace{i\omega_2}_{m_4+N};\underbrace{-i\omega_2}_{m_4}) \tag{44}$$

where n has been replaced by the appropriate sum of subscripts, and the constraints on the summations have been removed.

Example 5.5 As an illustration of the use of (44), I will list the terms in the response of a degree-3 polynomial system to the input (38). The complex conjugate terms will be omitted since they add no information. The contribution of the degree-1 subsystem is found by imposing the restriction $2m_2+2m_4+M+N = 1$ in (44). In this case, there are no negative-frequency terms other than complex-conjugate terms, so the list with nonnegative M and N is complete as shown in Table 5.2.

Table 5.2
*Frequency-Response Terms: Degree-1 Subsystem**

Summation Indices				Frequency
m_2	m_4	M	N	Term
0	0	1	0	$A_1H_1(i\omega_1)e^{i\omega_1 t}$
0	0	0	1	$A_2H_1(i\omega_2)e^{i\omega_2 t}$

* *(plus complex—conjugate frequency terms)*

The contribution of the degree-2 subsystem involves essentially repeating Table 5.1. But the notation is different in the present context, so I will go ahead. Imposing the constraint $2m_2+2m_4+M+N = 2$ in (44) gives the list in Table 5.3. Notice in this case there is only one distinct frequency component generated by allowing M and/or N to become negative (ignoring complex conjugates). Such a term will be called a *sign switch* to indicate how it is obtained from previously computed terms.

Table 5.3
*Frequency-Response Terms: Degree-2 Subsystem**

Summation Indices				Frequency
m_2	m_4	M	N	Term
1	0	0	0	$2!A_1^2H_{2sym}(i\omega_1,-i\omega_1)$
0	1	0	0	$2!A_2^2H_{2sym}(i\omega_2,-i\omega_2)$
0	0	2	0	$A_1^2H_{2sym}(i\omega_1,i\omega_1)e^{i2\omega_1 t}$
0	0	0	2	$A_2^2H_{2sym}(i\omega_2,i\omega_2)e^{i2\omega_2 t}$
0	0	1	1	$2!A_1A_2H_{2sym}(i\omega_1,i\omega_2)e^{i(\omega_1+\omega_2)t}$
sign switch				$2!A_1A_2H_{2sym}(i\omega_1,-i\omega_2)e^{i(\omega_1-\omega_2)t}$

* *(plus complex—conjugate frequency terms)*

In a similar manner, setting $2m_2+2m_4+M+N = 3$ gives the contribution of the degree-3 subsystem as shown in Table 5.4.

Table 5.4
*Frequency-Response Terms: Degree-3 Subsystem**

Summation Indices				Frequency
m_2	m_4	M	N	Term
1	0	1	0	$\frac{3!}{2!}A_1^3H_{3sym}(i\omega_1,i\omega_1,-i\omega_1)e^{i\omega_1 t}$
1	0	0	1	$3!A_1^2A_2H_{3sym}(i\omega_1,-i\omega_1,i\omega_2)e^{i\omega_1 t}$
0	1	1	0	$3!A_1A_2^2H_{3sym}(i\omega_1,i\omega_2,-i\omega_2)e^{i\omega_1 t}$
0	1	0	1	$\frac{3!}{2!}A_2^3H_{3sym}(i\omega_2,i\omega_2,-i\omega_2)e^{i\omega_2 t}$
0	0	2	1	$\frac{3!}{2!}A_1^2A_2H_{3sym}(i\omega_1,i\omega_1,i\omega_2)e^{i(2\omega_1+\omega_2)t}$
0	0	1	2	$\frac{3!}{2!}A_1A_2^2H_{3sym}(i\omega_1,i\omega_2,i\omega_2)e^{i(\omega_1+2\omega_2)t}$
0	0	3	0	$A_1^3H_{3sym}(i\omega_1,i\omega_1,i\omega_1)e^{i3\omega_1 t}$
0	0	0	3	$A_2^3H_{3sym}(i\omega_2,i\omega_2,i\omega_2)e^{i3\omega_2 t}$
sign switch				$\frac{3!}{2!}A_1^2A_2H_{3sym}(i\omega_1,i\omega_1,-i\omega_2)e^{i(2\omega_1-\omega_2)t}$
sign switch				$\frac{3!}{2!}A_1A_2^2H_{3sym}(i\omega_1,-i\omega_2,-i\omega_2)e^{i(\omega_1-2\omega_2)t}$

* (plus complex—conjugate frequency terms)

Of course, to complete this example, all these terms should be combined—a task I leave to the reader.

To consider inputs that are sums of more than two sinusoidal terms, the same approach is followed. For example, it is straightforward although tedious to verify the following fact. For the input

$$u(t) = 2A_1\cos(\omega_1 t) + 2A_2 cos(\omega_2 t) + 2A_3 cos(\omega_3 t) \tag{45}$$

to a Volterra system, the coefficient of the exponential $e^{i[L\omega_1+M\omega_2+N\omega_3]t}$, $L,M,N \geq 0$, in the steady-state response is

$$\sum_{m_2=0}^{\infty}\sum_{m_4=0}^{\infty}\sum_{m_6=0}^{\infty} \frac{(2m_2+2m_4+2m_6+L+M+N)!A_1^{2m_2+L}A_2^{2m_4+M}A_3^{2m_6+N}}{(m_2+L)!m_2!(m_4+M)!m_4!(m_6+N)!m_6!}$$

$$H_{(2m_2+2m_4+2m_6+L+M+N)sym}(i\omega_1,-i\omega_1,i\omega_2,-i\omega_2,i\omega_3,-i\omega_3) \tag{46}$$

where the various numbers of arguments are entered into the transfer function the obvious number of times—to be pedantic, m_2+L, m_2, m_4+M, m_4, m_6+N, m_6, respectively. When L, M, or N are negative, the coefficient is found by changing the sign of the corresponding frequency arguments, just as before. Also just as before, the frequency components may not be distinct, depending on the relative values of ω_1, ω_2, and ω_3. I should emphasize that (46) gives the coefficient of just one complex exponential. So, what can be said about the total steady-state response? Not much more than that it is a jungle into which the prudent venture only with inkwell full.

5.4 Response to Random Inputs

Just as in the preceding sections, the linear theory to be generalized will be reviewed first. Suppose the input to the system

$$y(t) = \int_{-\infty}^{\infty} h(\sigma)u(t-\sigma)\, d\sigma \tag{47}$$

is a sample function from a real stochastic process with *expected value* $E[u(t)]$ and *autocorrelation*

$$R_{uu}(t_1,t_2) = E[u(t_1)u(t_2)] \tag{48}$$

Then the output is a sample function from a real stochastic process, and it is of interest to find the expected value of the output, $E[y(t)]$, the input/output *cross-correlation*

$$R_{yu}(t_1,t_2) = E[y(t_1)u(t_2)] \tag{49}$$

and the output autocorrelation

$$R_{yy}(t_1,t_2) = E[y(t_1)y(t_2)] \tag{50}$$

Proceeding by direct calculation, it is clear that since expectation can be interchanged with integration,

$$E[y(t)] = \int_{-\infty}^{\infty} h(\sigma)E[u(t-\sigma)]\, d\sigma \tag{51}$$

Furthermore,

$$y(t_1)u(t_2) = \int_{-\infty}^{\infty} h(\sigma)u(t_1-\sigma)u(t_2)\, d\sigma$$

so that taking expected values on both sides gives

$$R_{yu}(t_1,t_2) = \int_{-\infty}^{\infty} h(\sigma) R_{uu}(t_1-\sigma,t_2)\, d\sigma \tag{52}$$

Similarly,

$$y(t_1)y(t_2) = \int_{-\infty}^{\infty}\int_{-\infty}^{\infty} h(\sigma_1)h(\sigma_2)u(t_1-\sigma_1)u(t_2-\sigma_2)\, d\sigma_1 d\sigma_2$$

and thus

$$R_{yy}(t_1,t_2) = \int_{-\infty}^{\infty}\int_{-\infty}^{\infty} h(\sigma_1)h(\sigma_2)R_{uu}(t_1-\sigma_1,t_2-\sigma_2)\, d\sigma_1 d\sigma_2 \tag{53}$$

Notice that a number of technical matters again are being ignored. For example, it is assumed implicitly that $E[u(t)]$ and $R_{uu}(t_1,t_2)$ are sufficiently well behaved to permit the integrations indicated above. Such considerations are not too difficult to fill in, and that task is left to the reader, as usual.

The correlation relationships often are expressed in terms of a *multivariable Fourier transform.* In strict analogy to the usual single-variable Fourier transform

$$F(\omega) = F[f(t)] = \int_{-\infty}^{\infty} f(t)e^{-i\omega t}\, dt \tag{54}$$

the multivariable Fourier transform of a function $f(t_1,\ldots,t_n)$ is defined by

$$F(\omega_1,\ldots,\omega_n) = \int_{-\infty}^{\infty} f(t_1,\ldots,t_n)e^{-i\omega_1 t_1}\cdots e^{-i\omega_n t_n}\, dt_1 \cdots dt_n \tag{55}$$

Of course, this is no surprise, given the discussion of the multivariable Laplace transform in Chapter 2. Furthermore, the multivariable Fourier transform exhibits all the properties that might reasonably be expected after a review of the properties of the Laplace transform in Chapter 2. The inverse Fourier transform is given by

$$f(t_1,\ldots,t_n) = \frac{1}{(2\pi)^n}\int_{-\infty}^{\infty} F(\omega_1,\ldots,\omega_n)e^{i\omega_1 t_1}\cdots e^{i\omega_n t_n}\, d\omega_1 \cdots d\omega_n \tag{56}$$

For the purposes of this chapter, the Fourier transform of $h(t)$ is called the *system function,* and it is written as $H(\omega)$. In this context, perhaps I should remind the reader of the common notational collision between Laplace and Fourier transforms. If a (Laplace) transfer function $H(s)$ exists for $Re[s] = 0$, then the system function is given by $H(s)\,|_{s=i\omega} = H(i\omega)$. However, Laplace aside, it is more convenient to use the notation $H(\omega)$ for the system function. Since the Laplace transform will be set aside for the material dealing with random input signals, I will use the $H(\omega)$ notation for the system function, and for all single- or multi-variable Fourier transforms. Incidentally, the hypotheses needed to insure the existence of Fourier transforms will be assumed. For example, the system stability property corresponding to

$$\int_{-\infty}^{\infty} |h(t)|\, dt < \infty$$

can be assumed to guarantee the existence of the system function $H(\omega)$.

Letting

$$S_{uu}(\omega_1, \omega_2) = F[R_{uu}(t_1,t_2)] \tag{57}$$

with similar definitions for the transforms of the other correlation functions, a straightforward calculation shows that (52) and (53) can be represented by

$$\begin{aligned} S_{yu}(\omega_1, \omega_2) &= H(\omega_1) S_{uu}(\omega_1, \omega_2) \\ S_{yy}(\omega_1, \omega_2) &= H(\omega_1) H(\omega_2) S_{uu}(\omega_1, \omega_2) \end{aligned} \tag{58}$$

These concepts are of most interest in the case where the real random process $u(t)$ is (strict-sense) stationary. For then, assuming the input signal was applied at $t = -\infty$, the output also is a real, stationary random process. In other words, the steady-state output is a real, stationary random process. Of course, there is an implicit stability assumption here. (The astute reader will notice that the stationarity condition is stronger than necessary for the linear case, and that only wide-sense stationarity is needed. However, in the nonlinear case strict-sense stationarity is required.)

In the case of stationary input, $E[u(t)]$ is a constant, so that

$$E[y(t)] - \int_{-\infty}^{\infty} h(\sigma)\, d\sigma\; E[u(t)] \tag{59}$$

Also, the autocorrelation function $R_{uu}(t_1,t_2)$ depends only on the difference $t_1 - t_2$. Following the usual notation by changing to the variables $t_2 = t$, $t_1 = t + \tau$, the autocorrelation $R_{uu}(t+\tau,t)$ is a function of τ only, and thus is written as $R_{uu}(\tau)$. To determine the input/output cross-correlation in terms of the new variables, (52) can be written as

$$R_{yu}(t+\tau,t) = \int_{-\infty}^{\infty} h(\sigma) R_{uu}(t+\tau-\sigma,t)\, d\sigma$$

and since the right side is independent of t, this is written in the form

$$R_{yu}(\tau) = \int_{-\infty}^{\infty} h(\sigma) R_{uu}(\tau-\sigma)\, d\sigma \tag{60}$$

Similarly, the output autocorrelation can be written as

$$R_{yy}(\tau) = \int_{-\infty}^{\infty}\int_{-\infty}^{\infty} h(\sigma_1) h(\sigma_2) R_{uu}(\tau-\sigma_1+\sigma_2)\, d\sigma_1 d\sigma_2 \tag{61}$$

These relationships can be expressed in the frequency domain using the single-variable Fourier transform. This can be accomplished directly in an easy fashion. However, to warm up for later developments, I will derive the expressions from the 2-variable Fourier transform formulas in (58). Using the new variables introduced above,

$$S_{uu}(\omega_1,\omega_2) = \int_{-\infty}^{\infty}\int_{-\infty}^{\infty} R_{uu}(t_1,t_2)e^{-i\omega_1 t_1}e^{-i\omega_2 t_2}\,dt_1 dt_2$$

$$= \int_{-\infty}^{\infty}\int_{-\infty}^{\infty} R_{uu}(t+\tau,t)e^{-i\omega_1(t+\tau)}e^{-i\omega_2 t}\,d\tau\,dt$$

$$= \int_{-\infty}^{\infty}\int_{-\infty}^{\infty} R_{uu}(\tau)e^{-i\omega_1\tau}e^{-i(\omega_1+\omega_2)t}\,d\tau\,dt$$

Integrating with respect to τ gives the Fourier transform $S_{uu}(\omega_1) = F[R_{uu}(\tau)]$, which is the *power spectral density* of the stationary random process. Then using the well known transform

$$\int_{-\infty}^{\infty} e^{-i\omega t}\,dt = 2\pi\delta_0(\omega)$$

leads to

$$S_{uu}(\omega_1,\omega_2) = 2\pi S_{uu}(\omega_1)\delta_0(\omega_1+\omega_2)$$

Integrating both sides with respect to ω_2 gives

$$S_{uu}(\omega_1) = \frac{1}{2\pi}\int_{-\infty}^{\infty} S_{uu}(\omega_1,\omega_2)\,d\omega_2 \tag{62}$$

This formula expresses the power spectral density of a stationary random process in terms of the 2-variable Fourier transform of the general autocorrelation function of that process. Of course a similar relationship is obtained for the *cross-spectral density* $S_{yu}(\omega_1)$ in terms of $S_{yu}(\omega_1,\omega_2)$ given in (58). Thus the first equation in (58) becomes

$$S_{yu}(\omega_1,\omega_2) = 2\pi H(\omega_1)S_{uu}(\omega_1)\delta_0(\omega_1+\omega_2)$$

so that the input/output cross-spectral density is given in terms of the input power spectral density by

$$S_{yu}(\omega_1) = \frac{1}{2\pi}\int_{-\infty}^{\infty} 2\pi H(\omega_1)S_{uu}(\omega_1)\delta_0(\omega_1+\omega_2)\,d\omega_2$$

$$= H(\omega_1)S_{uu}(\omega_1) \tag{63}$$

Proceeding in a similar fashion for the second relation in (58) gives the output power spectral density in terms of the input power spectral density as

$$S_{yy}(\omega_1) = H(\omega_1)H(-\omega_1)S_{uu}(\omega_1)$$
$$= |H(\omega_1)|^2 S_{uu}(\omega_1) \tag{64}$$

I should note at this point that under appropriate ergodicity assumptions, the various correlations and spectral densities in the stationary case can be expressed as time averages. This fact will be crucial in Chapter 7, when identification techniques are discussed. Also note that, in terms of the system function, the expected value of the output given in (59) can be written in the form

$$E[y(t)] = H(0)E[u(t)] \tag{65}$$

Now consider the generalization of the ideas just reviewed to nonlinear systems described by

$$y(t) = \int_{-\infty}^{\infty} h(\sigma_1,\ldots,\sigma_n)u(t-\sigma_1)\cdots u(t-\sigma_n)\,d\sigma_1\cdots d\sigma_n \tag{66}$$

The discussion of polynomial or Volterra systems will be postponed until this homogeneous case is treated, as usual.

When $u(t)$ is a real random process, direct calculation gives

$$E[y(t)] = \int_{-\infty}^{\infty} h(\sigma_1,\ldots,\sigma_n)E[u(t-\sigma_1)\cdots u(t-\sigma_n)]\,d\sigma_1\cdots d\sigma_n$$

$$= \int_{-\infty}^{\infty} h(\sigma_1,\ldots,\sigma_n)R_{uu}^{(n)}(t-\sigma_1,\ldots,t-\sigma_n)\,d\sigma_1\cdots d\sigma_n \tag{67}$$

where the nth*-order autocorrelation function* of the input is defined by

$$R_{uu}^{(n)}(t_1,\ldots,t_n) = E[u(t_1)\cdots u(t_n)]$$

In a similar fashion the input/output cross-correlation, and the output autocorrelation can be written in the forms

$$R_{yu}(t_1,t_2) = \int_{-\infty}^{\infty} h(\sigma_1,\ldots,\sigma_n)R_{uu}^{(n+1)}(t_1-\sigma_1,\ldots,t_1-\sigma_n,t_2)\,d\sigma_1\cdots d\sigma_n \tag{68}$$

$$R_{yy}(t_1,t_2) = \int_{-\infty}^{\infty} h(\sigma_1,\ldots,\sigma_n)h(\sigma_{n+1},\ldots,\sigma_{2n})$$

$$R_{uu}^{(2n)}(t_1-\sigma_1,\ldots,t_1-\sigma_n,t_2-\sigma_{n+1},\ldots,t_2-\sigma_{2n})\,d\sigma_1\cdots\sigma_{2n} \tag{69}$$

For $n = 1$ these expressions are just those discussed previously. But for $n > 1$ the expected value of the output and the (order-2) output correlations depend on higher-order input autocorrelations. In other words, as n increases more statistical information about the input is needed to characterize, for example, the output autocorrelation.

The expressions (67), (68), and (69) can be written in the form of convolutions followed by variable associations, a form that is reminiscent of the convolutions and variable associations that arise in considering the input/output representation of a homogeneous system using the multivariable Laplace transform. Since it is of interest to express (68) and (69) in terms of Fourier transforms, it is convenient to separate the convolution aspect from the association aspect, as was done in Chapter 2. To do this, define the *multivariable input/output cross-correlation* by

$$\hat{R}_{yu}(t_1,\ldots,t_{n+1}) =$$
$$\int_{-\infty}^{\infty} h(\sigma_1,\ldots,\sigma_n) R_{uu}^{(n+1)}(t_1-\sigma_1,\ldots,t_n-\sigma_n,t_{n+1})\, d\sigma_1 \cdots d\sigma_n \tag{70}$$

so that

$$R_{yu}(t_1,t_2) = \hat{R}_{yu}(t_1,\ldots,t_{n+1}) \Big|_{\substack{t_1=\cdots=t_n=t_1 \\ t_{n+1}=t_2}} \tag{71}$$

In a similar manner, the *multivariable output autocorrelation* is defined by

$$\hat{R}_{yy}(t_1,\ldots,t_{2n}) =$$
$$\int_{-\infty}^{\infty} h(\sigma_1,\ldots,\sigma_n) h(\sigma_{n+1},\ldots,\sigma_{2n}) R_{uu}^{(2n)}(t_1-\sigma_1,\ldots,t_{2n}-\sigma_{2n})\, d\sigma_1 \cdots d\sigma_{2n} \tag{72}$$

so that

$$R_{yy}(t_1,t_2) = \hat{R}_{yy}(t_1,\ldots,t_{2n}) \Big|_{\substack{t_1=\cdots=t_n=t_1 \\ t_{n+1}=\cdots=t_{2n}=t_2}} \tag{73}$$

These intermediate multivariable quantities have no significance other than to facilitate the representation via Fourier transforms. Let the Fourier transform of the order-n autocorrelation function of the input be

$$S_{uu}^{(n)}(\omega_1,\ldots,\omega_n) = F[R_{uu}^{(n)}(t_1,\ldots,t_n)] \tag{74}$$

and the Fourier transforms of the multivariable cross- and autocorrelations of the output be

$$\hat{S}_{yu}(\omega_1,\ldots,\omega_{n+1}) = F[\hat{R}_{yu}(t_1,\ldots,t_{n+1})]$$
$$\hat{S}_{yy}(\omega_1,\ldots,\omega_{2n}) = F[\hat{R}_{yy}(t_1,\ldots,t_{2n})] \tag{75}$$

These will be called multivariable spectral densities, though they have little or nothing to do with spectral density. It follows from the readily established convolution property of Fourier transforms that in terms of the system function,

$$\hat{S}_{yu}(\omega_1,\ldots,\omega_{n+1}) = H(\omega_1,\ldots,\omega_n) S_{uu}^{(n+1)}(\omega_1,\ldots,\omega_{n+1}) \tag{76}$$

$$\hat{S}_{yy}(\omega_1,\ldots,\omega_{2n}) = H(\omega_1,\ldots,\omega_n) H(\omega_{n+1},\ldots,\omega_{2n}) S_{uu}^{(2n)}(\omega_1,\ldots,\omega_{2n}) \tag{77}$$

For $n = 1$ the circumflexes can be removed from the left side, and then these expressions agree with those in (58). The problem of interest now is to express $S_{yy}(\omega_1,\omega_2)$ and $S_{yu}(\omega_1,\omega_2)$ in terms of the multivariable spectral densities for $n > 1$. That is, to express the variable associations in (71) and (73) in terms of Fourier transforms. It takes a little bit of maneuvering to accomplish this, but the maneuvers should be familiar from the proof of the association-of-variables formula in Chapter 2.

The inverse Fourier transform relationship can be written for the multivariable cross-correlation as

$$\hat{R}_{yu}(t_1,\ldots,t_{n+1}) = \frac{1}{(2\pi)^{n+1}} \int_{-\infty}^{\infty} \hat{S}_{yu}(\gamma_1,\ldots,\gamma_{n+1}) e^{i\gamma_1 t_1} \cdots e^{i\gamma_{n+1}t_{n+1}}\, d\gamma_1 \cdots d\gamma_{n+1}$$

from which

$$R_{yu}(t_1,t_2) = \frac{1}{(2\pi)^{n+1}} \int_{-\infty}^{\infty} \hat{S}_{yu}(\gamma_1,\ldots,\gamma_{n+1}) e^{i(\gamma_1+\cdots+\gamma_n)t_1} e^{i\gamma_{n+1}t_2}\, d\gamma_1 \cdots d\gamma_{n+1}$$

Taking the Fourier transform of both sides gives

$$S_{yu}(\omega_1,\omega_2) =$$

$$\frac{1}{(2\pi)^{n+1}} \int_{-\infty}^{\infty} \hat{S}_{yu}(\gamma_1,\ldots,\gamma_{n+1}) e^{-i(\omega_1-\gamma_1-\cdots-\gamma_n)t_1} e^{-i(\omega_2-\gamma_{n+1})t_2}\, d\gamma_1 \cdots d\gamma_{n+1} dt_1 dt_2$$

and integrating with respect to t_1 and t_2,

$$S_{yu}(\omega_1,\omega_2) = \frac{1}{(2\pi)^{n-1}} \int_{-\infty}^{\infty} \hat{S}_{yu}(\gamma_1,\ldots,\gamma_{n+1}) \delta_0(\omega_1-\gamma_1-\cdots-\gamma_n)$$
$$\delta_0(\omega_2-\gamma_{n+1})\, d\gamma_1 \cdots d\gamma_{n+1}$$
$$= \frac{1}{(2\pi)^{n-1}} \int_{-\infty}^{\infty} H(\gamma_1,\ldots,\gamma_n) S_{uu}^{(n+1)}(\gamma_1,\ldots,\gamma_{n+1}) \delta_0(\omega_1-\gamma_1-\cdots-\gamma_n)$$
$$\delta_0(\omega_2-\gamma_{n+1}) d\gamma_1 \cdots d\gamma_{n+1} \tag{78}$$

Repeating this procedure for the output autocorrelation gives

$$S_{yy}(\omega_1,\omega_2) = \frac{1}{(2\pi)^{2n-2}} \int_{-\infty}^{\infty} \hat{S}_{yy}(\gamma_1,\ldots,\gamma_{2n}) \delta_0(\omega_1-\gamma_1-\cdots-\gamma_n)$$
$$\delta_0(\omega_2-\gamma_{n+1}-\cdots-\gamma_{2n}) d\gamma_1 \cdots d\gamma_{2n}$$
$$= \frac{1}{(2\pi)^{2n-2}} \int_{-\infty}^{\infty} H(\gamma_1,\ldots,\gamma_n) H(\gamma_{n+1},\ldots,\gamma_{2n}) S_{uu}^{(2n)}(\gamma_1,\ldots,\gamma_{2n})$$
$$\delta_0(\omega_1-\gamma_1-\cdots-\gamma_n) \delta_0(\omega_2-\gamma_{n+1}-\cdots-\gamma_{2n}) d\gamma_1 \cdots d\gamma_{2n} \tag{79}$$

The similarities here with the association-of-variables formulas in Chapter 2 may not be apparent yet, but I will discuss that shortly.

There is no question that these expressions for the output spectral density and cross-spectral density are formidable when actual computations or applications are contemplated. But they can be simplified somewhat by the process of imposing further assumptions on the input random process. Just as in the linear case, the first of these is (strict-sense) stationarity. When a stationary input is applied at $t = -\infty$ to a stationary homogeneous system, the usual and rather simple time-shift argument shows that the output random process is stationary. Thus, the output autocorrelation (and power spectral density) and the input/output cross-correlation (and cross-spectral density) can be expressed as functions of a single variable using techniques reviewed earlier. I will do the calculations for the spectral densities and leave the correlations to the Problems.

For the cross-spectral density, a relationship of the form (62) can be written, giving

$$S_{yu}(\omega_1) = \frac{1}{2\pi} \int_{-\infty}^{\infty} S_{yu}(\omega_1, \omega_2) \, d\omega_2$$

$$= \frac{1}{(2\pi)^n} \int_{-\infty}^{\infty} \hat{S}_{yu}(\gamma_1, \ldots, \gamma_{n+1}) \delta_0(\omega_1 - \gamma_1 - \cdots - \gamma_n)$$

$$\delta_0(\omega_2 - \gamma_{n+1}) \, d\gamma_1 \cdots d\gamma_{n+1} d\omega_2 \tag{80}$$

Integrating first with respect to ω_2 yields

$$S_{yu}(\omega_1) = \frac{1}{(2\pi)^n} \int_{-\infty}^{\infty} \hat{S}_{yu}(\gamma_1, \ldots, \gamma_{n+1}) \delta_0(\omega_1 - \gamma_1 - \cdots - \gamma_n) \, d\gamma_1 \cdots d\gamma_{n+1} \tag{81}$$

or, in terms of the system function and input spectral density of order $n+1$,

$$S_{yu}(\omega_1) = \frac{1}{(2\pi)^n} \int_{-\infty}^{\infty} H(\gamma_1, \ldots, \gamma_n) S_{uu}^{(n+1)}(\gamma_1, \ldots, \gamma_{n+1})$$

$$\delta_0(\omega_1 - \gamma_1 - \cdots - \gamma_n) \, d\gamma_1 \cdots d\gamma_{n+1} \tag{82}$$

Notice that integrating with respect to γ_1 in (82) gives

$$S_{yu}(\omega_1) = \frac{1}{(2\pi)^n} \int_{-\infty}^{\infty} \hat{S}_{yu}(\omega_1 - \gamma_2 - \cdots - \gamma_n, \gamma_2, \ldots, \gamma_{n+1}) \, d\gamma_2 \cdots d\gamma_{n+1} \tag{83}$$

an expression that is very much like an association-of-variables formula in Section 2.3. However, the unintegrated form in (82) will be more efficient for further developments.

A similar calculation for the output power spectral density gives

$$S_{yy}(\omega_1) = \frac{1}{(2\pi)^{2n-1}} \int_{-\infty}^{\infty} \hat{S}_{yy}(\gamma_1, \ldots, \gamma_{2n}) \delta_0(\omega_1 - \gamma_1 - \cdots - \gamma_n) d\gamma_1 \cdots d\gamma_{2n}$$

$$= \frac{1}{(2\pi)^{2n-1}} \int_{-\infty}^{\infty} H(\gamma_1, \ldots, \gamma_n) H(\gamma_{n+1}, \ldots, \gamma_{2n}) S_{uu}^{(2n)}(\gamma_1, \ldots, \gamma_{2n})$$

$$\delta_0(\omega_1 - \gamma_1 - \cdots - \gamma_n) \, d\gamma_1 \cdots d\gamma_{2n} \tag{84}$$

Again, this can be interpreted as an association-of-variables formula.

To achieve further simplification, it is assumed that the real, stationary, random-process input is zero-mean and Gaussian. For in this case the higher-order autocorrelations (spectral densities) of the input process can be expressed in terms of the order-2 autocorrelation (power spectral density). The derivation of this fact will not be given, rather, I simply will present the formulas.

The order-n autocorrelation function of a stationary, zero-mean, Gaussian random process $u(t)$ can be written as

$$R_{uu}^{(n)}(t_1,\ldots,t_n) = \begin{cases} \sum\limits_p \prod\limits_{j,k}^{n} R_{uu}(t_j-t_k), & n \text{ even} \\ 0, & n \text{ odd} \end{cases} \tag{85}$$

where $\prod\limits_{j,k}^{n}$ is a product over a set of $n/2$ (unordered) pairs of integers from $1, 2,\ldots,n$, and $\sum\limits_p$ is a sum over all

$$(n-1)(n-3)(n-5)\cdots(1) = \frac{n!}{(n/2)!2^{n/2}}$$

such products. While a more explicit notation can be adopted, it is so complicated that I will use (85) and further explain with examples.

Example 5.6 For $n = 2$ there is only one pair, namely $(1, 2)$. Thus

$$R_{uu}^{(2)}(t_1,t_2) = R_{uu}(t_1-t_2)$$

that is, the usual order-2 autocorrelation. For $n = 4$ there are three sets of two pairs, namely $(1, 2), (3, 4)$; $(1, 3), (2, 4)$; and $(1, 4), (2, 3)$. Thus

$$\begin{aligned} R_{uu}^{(4)}(t_1,t_2,t_3,t_4) = {} & R_{uu}(t_1-t_2)R_{uu}(t_3-t_4) \\ & + R_{uu}(t_1-t_3)R_{uu}(t_2-t_4) + R_{uu}(t_1-t_4)R_{uu}(t_2-t_3) \end{aligned} \tag{86}$$

In a similar fashion the higher-order spectral densities can be expressed in terms of the order-2 power spectral density. Taking the n-variable Fourier transform of (85) gives

$$S_{uu}^{(n)}(\omega_1,\ldots,\omega_n) = \begin{cases} (2\pi)^{n/2} \sum\limits_p \prod\limits_{j,k}^{n} S_{uu}(\omega_j)\delta_0(\omega_j+\omega_k), & n \text{ even} \\ 0, & n \text{ odd} \end{cases} \tag{87}$$

Example 5.7 For $n = 2$ this formula gives

$$S_{uu}^{(2)}(\omega_1, \omega_2) = 2\pi S_{uu}(\omega_1)\delta_0(\omega_1+\omega_2)$$

an expression that was derived at the beginning of this section. For $n = 4$ I leave the calculation as an exercise and provide the result:

$$S_{uu}^{(4)}(\omega_1,\omega_2,\omega_3,\omega_4) = (2\pi)^2 S_{uu}(\omega_1)S_{uu}(\omega_3)\delta_0(\omega_1+\omega_2)\delta_0(\omega_3+\omega_4)$$
$$+ (2\pi)^2 S_{uu}(\omega_1)S_{uu}(\omega_2)\delta_0(\omega_1+\omega_3)\delta_0(\omega_2+\omega_4)$$
$$+ (2\pi)^2 S_{uu}(\omega_1)S_{uu}(\omega_2)\delta_0(\omega_1+\omega_4)\delta_0(\omega_2+\omega_3) \tag{88}$$

Example 5.8 To illustrate the use of these formulas, the expected value of the output will be computed for the case where the input random process is real, stationary, zero-mean, Gaussian and white with unit intensity. That is, $R_{uu}(\tau) = \delta_0(\tau)$. Also, it will be assumed that the system is described in terms of the symmetric kernel or symmetric system function. In this case substitution of (85) into (67) gives, for $n \geq 1$, $E[y(t)] = 0$ when n is odd, and

$$E[y(t)] = \int_{-\infty}^{\infty} h_{sym}(\sigma_1,\ldots,\sigma_n) \sum_p \prod_{j,k}^{n} \delta_0(\sigma_k-\sigma_j)\, d\sigma_1 \cdots d\sigma_n$$
$$= \sum_p \int_{-\infty}^{\infty} h_{sym}(\sigma_1,\ldots,\sigma_n) \prod_{j,k}^{n} \delta_0(\sigma_k-\sigma_j)\, d\sigma_1 \cdots d\sigma_n,\quad n \text{ even}$$

Now in each term of the sum, the $n/2$ impulses can be integrated out, and this will leave the kernel with only $n/2$ distinct arguments. By symmetry of the kernel, the like arguments can be arranged in pairs, and since they are just variables of integration, they can be labeled in the form $h_{sym}(\sigma_1,\sigma_1,\ldots,\sigma_{n/2},\sigma_{n/2})$. There will be $(n-1)(n-3)\cdots(1)$ terms of this type, so the result is

$$E[y(t)] = \frac{n!}{(n/2)!2^{n/2}} \int_{-\infty}^{\infty} h_{sym}(\sigma_1,\sigma_1,\ldots,\sigma_{n/2},\sigma_{n/2})\, d\sigma_1 \cdots d\sigma_{n/2},\quad n \text{ even}$$

Using (82), the cross-spectral density and output power spectral density will now be computed for a degree-n system described by the symmetric transfer function with a real, stationary, zero-mean, Gaussian-random-process input. For the cross-spectral density, it is clear from (87) that $S_{yu}(\omega_1) = 0$ for $n + 1$ odd, that is, for a homogeneous system of even degree. When $n + 1$ is even, a simple substitution gives

$$S_{yu}(\omega_1) = \frac{1}{(2\pi)^{(n-1)/2}} \int_{-\infty}^{\infty} H_{sym}(\gamma_1,\ldots,\gamma_n)\delta_0(\omega_1-\gamma_1-\cdots-\gamma_n)$$
$$\sum_p \prod_{j,k}^{n+1} S_{uu}(\gamma_j)\delta_0(\gamma_j+\gamma_k)\, d\gamma_1 \cdots d\gamma_{n+1}$$
$$= \frac{1}{(2\pi)^{(n-1)/2}} \sum_p \int_{-\infty}^{\infty} H_{sym}(\gamma_1,\ldots,\gamma_n)\delta_0(\omega_1-\gamma_1-\cdots-\gamma_n)$$
$$\prod_{j,k}^{n+1} S_{uu}(\gamma_j)\delta_0(\gamma_j+\gamma_k)\, d\gamma_1 \cdots d\gamma_{n+1} \tag{89}$$

Before working on this expression in the general case, an example is instructive. And, of course, the $n+1 = 2$ case is too simple, giving just what was derived for linear systems.

Example 5.9 For $n+1 = 4$, (89) yields

$$S_{yu}(\omega_1) = \frac{1}{2\pi}\int_{-\infty}^{\infty} H_{sym}(\gamma_1,\gamma_2,\gamma_3)\delta_0(\omega_1-\gamma_1-\gamma_2-\gamma_3)$$
$$S_{uu}(\gamma_1)S_{uu}(\gamma_3)\delta_0(\gamma_1+\gamma_2)\delta_0(\gamma_3+\gamma_4)\ d\gamma_1 d\gamma_2 d\gamma_3 d\gamma_4$$
$$+\frac{1}{2\pi}\int_{-\infty}^{\infty} H_{sym}(\gamma_1,\gamma_2,\gamma_3)\delta_0(\omega_1-\gamma_1-\gamma_2-\gamma_3)$$
$$S_{uu}(\gamma_1)S_{uu}(\gamma_2)\delta_0(\gamma_1+\gamma_3)\delta_0(\gamma_2+\gamma_4)\ d\gamma_1 d\gamma_2 d\gamma_3 d\gamma_4$$
$$+\frac{1}{2\pi}\int_{-\infty}^{\infty} H_{sym}(\gamma_1,\gamma_2,\gamma_3)\delta_0(\omega_1-\gamma_1-\gamma_2-\gamma_3)$$
$$S_{uu}(\gamma_1)S_{uu}(\gamma_2)\delta_0(\gamma_1+\gamma_4)\delta_0(\gamma_2+\gamma_3)\ d\gamma_1 d\gamma_2 d\gamma_3 d\gamma_4$$

Integrating with respect to γ_4 in each of these terms gives

$$S_{yu}(\omega_1) = \frac{1}{2\pi}\int_{-\infty}^{\infty} H_{sym}(\gamma_1,\gamma_2,\gamma_3)\delta_0(\omega_1-\gamma_1-\gamma_2-\gamma_3)$$
$$S_{uu}(\gamma_1)S_{uu}(\gamma_3)\delta_0(\gamma_1+\gamma_2)\ d\gamma_1 d\gamma_2 d\gamma_3$$
$$+\frac{1}{2\pi}\int_{-\infty}^{\infty} H_{sym}(\gamma_1,\gamma_2,\gamma_3)\delta_0(\omega_1-\gamma_1-\gamma_2-\gamma_3)$$
$$S_{uu}(\gamma_1)S_{uu}(\gamma_2)\delta_0(\gamma_1+\gamma_3)\ d\gamma_1 d\gamma_2 d\gamma_3$$
$$+\frac{1}{2\pi}\int_{-\infty}^{\infty} H_{sym}(\gamma_1,\gamma_2,\gamma_3)\delta_0(\omega_1-\gamma_1-\gamma_2-\gamma_3)$$
$$S_{uu}(\gamma_1)S_{uu}(\gamma_2)\delta_0(\gamma_2+\gamma_3)\ d\gamma_1 d\gamma_2 d\gamma_3$$

Now integrate the first term with respect to γ_3, the second term with respect to γ_2, and the third term with respect to γ_1 to obtain

$$S_{yu}(\omega_1) = \frac{1}{2\pi}\int_{-\infty}^{\infty} H_{sym}(\gamma_1,\gamma_2,\omega_1-\gamma_1-\gamma_2)S_{uu}(\gamma_1)S_{uu}(\omega_1-\gamma_1-\gamma_2)\delta_0(\gamma_1+\gamma_2)\ d\gamma_1 d\gamma_2$$
$$+\frac{1}{2\pi}\int_{-\infty}^{\infty} H_{sym}(\gamma_1,\omega_1-\gamma_1-\gamma_3,\gamma_3)S_{uu}(\gamma_1)S_{uu}(\omega_1-\gamma_1-\gamma_3)\delta_0(\gamma_1+\gamma_3)\ d\gamma_1 d\gamma_3$$
$$+\frac{1}{2\pi}\int_{-\infty}^{\infty} H_{sym}(\omega_1-\gamma_2-\gamma_3,\gamma_2,\gamma_3)S_{uu}(\omega_1-\gamma_2-\gamma_3)S_{uu}(\gamma_2)\delta_0(\gamma_2+\gamma_3)\ d\gamma_2 d\gamma_3$$

Finally, integrating with respect to γ_2 in the first term and γ_3 in the remaining two terms gives

$$S_{yu}(\omega_1) = \frac{1}{2\pi} \int_{-\infty}^{\infty} H_{sym}(\gamma_1, -\gamma_1, \omega_1) S_{uu}(\gamma_1) S_{uu}(\omega_1)\, d\gamma_1$$

$$+ \frac{1}{2\pi} \int_{-\infty}^{\infty} H_{sym}(\gamma_1, \omega_1, -\gamma_1)) S_{uu}(\gamma_1) S_{uu}(\omega_1)\, d\gamma_1$$

$$+ \frac{1}{2\pi} \int_{-\infty}^{\infty} H_{sym}(\omega_1, \gamma_2, -\gamma_2) S_{uu}(\omega_1) S_{uu}(\gamma_2)\, d\gamma_2$$

But from this expression it is clear that since $H_{sym}(\omega_1, \omega_2, \omega_3)$ is symmetric, the input/output cross-spectral density is

$$S_{yu}(\omega) = \frac{3}{2\pi} S_{uu}(\omega) \int_{-\infty}^{\infty} H_{sym}(\omega, \gamma, -\gamma) S_{uu}(\gamma)\, d\gamma \tag{90}$$

This example illustrates the fact that all terms in the summation in (89) are identical since $H_{sym}(\omega_1, . . . \omega_n)$ is symmetric. Thus to get the general expression for the cross-spectral density when $n+1$ is even, it is necessary to work only with a single term and multiply the result by the number of terms in the summation. Choosing the term corresponding to the set of pairs $(1,2), (3,4), ..., (n, n+1)$ gives

$$S_{yu}(\omega_1) = \frac{n(n-2) \cdots (1)}{(2\pi)^{(n-1)/2}} \int_{-\infty}^{\infty} H_{sym}(\gamma_1, ..., \gamma_n) \delta_0(\omega_1 - \gamma_1 - \cdots - \gamma_n) S_{uu}(\gamma_1)$$

$$S_{uu}(\gamma_3) \cdots S_{uu}(\gamma_n) \delta_0(\gamma_1+\gamma_2) \delta_0(\gamma_3+\gamma_4) \cdots \delta_0(\gamma_n+\gamma_{n+1})\, d\gamma_1 \cdots d\gamma_{n+1}$$

Integrating with respect to γ_2, then γ_4, and so on, gives, with a relabeling of variables,

$$S_{yu}(\omega) = \frac{n(n-2)(n-4) \cdots (1)}{(2\pi)^{(n-1)/2}} S_{uu}(\omega) \int_{-\infty}^{\infty} H_{sym}(\omega, \gamma_1, -\gamma_1, \gamma_2, -\gamma_2,$$

$$. . . ., \gamma_{\frac{n-1}{2}}, -\gamma_{\frac{n-1}{2}}) S_{uu}(\gamma_1) S_{uu}(\gamma_2) \cdots S_{uu}(\gamma_{\frac{n-1}{2}})\, d\gamma_1$$

$$\cdots d\gamma_{\frac{n-1}{2}}, \quad n+1 \text{ even} \tag{91}$$

while $S_{yu}(\omega) = 0$, for $n+1$ odd.

Now I begin what starts out appearing to be a similar calculation for the output power spectral density for a homogeneous system with real, stationary, zero-mean, Gaussian input. Again the symmetric kernel and transfer function representations are used for the system. Substituting (87) into (84) gives

$$S_{yy}(\omega) = \frac{1}{(2\pi)^{n-1}} \sum_p \int_{-\infty}^{\infty} H_{sym}(\gamma_1, ..., \gamma_n) H_{sym}(\gamma_{n+1}, ..., \gamma_{2n})$$

$$\delta_0(\omega-\gamma_1- \cdots -\gamma_n)\prod_{j,k}^{2n} S_{uu}(\gamma_j)\delta_0(\gamma_j+\gamma_k)\ d\gamma_1 \cdots d\gamma_{2n} \tag{92}$$

But this situation is considerably more complex than the cross-spectral density case because $H_{sym}(\gamma_1, \ldots, \gamma_n)H_{sym}(\gamma_{n+1}, \ldots, \gamma_{2n})$ is in general not symmetric for a symmetric system function. Thus, different types of terms will arise in the summation. Indeed, the general form for $S_{yy}(\omega)$ is extremely complicated. I will derive the result for $n = 2$ and state the result for $n = 3$, leaving further considerations to the assiduous reader, or to the literature.

Example 5.10 For $n = 2$, (92) becomes

$$S_{yy}(\omega) = \frac{1}{2\pi}\int_{-\infty}^{\infty} H_{sym}(\gamma_1,\gamma_2)H_{sym}(\gamma_3,\gamma_4)\delta_0(\omega-\gamma_1-\gamma_2)S_{uu}(\gamma_1)S_{uu}(\gamma_3)$$

$$\delta_0(\gamma_1+\gamma_2)\delta_0(\gamma_3+\gamma_4)\ d\gamma_1 d\gamma_2 d\gamma_3 d\gamma_4$$

$$+\frac{1}{2\pi}\int_{-\infty}^{\infty} H_{sym}(\gamma_1,\gamma_2)H_{sym}(\gamma_3,\gamma_4)\delta_0(\omega-\gamma_1-\gamma_2)S_{uu}(\gamma_1)S_{uu}(\gamma_2)$$

$$\delta_0(\gamma_1+\gamma_3)\delta_0(\gamma_2+\gamma_4)\ d\gamma_1 d\gamma_2 d\gamma_3 d\gamma_4$$

$$+\frac{1}{2\pi}\int_{-\infty}^{\infty} H_{sym}(\gamma_1,\gamma_2)H_{sym}(\gamma_3,\gamma_4)\delta_0(\omega-\gamma_1-\gamma_2)S_{uu}(\gamma_1)S_{uu}(\gamma_2)$$

$$\delta_0(\gamma_1+\gamma_4)\delta_0(\gamma_2+\gamma_3)\ d\gamma_1 d\gamma_2 d\gamma_3 d\gamma_4$$

Integrating each term with respect to γ_4 gives

$$S_{yy}(\omega) = \frac{1}{2\pi}\int_{-\infty}^{\infty} H_{sym}(\gamma_1,\gamma_2)H_{sym}(\gamma_3,-\gamma_3)\delta_0(\omega-\gamma_1-\gamma_2)S_{uu}(\gamma_1)S_{uu}(\gamma_3)$$

$$\delta_0(\gamma_1+\gamma_2)\ d\gamma_1 d\gamma_2 d\gamma_3$$

$$+\frac{1}{2\pi}\int_{-\infty}^{\infty} H_{sym}(\gamma_1,\gamma_2)H_{sym}(\gamma_3,-\gamma_2)\delta_0(\omega-\gamma_1-\gamma_2)S_{uu}(\gamma_1)S_{uu}(\gamma_2)$$

$$\delta_0(\gamma_1+\gamma_3)\ d\gamma_1 d\gamma_2 d\gamma_3$$

$$+\frac{1}{2\pi}\int_{-\infty}^{\infty} H_{sym}(\gamma_1,\gamma_2)H_{sym}(\gamma_3,-\gamma_1)\delta_0(\omega-\gamma_1-\gamma_2)S_{uu}(\gamma_1)S_{uu}(\gamma_2)$$

$$\delta_0(\gamma_2+\gamma_3)\ d\gamma_1 d\gamma_2 d\gamma_3$$

It should be fairly clear how to proceed. Integrating the first term with respect to γ_2 and the last two terms with respect to both γ_2 and γ_3 gives

$$S_{yy}(\omega) = \frac{1}{2\pi}\,\delta_0(\omega)\int_{-\infty}^{\infty} H_{sym}(\gamma_1,-\gamma_1)H_{sym}(\gamma_3,-\gamma_3)S_{uu}(\gamma_1)S_{uu}(\gamma_3)\;d\gamma_1 d\gamma_3$$

$$+\frac{1}{2\pi}\int_{-\infty}^{\infty} H_{sym}(\gamma_1,\omega-\gamma_1)H_{sym}(-\gamma_1,-\omega+\gamma_1)S_{uu}(\gamma_1)S_{uu}(\omega-\gamma_1)\;d\gamma_1$$

$$+\frac{1}{2\pi}\int_{-\infty}^{\infty} H_{sym}(\gamma_1,\omega-\gamma_1)H_{sym}(-\omega+\gamma_1,-\gamma_1)S_{uu}(\gamma_1)S_{uu}(\omega-\gamma_1)\;d\gamma_1$$

Using the symmetry of the system function to combine the last two terms allows the output power spectral density for the $n = 2$ case to be written in the form

$$S_{yy}(\omega) = \frac{1}{2\pi}\,\delta_0(\omega)\int_{-\infty}^{\infty} H_{sym}(\gamma_1,-\gamma_1)H_{sym}(\gamma_2,-\gamma_2)S_{uu}(\gamma_1)S_{uu}(\gamma_2)d\gamma_1 d\gamma_2$$

$$+\frac{1}{\pi}\int_{-\infty}^{\infty} H_{sym}(\omega-\gamma,\gamma)H_{sym}(-\omega+\gamma,-\gamma)S_{uu}(\gamma)S_{uu}(\omega-\gamma)\;d\gamma \tag{93}$$

This example illustrates the different types of terms that can arise in the general formula. For the record, I also list the result for degree-3 homogeneous systems.

$$S_{yy}(\omega) = \frac{6}{(2\pi)^2}\int_{-\infty}^{\infty} H_{sym}(\omega-\gamma_1-\gamma_2,\gamma_1,\gamma_2)H_{sym}(-\omega+\gamma_1+\gamma_2,-\gamma_1,-\gamma_2)$$

$$S_{uu}(\omega-\gamma_1-\gamma_2)S_{uu}(\gamma_1)S_{uu}(\gamma_2)\;d\gamma_1 d\gamma_2$$

$$+\frac{9}{(2\pi)^2}\,S_{uu}(\omega)\int_{-\infty}^{\infty} H_{sym}(\omega,\gamma_1,-\gamma_1)H_{sym}(-\omega,\gamma_2,-\gamma_2)S_{uu}(\gamma_1)S_{uu}(\gamma_2)\;d\gamma_1 d\gamma_2 \tag{94}$$

Example 5.11 Suppose that the input to the system shown in Figure 5.1 is a real, stationary, zero-mean, Gaussian random process with power spectral density

$$S_{uu}(\omega) = \frac{A^2}{\omega^2+\alpha^2}$$

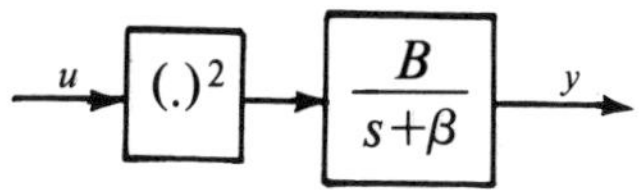

Figure 5.1. A degree-2 homogeneous system.

To find the power spectral density of the output, first note that the symmetric system function in this case is

$$H_{sym}(\omega_1,\omega_2) = \frac{B}{i\omega_1 + i\omega_2 + \beta}$$

Thus

$$S_{yy}(\omega) = \frac{1}{2\pi}\,\delta_0(\omega)\int_{-\infty}^{\infty}\int_{-\infty}^{\infty}\frac{B^2}{\beta^2}\,\frac{A^4}{(\gamma_1^2+\alpha^2)(\gamma_2^2+\alpha^2)}\,d\gamma_1 d\gamma_2$$

$$+\frac{1}{\pi}\int_{-\infty}^{\infty}\frac{B^2}{(i\omega+\beta)(-i\omega+\beta)}\,\frac{A^4}{(\gamma^2+\alpha^2)[(\omega-\gamma)^2+\alpha^2]}\,d\gamma$$

$$= \frac{A^4B^2}{2\pi\beta^2}\,\delta_0(\omega)[\int_{-\infty}^{\infty}\frac{1}{\gamma^2+\alpha^2}\,d\gamma]^2 + \frac{B^2A^4/\pi}{\omega^2+\beta^2}\int_{-\infty}^{\infty}\frac{1}{(\gamma^2+\alpha^2)[(\omega-\gamma)^2+\alpha^2]}\,d\gamma$$

Performing the integrations (tables are allowed) gives

$$\int_{-\infty}^{\infty}\frac{1}{\gamma^2+\alpha^2}\,d\gamma = \frac{\pi}{\alpha}$$

$$\int_{-\infty}^{\infty}\frac{1}{(\gamma^2+\alpha^2)[(\omega-\gamma)^2+\alpha^2]}\,d\gamma = \frac{4\pi^2}{\omega^2+4\alpha^2}$$

Thus the output power spectral density is

$$S_{yy}(\omega) = \frac{A^4B^2\pi}{2\alpha^2\beta^2}\,\delta_0(\omega) + \frac{4\pi A^4B^2}{(\omega^2+\beta^2)(\omega^2+4\alpha^2)}$$

Now consider the case of Volterra and polynomial systems with random inputs. Since the expressions are formidable, the results will be given only for the first few terms - at least in the output power spectral density calculation.

It will be convenient to use the notation

$$y(t) = \sum_{n=1}^{\infty} y_n(t) \tag{95}$$

where $y_n(t)$ is the output of the degree-n homogeneous term. And to be absolutely specific, a subscript will be added to the kernels and transfer functions to indicate the degree. When the input is a sample function from an arbitrary, real random process, the input/output cross-correlation can be written as

$$\begin{aligned} R_{yu}(t_1,t_2) &= E[y(t_1)u(t_2)] \\ &= \sum_{n=1}^{\infty} E[y_n(t_1)u(t_2)] = \sum_{n=1}^{\infty} R_{y_nu}(t_1,t_2) \end{aligned} \tag{96}$$

where $R_{y_nu}(t_1,t_2)$ denotes the cross-correlation for the degree-n homogeneous case just considered. Thus the cross-correlation and cross-spectral density expressions for polynomial or Volterra systems are found simply by summing the previously derived expressions over n, the subsystem degree. For example, when the input is real, stationary, zero-mean, and Gaussian, the input/output cross-spectral density is given by

$$S_{yu}(\omega) = S_{uu}(\omega) \sum_{odd\ n \geq 1} \frac{(n)(n-2)\cdots(1)}{(2\pi)^{(n-1)/2}} \int_{-\infty}^{\infty} H_{nsym}(\omega,\gamma_1,-\gamma_1,\ldots,\gamma_{\frac{n-1}{2}},-\gamma_{\frac{n-1}{2}})$$

$$S_{uu}(\gamma_1)\cdots S_{uu}(\gamma_{\frac{n-1}{2}})\, d\gamma_1 \cdots d\gamma_{\frac{n-1}{2}} \tag{97}$$

Returning to the case of an arbitrary real random input, the calculation of the output autocorrelation or power spectral density is considerably more complex. To see why, note that

$$R_{yy}(t_1,t_2) = E[y(t_1)y(t_2)] = \sum_{n=1}^{\infty}\sum_{m=1}^{\infty} E[y_n(t_1)y_m(t_2)] \tag{98}$$

This expression can be written in the notation

$$R_{yy}(t_1,t_2) = \sum_{n=1}^{\infty}\sum_{m=1}^{\infty} R_{y_ny_m}(t_1,t_2) \tag{99}$$

where $R_{y_ny_m}(t_1,t_2) = E[y_n(t_1)y_m(t_2)]$ is called the *partial output autocorrelation.* The computation of this term is only slightly different from the computations considered earlier, and the tools are quite familiar by now.

Performing the obvious calculations gives an expression for the partial output autocorrelation,

$$R_{y_ny_m}(t_1,t_2) = \int_{-\infty}^{\infty} h_{nsym}(\sigma_1,\ldots,\sigma_n)h_{msym}(\sigma_{n+1},\ldots,\sigma_{n+m})$$

$$R_{uu}^{(n+m)}(t_1-\sigma_1,\ldots,t_1-\sigma_n,t_2-\sigma_{n+1},\ldots,t_2-\sigma_{n+m})\, d\sigma_1 \cdots d\sigma_{n+m} \tag{100}$$

Just as before when faced with expressions of this form, it is convenient to define a *multivariable partial output autocorrelation* by

$$\hat{R}_{y_ny_m}(t_1,\ldots,t_{n+m}) = \int_{-\infty}^{\infty} h_{nsym}(\sigma_1,\ldots,\sigma_n)h_{msym}(\sigma_{n+1},\ldots,\sigma_{n+m})$$

$$R_{uu}^{(n+m)}(t_1-\sigma_1,\ldots,t_{n+m}-\sigma_{n+m})\, d\sigma_1 \cdots d\sigma_{n+m} \tag{101}$$

so that

$$R_{y_ny_m}(t_1,t_2) = \hat{R}_{y_ny_m}(t_1,\ldots,t_{n+m}) \Bigg|_{\substack{t_1=\cdots=t_n=t_1 \\ t_{n+1}=\cdots=t_{n+m}=t_2}} \tag{102}$$

Again, the advantage of this notation is that the convolution properties of the Fourier transform can be applied in a direct fashion, and the variable associations can be handled separately. Letting

$$S_{y_n y_m}(\omega_1,\omega_2) = F[R_{y_n y_m}(t_1,t_2)]$$

$$\hat{S}_{y_n y_m}(\omega_1,\ldots,\omega_{n+m}) = F[\hat{R}_{y_n y_m}(t_1,\ldots,t_{n+m})]$$

$$S_{uu}^{(n+m)}(\omega_1,\ldots,\omega_{n+m}) = F[R_{uu}^{(n+m)}(t_1,\ldots,t_{n+m})] \tag{103}$$

and using the system function defined previously gives

$$\hat{S}_{y_n y_m}(\omega_1,\ldots,\omega_{n+m}) = H_{nsym}(\omega_1,\ldots,\omega_n)$$
$$H_{msym}(\omega_{n+1},\ldots,\omega_{n+m})S_{uu}^{(n+m)}(\omega_1,\ldots,\omega_{n+m}) \tag{104}$$

Repeating the derivation leading to (79) gives, in the present setting

$$S_{y_n y_m}(\omega_1,\omega_2) = \frac{1}{(2\pi)^{n+m-2}}\int_{-\infty}^{\infty} \hat{S}_{y_n y_m}(\gamma_1,\ldots,\gamma_{n+m})\delta_0(\omega_1-\gamma_1-\cdots-\gamma_n)$$
$$\delta_0(\omega_2-\gamma_{n+1}-\cdots-\gamma_{n+m})\,d\gamma_1\cdots d\gamma_{n+m} \tag{105}$$

Furthermore, for stationary random process inputs, the partial output power spectral density is given by

$$S_{y_n y_m}(\omega) = \frac{1}{(2\pi)^{n+m-1}}\int_{-\infty}^{\infty} \hat{S}_{y_n y_m}(\gamma_1,\ldots,\gamma_{n+m})$$
$$\delta_0(\omega-\gamma_1-\cdots-\gamma_n)\,d\gamma_1\cdots d\gamma_{n+m} \tag{106}$$

Of course, this formula checks with (84) for the case $m = n$.

Now assume that the input is real, stationary, zero-mean, Gaussian, and with power spectral density $S_{uu}(\omega)$. Substituting (87) and (104) into (106) gives

$$S_{y_n y_m}(\omega) = 0,\quad n+m\ odd \tag{107}$$

and

$$S_{y_n y_m}(\omega) = \frac{1}{(2\pi)^{(n+m-2)/2}}\sum_p \int_{-\infty}^{\infty} H_{nsym}(\gamma_1,\ldots,\gamma_n)H_{msym}(\gamma_{n+1},\ldots,\gamma_{n+m})$$
$$\delta_0(\omega-\gamma_1-\cdots-\gamma_n)\prod_{j,k}^{n+m} S_{uu}(\gamma_j)\delta_0(\gamma_j+\gamma_k)\,d\gamma_1\cdots d\gamma_{n+m},\quad n+m\ even \tag{108}$$

The reduction of this expression to more explicit form is a combinatorial problem of some complexity because the integrand lacks symmetry. I will be content to work out the terms that give the output power spectral density for polynomial systems of degree 3.

Example 5.12 To compute $S_{yy}(\omega)$ for the case of a degree-3 polynomial system, the terms $S_{y_ny_m}(\omega)$ must be computed for $n,m = 1,2,3$. But it is evident that

$$S_{y_1y_2}(\omega) = S_{y_2y_1}(\omega) = S_{y_2y_3}(\omega) = S_{y_3y_2}(\omega) = 0$$

For $n = m = 1,2,3$ the partial output power spectral densities have been calculated previously, and are given in (64), (93), and (94). For $n = 1$ and $m = 3$, (108) gives

$$S_{y_1y_3}(\omega) = \frac{1}{2\pi} \sum_p \int_{-\infty}^{\infty} H_1(\gamma_1) H_{3sym}(\gamma_2,\gamma_3,\gamma_4)\delta_0(\omega-\gamma_1)$$

$$\prod_{j,k}^{4} S_{uu}(\gamma_j)\delta_0(\gamma_j+\gamma_k)\; d\gamma_1 \cdots d\gamma_4$$

$$= \frac{1}{2\pi} \int_{-\infty}^{\infty} H_1(\gamma_1) H_{3sym}(\gamma_2,\gamma_3,\gamma_4)\delta_0(\omega-\gamma_1) S_{uu}(\gamma_1) S_{uu}(\gamma_3)$$

$$\delta_0(\gamma_1+\gamma_2)\delta_0(\gamma_3+\gamma_4)\; d\gamma_1 \cdots d\gamma_4$$

$$+ \frac{1}{2\pi} \int_{-\infty}^{\infty} H_1(\gamma_1) H_{3sym}(\gamma_2,\gamma_3,\gamma_4)\delta_0(\omega-\gamma_1) S_{uu}(\gamma_1) S_{uu}(\gamma_2)$$

$$\delta_0(\gamma_1+\gamma_3)\delta_0(\gamma_2+\gamma_4)\; d\gamma_1 \cdots d\gamma_4$$

$$+ \frac{1}{2\pi} \int_{-\infty}^{\infty} H_1(\gamma_1) H_{3sym}(\gamma_2,\gamma_3,\gamma_4)\delta_0(\omega-\gamma_1) S_{uu}(\gamma_1) S_{uu}(\gamma_2)$$

$$\delta_0(\gamma_1+\gamma_4)\delta_0(\gamma_2+\gamma_3)\; d\gamma_1 \cdots d\gamma_4$$

Performing the integrations gives

$$S_{y_1y_3}(\omega) = \frac{3}{2\pi} H_1(\omega) S_{uu}(\omega) \int_{-\infty}^{\infty} H_{3sym}(-\omega,\gamma,-\gamma) S_{uu}(\gamma)\; d\gamma$$

In a similar fashion $S_{y_3y_1}(\omega)$ can be computed. Alternatively, the easily proved fact that $S_{y_3y_1}(\omega) = S_{y_1y_3}(-\omega)$ can be used to obtain

$$S_{y_3y_1}(\omega) = \frac{3}{2\pi} H_1(-\omega) S_{uu}(\omega) \int_{-\infty}^{\infty} H_{3sym}(\omega,\gamma,-\gamma) S_{uu}(\gamma)\; d\gamma$$

Now, collecting together all the terms gives the expression

$$S_{yy}(\omega) = H_1(\omega)H_1(-\omega)S_{uu}(\omega) + \frac{3}{2\pi} H_1(\omega) S_{uu}(\omega) \int_{-\infty}^{\infty} H_{3sym}(-\omega,\gamma,-\gamma) S_{uu}(\gamma)\; d\gamma$$

$$+ \frac{3}{2\pi} H_1(-\omega) S_{uu}(\omega) \int_{-\infty}^{\infty} H_{3sym}(\omega,\gamma,-\gamma) S_{uu}(\gamma)\; d\gamma$$

$$+ \frac{1}{2\pi} \delta_0(\omega) \int_{-\infty}^{\infty}\int_{-\infty}^{\infty} H_{2sym}(\gamma_1,-\gamma_1) H_{2sym}(\gamma_2,-\gamma_2) S_{uu}(\gamma_1) S_{uu}(\gamma_2)\; d\gamma_1 d\gamma_2$$

$$+ \frac{1}{\pi} \int_{-\infty}^{\infty} H_{2sym}(\omega-\gamma,\gamma) H_{2sym}(-\omega+\gamma,-\gamma) S_{uu}(\gamma) S_{uu}(\omega-\gamma)\, d\gamma$$

$$+ \frac{6}{(2\pi)^2} \int_{-\infty}^{\infty} \int_{-\infty}^{\infty} H_{3sym}(\omega-\gamma_1-\gamma_2,\gamma_1,\gamma_2) H_{3sym}(-\omega+\gamma_1+\gamma_2,-\gamma_1,-\gamma_2)$$

$$S_{uu}(\omega-\gamma_1-\gamma_2) S_{uu}(\gamma_1) S_{uu}(\gamma_2)\, d\gamma_1 d\gamma_2$$

$$+ \frac{9}{(2\pi)^2} S_{uu}(\omega) \int_{-\infty}^{\infty} \int_{-\infty}^{\infty} H_{3sym}(\omega,\gamma_1,-\gamma_1) H_{3sym}(-\omega,\gamma_2,-\gamma_2)$$

$$S_{uu}(\gamma_1) S_{uu}(\gamma_2)\, d\gamma_1 d\gamma_2 \qquad (109)$$

Example 5.13 For the phase-locked loop introduced in Example 3.3, the first three symmetric transfer functions are shown in Example 3.8 to be

$$H(s) = \frac{1}{s+K}, \quad H_{2sym}(s_1,s_2) = 0,$$

$$H_{3sym}(s_1,s_2,s_3) = \frac{K/6}{(s_1+s_2+s_3+K)(s_1+K)(s_2+K)(s_3+K)}$$

Suppose the message signal is real, stationary, zero-mean, Gaussian white noise with intensity A. Thus,

$$R_{uu}(\tau) = A\delta_0(\tau), \quad S_{uu}(\omega) = A$$

Then it is straightforward to verify that, through degree 3, the loop error signal $x(t)$ also is zero mean. In particular, from (65) the degree-1 component of $x(t)$ is zero mean, and from Example 5.8 the degree-2 and degree-3 components are also.

To illustrate the calculation of the power spectral density for $x(t)$, I will evaluate the terms in (109) that are of degree $\leqslant 2$ in the noise intensity A. These terms give

$$S_{xx}(\omega) = AH(\omega)H(-\omega) + \frac{3A^2}{2\pi} H(\omega) \int_{-\infty}^{\infty} H_{3sym}(-\omega,\gamma,-\gamma)\, d\gamma$$

$$+ \frac{3A^2}{2\pi} H(-\omega) \int_{-\infty}^{\infty} H_{3sym}(\omega,\gamma,-\gamma)\, d\gamma$$

Clearly,

$$AH(\omega)H(-\omega) = \frac{A}{\omega^2 + K^2}$$

and a simple calculation gives

$$\frac{3A^2}{2\pi} H(\omega) \int_{-\infty}^{\infty} H_{3sym}(-\omega,\gamma,-\gamma)\, d\gamma = \frac{KA^2/4\pi}{(\omega^2+K_2)(-i\omega+K)} \int_{-\infty}^{\infty} \frac{1}{\gamma^2+K^2}\, d\gamma$$

$$= \frac{A^2/4}{(\omega^2+K^2)(-i\omega+K)}$$

Since the third term can be obtained form the second by replacing ω by $-\omega$, the power spectral density of the loop error is, through degree 2 in A,

$$S_{xx}(\omega) = \frac{A}{\omega^2 + K^2} + \frac{KA^2/2}{(\omega^2 + K^2)^2}$$

5.5 The Wiener Orthogonal Representation

A major difficulty in computing the output power spectral density, or autocorrelation, for a polynomial or Volterra system is the profusion of partial output spectral densities or autocorrelations (cross-terms). For this reason, and for other reasons that will be discussed in Chapter 7, I will now consider a series representation that has certain orthogonality properties with respect to the statistical characterization of the response. Under appropriate convergence conditions, this Wiener series representation can be viewed as a rearrangement of the terms in a Volterra series representation. However, this viewpoint can be confusing, and it probably is best to regard the Wiener representation as a separate topic, at least at the outset.

Throughout this section it is assumed that the input signal is a sample function from a real, stationary, zero-mean, white Gaussian random process with intensity A. The exposition will be in terms of infinite series, with the usual avoidance of convergence issues. Actually, the convergence properties of the Wiener representation are naturally addressed in the mean square sense, and it can be shown that the resulting conditions are less restrictive than those for the Volterra series. These issues will be left to the literature cited in Section 5.6.

The Wiener representation for a system takes the form

$$y(t) = \sum_{n=0}^{\infty} G_n[k_n,u(t)] \tag{110}$$

where each *Wiener operator* $G_n[k_n,u(t)]$ is a degree-n polynomial operator that is specified (in a yet to be determined, and at this point nonobvious, manner) by a symmetric *Wiener kernel* $k_n(t_1,\ldots,t_n)$. Notice that the operator notation from Chapter 1 has been changed slightly in that the subscript indicates now the polynomial degree of the operator, and the dependence on k_n is displayed. Also there is no subscript "sym" on the Wiener kernel, even though it is symmetric. This conforms with the traditional notation, and helps to distinguish the Wiener kernel from the symmetric Volterra series kernel $h_{nsym}(t_1,\ldots,t_n)$.

The important condition to be imposed is that what might be called the partial output autocorrelations in this new representation satisfy

$$E[G_n[k_n,u(t+\tau)]G_m[k_m,u(t)]] = 0, \quad \text{for } all\ \tau,\ m \neq n \tag{111}$$

Of course, when this condition is satisfied, the output autocorrelation is given by

$$R_{yy}(\tau) = \sum_{n=0}^{\infty} E[G_n[k_n,u(t+\tau)]G_n[k_n,u(t)]] \tag{112}$$

Although the Wiener representation can be determined through an elegant, general argument, it is instructive to begin in an elementary fashion. (A more elegant derivation will be used for the discrete-time case in Chapter 6.) The approach is to find $G_n[k_n,u(t)]$ by requiring that it be a degree-n polynomial operator that satisfies

$$E[G_n[k_n,u(t+\tau)]F_j[u(t)]] = 0, \quad \text{for } all\ \tau,\ j = 0,1,\ldots,n-1 \tag{113}$$

where $F_j[u(t)]$ is any homogeneous operator of degree j. Of course this condition guarantees that $G_n[k_n,u(t)]$ is orthogonal to any polynomial operator of degree $n-1$ or less. In the following development, the symmetric kernel corresponding to $F_j[u(t)]$ will be denoted by $f_{jsym}(t_1,\ldots,t_j)$, except that the "sym" is superfluous when $j = 0,1$.

The degree-0 Wiener operator is defined to be simply $G_0[k_0,u(t)] = k_0$. The degree-1 Wiener operator is assumed to take the general form

$$G_1[k_1,u(t)] = \int_{-\infty}^{\infty} k_1(\sigma)u(t-\sigma)\,d\sigma + k_{1,0} \tag{114}$$

where $k_1(t)$ is the degree-1 Wiener kernel, and $k_{1,0}$ is a constant that remains to be chosen. This operator must be orthogonal to any degree-0 homogeneous operator $F_0[u(t)] = f_0$, that is

$$\begin{aligned} 0 &= E[G_1[k_1,u(t+\tau)]F_0[u(t)]] \\ &= \int_{-\infty}^{\infty} f_0k_1(\sigma)E[u(t+\tau-\sigma)]\,d\sigma + f_0k_{1,0}, \quad \text{for } all\ \tau \end{aligned}$$

for any f_0. And since the expected value in the first term is 0, this condition can be satisfied by taking $k_{1,0} = 0$. Thus the degree-1 Wiener operator takes the (familiar) form

$$G_1[k_1,u(t)] = \int_{-\infty}^{\infty} k_1(\sigma)u(t-\sigma)\,d\sigma \tag{115}$$

So far the Wiener representation looks just like a Volterra series representation, except that the kernels may be different.

Now I proceed to degree 2, where more interesting things begin to happen. The general form of $G_2[k_2,u(t)]$ is

$$G_2[k_2,u(t)] = \int_{-\infty}^{\infty} k_2(\sigma_1,\sigma_2)u(t-\sigma_1)u(t-\sigma_2)\,d\sigma_1 d\sigma_2$$

$$+ \int_{-\infty}^{\infty} k_{2,1}(\sigma_1)u(t-\sigma_1)\, d\sigma_1 + k_{2,0} \tag{116}$$

where $k_2(t_1,t_2)$ is symmetric, and where the conditions to be satisfied are

$$E[G_2[k_2,u(t+\tau)]F_1[u(t)]] = 0$$
$$E[G_2[k_2,u(t+\tau)]F_0[u(t)]] = 0 \tag{117}$$

for all τ. The first condition gives

$$0 = \int_{-\infty}^{\infty} k_2(\sigma_1,\sigma_2)f_1(\sigma)E[u(t+\tau-\sigma_1)u(t+\tau-\sigma_2)u(t-\sigma)]\, d\sigma_1 d\sigma_2 d\sigma$$

$$+ \int_{-\infty}^{\infty} k_{2,1}(\sigma_1)f_1(\sigma)E[u(t+\tau-\sigma_1)u(t-\sigma)]\, d\sigma d\sigma_1$$

$$+ \int_{-\infty}^{\infty} k_{2,0}f_1(\sigma)E[u(t-\sigma)]\, d\sigma$$

$$= A \int_{-\infty}^{\infty} k_{2,1}(\sigma+\tau)f_1(\sigma)\, d\sigma$$

To guarantee that this is satisfied regardless of $f_1(t)$, take $k_{2,1}(t) = 0$. Now the second condition in (117) gives

$$0 = \int_{-\infty}^{\infty} k_2(\sigma_1,\sigma_2)f_0E[u(t+\tau-\sigma_1)u(t+\tau-\sigma_2)]\, d\sigma_1 d\sigma_2 + k_{2,0}f_0$$

$$= A \int_{-\infty}^{\infty} k_2(\sigma,\sigma)f_0\, d\sigma + k_{2,0}f_0$$

This can be satisfied by taking

$$k_{2,0} = -A \int_{-\infty}^{\infty} k_2(\sigma,\sigma)\, d\sigma$$

so that the degree-2 Wiener operator is

$$G_2[k_2,u(t)] = \int_{-\infty}^{\infty} k_2(\sigma_1,\sigma_2)u(t-\sigma_1)u(t-\sigma_2)d\sigma_1 d\sigma_2 - A \int_{-\infty}^{\infty} k_2(\sigma,\sigma)d\sigma \tag{118}$$

This is the first illustration of how the Wiener polynomial operators are specified by a single kernel. Also, notice that there is an implicit technical assumption here, namely, that the integral of $k_2(t,t)$ is finite.

I will work out one more just to show that nothing surprising happens. The degree-3 Wiener operator will take the general form

$$G_3[k_3,u(t)] = \int_{-\infty}^{\infty} k_3(\sigma_1,\sigma_2,\sigma_3)u(t-\sigma_1)u(t-\sigma_2)u(t-\sigma_3)\, d\sigma_1 d\sigma_2 d\sigma_3$$
$$+ \int_{-\infty}^{\infty} k_{3,2}(\sigma_1,\sigma_2)u(t-\sigma_1)u(t-\sigma_2)\, d\sigma_1 d\sigma_2$$
$$+ \int_{-\infty}^{\infty} k_{3,1}(\sigma_1)u(t-\sigma_1)\, d\sigma_1 + k_{3,0} \tag{119}$$

where the degree-3 Wiener kernel $k_3(t_1,t_2,t_3)$ is symmetric. Imposing the condition of orthogonality to degree-0 homogeneous operators gives that

$$A \int_{-\infty}^{\infty} k_{3,2}(\sigma,\sigma)f_0\, d\sigma + k_{3,0}f_0 = 0$$

must hold for all f_0. To satisfy this condition, set

$$k_{3,0} = -A \int_{-\infty}^{\infty} k_{3,2}(\sigma,\sigma)\, d\sigma$$

Orthogonality with respect to degree-1 homogeneous operators gives the condition

$$0 = 3A^2 \int_{-\infty}^{\infty} k_3(\sigma_1,\sigma_1,\sigma+\tau)f_1(\sigma)\, d\sigma_1 d\sigma + A \int_{-\infty}^{\infty} k_{3,1}(\sigma+\tau)f_1(\sigma)\, d\sigma$$

for all $f_1(t)$. To satisfy this, set

$$k_{3,1}(t) = -3A \int_{-\infty}^{\infty} k_3(\sigma,\sigma,t)\, d\sigma$$

Up to this point the degree-3 Wiener operator has been specialized to the form

$$G_3[k_3,u(t)] = \int_{-\infty}^{\infty} k_3(\sigma_1,\sigma_2,\sigma_3)u(t-\sigma_1)u(t-\sigma_2)u(t-\sigma_3)\, d\sigma_1 d\sigma_2 d\sigma_3$$
$$+ \int_{-\infty}^{\infty} k_{3,2}(\sigma_1,\sigma_2)u(t-\sigma_1)u(t-\sigma_2)\, d\sigma_1 d\sigma_2$$
$$- 3A \int_{-\infty}^{\infty} k_3(\sigma_1,\sigma_1,\sigma)u(t-\sigma)\, d\sigma_1 d\sigma - A \int_{-\infty}^{\infty} k_{3,2}(\sigma,\sigma)\, d\sigma \tag{120}$$

Imposing the (final) condition that (120) be orthogonal to all degree-2 homogeneous operators leads, after some calculation, to the choice $k_{3,2}(t_1,t_2) = 0$. Thus the degree-3 Wiener operator in (110) is

$$G_3[k_3,u(t)] = \int_{-\infty}^{\infty} k_3(\sigma_1,\sigma_2,\sigma_3)u(t-\sigma_1)u(t-\sigma_2)u(t-\sigma_3)\, d\sigma_1 d\sigma_2 d\sigma_3$$

$$- 3A \int_{-\infty}^{\infty} k_3(\sigma_1,\sigma_1,\sigma)u(t-\sigma)\, d\sigma_1 d\sigma \tag{121}$$

The general result can be stated as follows.

Theorem 5.1 The degree-n Wiener operator is given by

$$G_n[k_n,u(t)] = \sum_{i=0}^{[n/2]} \frac{(-1)^i n! A^i}{2^i (n-2i)! i!} \int_{-\infty}^{\infty} k_n(\sigma_1,\ldots,\sigma_{n-2i},\tau_1,\tau_1,\ldots,\tau_i,\tau_i)$$

$$d\tau_1 \cdots d\tau_i u(t-\sigma_1) \cdots u(t-\sigma_{n-2i})\, d\sigma_1 \cdots d\sigma_{n-2i} \tag{122}$$

where $[n/2]$ indicates the greatest integer $\leqslant n/2$, the Wiener kernel $k_n(t_1,\ldots,t_n)$ is symmetric, and where A is the intensity of the real, stationary, zero-mean, Gaussian-white-noise input.

Proof (Sketch) Suppose n is an even integer for definiteness. (The proof for odd n is similar.) Then the square brackets can be erased from the upper limit of the summation sign. Also, retaining the notation $F_j[u(t)]$ for an arbitrary degree-j homogeneous operator with symmetric kernel $f_{jsym}(t_1,\ldots,t_j)$,

$$E[G_n[k_n,u(t+\tau)]F_j[u(t)]] = 0, \quad odd\ j < n$$

This is because all terms will involve the expected value of a product of an odd number of zero-mean Gaussian random variables. Thus, it remains to show that

$$E[G_n[k_n,u(t+\tau)]F_{2j}[u(t)]] = 0, \quad j = 0,1,\ldots,\frac{n-2}{2}$$

For $j = 0$ this condition reduces to showing that $E[G_n[k_n,u(t)]] = 0$ for $n > 0$. Direct calculation using (85) gives

$$E[G_n[k_n,u(t)]] = \sum_{i=0}^{n/2} \frac{(-1)^i n! A^i}{2^i (n-2i)! i!} \int_{-\infty}^{\infty} k_n(\sigma_1,\ldots,\sigma_{n-2i},\tau_1,\tau_1,\ldots,\tau_i,\tau_i)$$

$$d\tau_1 \cdots d\tau_i A^{(n-2i)/2} \sum_p \prod_{j,k}^{n-2i} \delta_0(\sigma_j-\sigma_k)\, d\sigma_1 \cdots d\sigma_{n-2i}$$

For each fixed i, integrating out the $(n-2i)/2$ impulses in each product will yield identical results because of the symmetry of the Wiener kernel. Furthermore, from (85) there are

$$\frac{(n-2i)!}{(\frac{n-2i}{2})!2^{(n-2i)/2}} = \frac{(n-2i)!}{(\frac{n}{2}-i)!2^{n/2}2^{-i}}$$

products in the summation in the ith term. Therefore, with considerable relabeling of variables, I can write

$$E[G_n[k_n,u(t)]] = \frac{n! A^{n/2}}{2^{n/2}} \sum_{i=0}^{n/2} \frac{(-1)^i}{(\frac{n}{2}-i)! i!} \int_{-\infty}^{\infty} k_n(\sigma_1,\sigma_1,\ldots,\sigma_{n/2},\sigma_{n/2})\, d\sigma_1 \cdots d\sigma_{n/2}$$

But

$$\sum_{i=0}^{n/2} \frac{(-1)^i}{(\frac{n}{2}-i)!i!} = \frac{1}{(n/2)!} \sum_{i=0}^{n/2} (-1)^i \binom{n/2}{i} = 0$$

so the result is

$$E[G_n[k_n,u(t)]] = 0, \quad n = 1,2,\ldots$$

For $j = 1$ it must be shown that, assuming $n > 2$,

$$E[G_n[k_n,u(t+\tau)]F_2[u(t)]] = 0$$

where the degree-2 operator $F_2[u(t)]$ is arbitrary. Again, proceeding by direct calculation gives

$$E[G_n[k_n,u(t+\tau)]F_2[u(t)]]$$

$$= \sum_{i=0}^{n/2} \frac{(-1)^i n! A^i}{2^i(n-2i)!i!} \int_{-\infty}^{\infty} k_n(\sigma_1,\ldots,\sigma_{n-2i},\tau_1,\tau_1,\ldots,\tau_i,\tau_i)$$

$$d\tau_1 \cdots d\tau_i f_{2sym}(\sigma_{n-2i+1},\sigma_{n-2i+2})E[u(t+\tau-\sigma_1)$$

$$\cdots u(t+\tau-\sigma_{n-2i})u(t-\sigma_{n-2i+1})u(t-\sigma_{n-2i+2})]\, d\sigma_1 \cdots d\sigma_{n-2i+2}$$

$$= \sum_{i=0}^{n/2} \frac{(-1)^i n! A^i}{(n-2i)!i!2^i} \int_{-\infty}^{\infty} k_n(\sigma_1,\ldots,\sigma_{n-2i},\tau_1,\tau_1,\ldots,\tau_i,\tau_i)\, d\tau_1 \cdots d\tau_i$$

$$f_{2sym}(\sigma_{n-2i+1},\sigma_{n-2i+2})A^{(n-2i+2)/2} \sum_p \prod_{j,k}^{n-2i+2} \delta_0(\sigma_j-\sigma_k)\, d\sigma_1 \cdots d\sigma_{n-2i+2}$$

First, for each fixed i consider all the product terms in $\sum_p$ that contain a factor of the form $\delta_0(\sigma_{n-2i+1}-\sigma_{n-2i+2})$. Integrating out this impulse gives an identical result in each term, and there are

$$\frac{(n-2i)!}{(\frac{n-2i}{2})!2^{(n-2i)/2}} = \frac{(n-2i)!}{(\frac{n}{2}-i)!2^{n/2}2^{-i}}$$

such terms for each i. The remaining products of $(n-2i)/2$ impulses will contain all possible pairs of arguments from $\sigma_1,\ldots,\sigma_{n-2i}$. Thus, these terms give

$$\sum_{i=0}^{n/2} \frac{(-1)^i n! A^{(n+2)/2}}{(\frac{n}{2}-i)!i!2^{n/2}} \int_{-\infty}^{\infty} k_n(\sigma_1,\ldots,\sigma_{n-2i},\tau_1,\tau_1,\ldots,\tau_i,\tau_i)\, d\tau_1 \cdots d\tau_i$$

$$f_{2sym}(\sigma_{n-2i+1},\sigma_{n-2i+1}) \sum_p \prod_{j,k}^{n-2i} \delta_0(\sigma_j-\sigma_k)\, d\sigma_1 \cdots d\sigma_{n-2i+1}$$

$$= \frac{n!A^{(n+2)/2}}{2^{n/2}} \sum_{i=0}^{n/2} \frac{(-1)^i}{(\frac{n}{2}-i)!i!} \int_{-\infty}^{\infty} f_{2sym}(\sigma_{n-2i+1},\sigma_{n-2i+1})\, d\sigma_{n-2i+1}$$

$$\int_{-\infty}^{\infty} k_n(\sigma_1,\sigma_1,\ldots,\sigma_{n/2},\sigma_{n/2})\, d\sigma_1 \cdots d\sigma_{n/2}$$

$$= 0$$

For emphasis, I will restate the key fact that has just been used. The set of all those terms in

$$\sum_p \prod_{j,k}^{n-2i+2} \delta_0(\sigma_j - \sigma_k)$$

that contain an impulse of the form $\delta_0(\sigma_{n-2i+1}-\sigma_{n-2i+2})$ can be written as

$$\delta_0(\sigma_{n-2i+1} - \sigma_{n-2i+2}) \sum_p \prod_{j,k}^{n-2i} \delta_0(\sigma_j - \sigma_k)$$

This result will be used in the sequel. Now, for each fixed i consider the remaining terms, all of which contain factors of the form $\delta_0(\sigma_j-\sigma_{n-2i+1})\delta_0(\sigma_k-\sigma_{n-2i+2})$ for $j,k \leqslant n-2i$. Of course, these terms occur only for $i < n/2$, and there are

$$\frac{(n-2i+2)!}{(\frac{n-2i+2}{2})!2^{(n-2i+2)/2}} - \frac{(n-2i)!}{(\frac{n}{2}-i)!2^{(n-2i)/2}} = \frac{(n-2i)!}{(\frac{n}{2}-i)!2^{n/2}2^{-i}}(n-2i)$$

such terms for each i. Because of symmetry, all these terms will be identical after the impulses are integrated out. Thus, these terms give

$$\sum_{i=0}^{n/2-1} \frac{(-1)^i n! A^{(n+2)/2}(n-2i)}{(\frac{n}{2}-i)!i!2^{n/2}} \int_{-\infty}^{\infty} k_n(\sigma_1,\sigma_1,\ldots,\sigma_{n/2},\sigma_{n/2})$$

$$f_{2sym}(\sigma_1,\sigma_2)\, d\sigma_1 \cdots d\sigma_{n/2} = 0$$

since

$$\sum_{i=0}^{n/2-1} (-1)^i \binom{n/2}{i}(n-2i) = n\sum_{i=0}^{n/2-1} (-1)^i \binom{n/2-1}{i} = 0$$

It now should be clear that to verify the orthogonality condition in general requires just this type of calculation for larger values of j. This tedious exercise is omitted.

There are some general features to notice about $G_n[k_n,u(t)]$. It is a degree-n polynomial operator that contains homogeneous terms of degree n, $n-2$, . . ., 1 (n *odd*) *or* 0(n *even*). However, all the homogeneous terms are specified by the degree-n, symmetric, Wiener kernel $k_n(t_1,\ldots,t_n)$ and by the input noise intensity A. Lest the reader be puzzled over the notational abuse in (122) when $i = 0$, that term is precisely

$$\int_{-\infty}^{\infty} k_n(\sigma_1,\ldots,\sigma_n)u(t-\sigma_1)\cdots u(t-\sigma_n)\,d\sigma_1\cdots d\sigma_n$$

Finally, it should be clear that certain integrability conditions on the Wiener kernel must hold if (122) is to make sense technically.

Now suppose a system is described by the Wiener representation in (111). Then, by the orthogonality property, the output autocorrelation is given by

$$R_{yy}(\tau) = E[y(t+\tau)y(t)] = \sum_{n=0}^{\infty} E[G_n[k_n,u(t+\tau)]G_n[k_n,u(t)]] \tag{123}$$

Before computing the general term, it is instructive to work out the first few. For $n = 0$, it is clear that

$$E[G_0[k_0,u(t+\tau)]G_0[k_0,u(t)] = E[k_0^2] = k_0^2 \tag{124}$$

For $n = 1$ the calculation is only slightly less trivial since this is the usual linear case:

$$\begin{aligned} E[G_1[k_1,u(t+\tau)]G_1[k_1,u(t)]] &= \int_{-\infty}^{\infty} k_1(\sigma_1)k_1(\sigma_2)E[u(t+\tau-\sigma_1)u(t-\sigma_2)]\,d\sigma_1 d\sigma_2 \\ &= A\int_{-\infty}^{\infty} k_1(\sigma+\tau)k_1(\sigma)\,d\sigma \end{aligned} \tag{125}$$

For $n = 2$ the calculation is a bit more involved, though all the steps have been done at least once before in this chapter.

$$\begin{aligned} &E[G_2[k_2,u(t+\tau)]G_2[k_2,u(t)]] \\ &\quad= \int_{-\infty}^{\infty} k_2(\sigma_1,\sigma_2)k_2(\sigma_3,\sigma_4)E[u(t+\tau-\sigma_1)u(t+\tau-\sigma_2) \\ &\quad\quad u(t-\sigma_3)u(t-\sigma_4)]\,d\sigma_1\cdots d\sigma_4 \\ &\quad - A\int_{-\infty}^{\infty} k_2(\sigma_1,\sigma_2)E[u(t+\tau-\sigma_1)u(t+\tau-\sigma_2)]\,d\sigma_1 d\sigma_2 \int_{-\infty}^{\infty} k_2(\sigma,\sigma)\,d\sigma \\ &\quad - A\int_{-\infty}^{\infty} k_2(\sigma_1,\sigma_2)E[u(t-\sigma_1)u(t-\sigma_2)]d\sigma_1 d\sigma_2 \int_{-\infty}^{\infty} k_2(\sigma,\sigma)\,d\sigma \\ &\quad + A^2\,[\int_{-\infty}^{\infty} k_2(\sigma,\sigma)\,d\sigma]^2 \end{aligned} \tag{126}$$

Upon expansion of the expectations, many of the terms add out, leaving the easily verified result:

$$E[G_2[k_2,u(t+\tau)]G_2[k_2,u(t)]] = 2A^2\int_{-\infty}^{\infty} k_2(\sigma_1+\tau,\sigma_2+\tau)k_2(\sigma_1,\sigma_2)\,d\sigma_1 d\sigma_2 \tag{127}$$

The general result will be presented more formally.

Theorem 5.2 For the Wiener polynomial operator $G_n[k_n,u(t)]$, where $u(t)$ is real, stationary, zero-mean, Gaussian white noise with intensity A,

$$E[G_n[k_n,u(t+\tau)]G_n[k_n,u(t)]]$$
$$= n!A^n \int_{-\infty}^{\infty} k_n(\sigma_1+\tau,\ldots,\sigma_n+\tau)k_n(\sigma_1,\ldots,\sigma_n)\,d\sigma_1\cdots d\sigma_n \qquad (128)$$

Proof (Sketch) To simplify the notation, the degree-n Wiener operator will be written in the general polynomial form shown below. (Recall that only homogeneous terms whose degree has the same parity as n occur in $G_n[k_n,u(t)]$.)

$$G_n[k_n,u(t)] = \sum_{\substack{k=0\\ p(k)=p(n)}}^{n} g_k(\sigma_1,\ldots,\sigma_k)u(t-\sigma_1)\cdots u(t-\sigma_k)\,d\sigma_1\cdots d\sigma_k$$

Then using the orthogonality property,

$$E[G_n[k_n,u(t+\tau)]G_n[k_n,u(t)]]$$
$$= E[\int_{-\infty}^{\infty} g_n(\sigma_1,\ldots,\sigma_n)u(t+\tau-\sigma_1)\cdots u(t+\tau-\sigma_n)\,d\sigma_1\cdots d\sigma_n\ G_n[k_n,u(t)]]$$
$$= \sum_{\substack{k=0\\ p(k)=p(n)}}^{n} \int_{-\infty}^{\infty} g_n(\sigma_1,\ldots,\sigma_n)g_k(\tau_1,\ldots,\tau_k)E[u(t+\tau-\sigma_1)$$
$$\cdots u(t+\tau-\sigma_n)u(t-\tau_1)\cdots u(t-\tau_k)]\,d\sigma_1\cdots d\sigma_n d\tau_1\cdots d\tau_k \qquad (129)$$

In the kth summand the expected value will contain a sum of products of $(n+k)/2$ impulses, and the argument of each impulse will be a difference of a pair of arguments chosen from $(t+\tau-\sigma_1),\ldots,(t+\tau-\sigma_n)$, $(t-\tau_1),\ldots,(t-\tau_k)$.

First consider the $k = n$ summand, and specifically those terms of the expected value wherein every impulse in the product has one of the σ_i variables and one of the τ_i variables in its argument. There will be $n!$ such products in the expected value, n choices of the τ_i variable to pair with the first σ_i variable, $n-1$ choices of the τ_i variable to pair with the second σ_i variable, and so on. (Since the product is over unordered pairs, the ordering of the σ_i's is immaterial.) Since $g_n(t_1,\ldots,t_n)$ is symmetrical, when the impulses are integrated out, each of the resulting terms will be identical. Thus this portion of the contribution of the $k = n$ term can be written as

$$n!A^n \int_{-\infty}^{\infty} g_n(\sigma_1+\tau,\ldots,\sigma_n+\tau)g_n(\sigma_1,\ldots,\sigma_n)\,d\sigma_1\cdots d\sigma_n$$

Of course, in the original notation this is precisely (128), and so the remainder of the argument is devoted to showing that all other terms in (129) are zero. The remaining products of impulses in the $k = n$ summand will contain at least one impulse with an argument that is a difference of two of the σ_i variables. This feature is shared by all the $k < n$ terms in (129) since there will be at least two more σ_i variables than τ_i variables. This kind of term will be discussed now for the general $k \leqslant n$ summand.

First, for fixed k in (129) consider each product-of-impulses term in the expected value that contains $\delta_0(\sigma_{n-1}-\sigma_n)$ as a factor. Using the key fact noted in the proof of Theorem 5.1, this collection of terms can be written as

$$A\delta_0(\sigma_{n-1}-\sigma_n)\; E[u(t+\tau-\sigma_1)\cdots u(t+\tau-\sigma_{n-2})u(t-\tau_1)\cdots u(t-\tau_k)]$$

Collecting these terms in (129) for each value of k, and integrating with respect to σ_n, gives that their contribution to (129) is

$$\sum_{\substack{k=0\\p(k)=p(n)}}^{n} A\int_{-\infty}^{\infty} g_n(\sigma_1,\ldots,\sigma_{n-2},\sigma_{n-1},\sigma_{n-1})g_k(\tau_1,\ldots,\tau_k)\; E[u(t+\tau-\sigma_1)$$

$$\cdots u(t+\tau-\sigma_{n-2})u(t-\tau_1)\cdots u(t-\tau_k)]\; d\sigma_1\cdots d\sigma_{n-1}d\tau_1\cdots d\tau_k$$

$$= A\; E[\int_{-\infty}^{\infty} g_n(\sigma_1,\ldots,\sigma_{n-2},\sigma_{n-1},\sigma_{n-1})u(t+\tau-\sigma_1)\cdots u(t+\tau-\sigma_{n-2})\; d\sigma_1\cdots d\sigma_{n-1}$$

$$\sum_{\substack{k=0\\p(k)=p(n)}}^{n} g_k(\tau_1,\ldots,\tau_k)u(t-\tau_1)\cdots u(t-\tau_k)\; d\tau_1\cdots d\tau_k]$$

$$= A\; E[\int_{-\infty}^{\infty} g_n(\sigma_1,\ldots,\sigma_{n-2},\sigma_{n-1},\sigma_{n-1})d\sigma_{n-1}u(t+\tau-\sigma_1)$$

$$\cdots u(t+\tau-\sigma_{n-2})\; d\sigma_1\cdots d\sigma_{n-2}\; G_n[k_n,u(t)]]$$

$$= 0$$

by the orthogonality property of $G_n[k_n,u(t)]$.

Now, for fixed k in (129), consider each product-of-impulses term in the expected value that contains $\delta_0(\sigma_{n-3}-\sigma_{n-2})$ as a factor, but that does *not* contain $\delta_0(\sigma_{n-1}-\sigma_n)$ as a factor. The collection of such terms can be written as the set of terms that contain a factor of the form $\delta_0(\sigma_{n-3}-\sigma_{n-2})$, minus the set of terms that contain a factor of $\delta_0(\sigma_{n-3}-\sigma_{n-2})\delta_0(\sigma_{n-1}-\sigma_n)$. That is,

$$A\delta_0(\sigma_{n-3}-\sigma_{n-2})\; E[u(t+\tau-\sigma_1)\cdots u(t+\tau-\sigma_{n-4})u(t+\tau-\sigma_{n-1})u(t+\tau-\sigma_n)$$

$$u(t-\tau_1)\cdots u(t-\tau_k)] - A^2\delta_0(\sigma_{n-3}-\sigma_{n-2})\delta_0(\sigma_{n-1}-\sigma_n)\; E[u(t+\tau-\sigma_1)$$

$$\cdots u(t+\tau-\sigma_{n-4})u(t-\tau_1)\cdots u(t-\tau_k)]$$

Collecting these terms in (129) for each value of k, and integrating out the common impulse factors, gives zero as follows. All terms corresponding to

$$A\delta_0(\sigma_{n-3}-\sigma_{n-2})E[u(t+\tau-\sigma_1)\cdots u(t+\tau-\sigma_{n-4})u(t+\tau-\sigma_{n-1})u(t+\tau-\sigma_n)]$$

give zero by the argument in the previous case. In a very similar fashion, the remaining set of terms gives

$$\sum_{\substack{k=0\\p(k)=p(n)}}^{n} -A^2 \int_{-\infty}^{\infty} g_n(\sigma_1,\ldots,\sigma_{n-4},\sigma_{n-3},\sigma_{n-3},\sigma_{n-1},\sigma_{n-1}) g_k(\tau_1,\ldots,\tau_k)$$

$$E[u(t+\tau-\sigma_1)\cdots u(t+\tau-\sigma_{n-4})u(t-\tau_1)\cdots u(t-\tau_k)]\, d\sigma_1\cdots d\sigma_{n-4}d\tau_1\cdots d\tau_k$$

$$= -A^2\, E[\int_{-\infty}^{\infty} g_n(\sigma_1,\ldots,\sigma_{n-4},\sigma_{n-3},\sigma_{n-3},\sigma_{n-1},\sigma_{n-1}) u(t+\tau-\sigma_1)$$

$$\cdots u(t+\tau-\sigma_{n-4})\, d\sigma_1\cdots d\sigma_{n-4}\; G_n[k_n,u(t)]]$$

$$= 0$$

The remainder of the proof continues in just this way, with the next step being to consider those terms in (129) that contain $\delta_0(\sigma_{n-5}-\sigma_{n-4})$ as a factor, but not $\delta_0(\sigma_{n-3}-\sigma_{n-2})$ or $\delta_0(\sigma_{n-1}-\sigma_n)$. It is left to the reader to compute the calculation of all the zeros.

Methods for determining the Wiener kernels $k_n(t_1,\ldots,t_n)$ for an unknown system will be a major topic in Chapter 7. However, one way to find the Wiener kernels for a known system is to establish the relationship between the Wiener kernels and the Volterra kernels. Suppose a system is described by the Wiener orthogonal representation, and also by the symmetric-kernel Volterra series representation

$$y(t) = \sum_{n=0}^{\infty} \int_{-\infty}^{\infty} h_{nsym}(\sigma_1,\ldots,\sigma_n) u(t-\sigma_1)\cdots u(t-\sigma_n)\, d\sigma_1\cdots d\sigma_n \tag{130}$$

Of course, it should be noted that strong convergence properties of both representations are required before the two can be related. Thus, the following developments are fine for the polynomial system case, but must be qualified by convergence hypotheses to be taken as rigorous in the infinite series case.

Theorem 5.3 Suppose a system is described by the Wiener orthogonal representation (110), (122), and by the symmetric Volterra system representation (130). Then the degree-N symmetric Volterra kernel is given by

$$h_{Nsym}(t_1,\ldots,t_N) = \sum_{j=0}^{\infty} \frac{(-1)^j (N+2j)!A^j}{N!j!2^j} \int_{-\infty}^{\infty} k_{N+2j}(t_1,\ldots,t_N,\tau_1,\tau_1,$$

$$\ldots,\tau_j,\tau_j)\, d\tau_1\cdots d\tau_j \tag{131}$$

Proof From (110) and (122), the Wiener representation for the system is

$$y(t) = \sum_{n=0}^{\infty}\sum_{m=0}^{[n/2]} \frac{(-1)^m n!A^m}{(n-2m)!m!2^m} \int_{-\infty}^{\infty} k_n(\sigma_1,\ldots,\sigma_{n-2m},\tau_1,\tau_1,\ldots,\tau_m,\tau_m)$$

$$d\tau_1 \cdots d\tau_m u(t-\sigma_1) \cdots u(t-\sigma_{n-2m})\, d\sigma_1 \cdots d\sigma_{n-2m}$$

To find an expression for $h_{Nsym}(t_1,\ldots,t_N)$, all terms of degree N must be extracted. These terms are precisely those with $n-2m = N$. Supposing first that N is even, if $n-2m = N$ then it is clear that n must be even, and $n \geqslant N$. Thus the degree-N terms in the Wiener representation are given by

$$\sum_{\substack{n=N \\ n\ even}}^{\infty} \frac{(-1)^{(n-N)/2} n!A^{(n-N)/2}}{N!((n-N)/2)!2^{(n-N)/2}} \int_{-\infty}^{\infty} k_n(\sigma_1,\ldots,\sigma_N,\tau_1,\tau_1,\ldots,\tau_{\frac{n-N}{2}},\tau_{\frac{n-N}{2}})$$

$$d\tau_1 \cdots d\tau_{\frac{n-N}{2}} u(t-\sigma_1) \cdots u(t-\sigma_N)\, d\sigma_1 \cdots d\sigma_N$$

To put this in a neater form, change the summation index n to $j = (n-N)/2$. This gives

$$\sum_{j=0}^{\infty} \frac{(-1)^j (N+2j)!A^j}{N!j!2^j} \int_{-\infty}^{\infty} k_{N+2j}(\sigma_1,\ldots,\sigma_N,\tau_1,\tau_1,\ldots,\tau_j,\tau_j)$$

$$d\tau_1 \cdots d\tau_j u(t-\sigma_1) \cdots u(t-\sigma_N)\, d\sigma_1 \cdots d\sigma_N$$

A very similar development for the case of N odd leads to exactly the same expression for the degree-N terms in the Wiener representation. Thus, it is clear that (131) gives the symmetric Volterra kernel for the system.

To express the Wiener kernels in terms of the symmetric Volterra kernels is a messier task. The approach used in the following proof is to write out (131) for the symmetric Volterra kernels $h_{Nsym}(t_1,\ldots,t_N)$, $h_{(N+2)sym}(t_1,\ldots,t_N,\sigma_1,\sigma_1)$, $h_{(N+4)sym}(t_1,\ldots,t_N,\sigma_1,\sigma_1,\sigma_2,\sigma_2)$, and so on. Then by tedious inspection it becomes clear that the Wiener kernel $k_N(t_1,\ldots,t_N)$ can be isolated using these expressions.

Theorem 5.4 Suppose a system is described by the symmetric Volterra system representation (130), and by the Wiener orthogonal representation (110), (122). Then the degree-N Wiener kernel is given by

$$k_N(t_1,\ldots,t_N) = \sum_{j=0}^{\infty} \frac{(N+2j)!A^j}{N!j!2^j} \int_{-\infty}^{\infty} h_{(N+2j)sym}(t_1,\ldots,t_N,\sigma_1,\sigma_1,$$

$$\ldots,\sigma_j,\sigma_j)\, d\sigma_1 \cdots d\sigma_j \tag{132}$$

Proof For convenience, let

$$a(N,j) = \frac{(N+2j)!A^j}{N!j!2^j}$$

Then (131) can be written as

$$h_{Nsym}(t_1,\ldots,t_N) = k_N(t_1,\ldots,t_N)$$

$$+ \sum_{j=1}^{\infty} (-1)^j a(N,j) \int_{-\infty}^{\infty} k_{N+2j}(t_1,\ldots,t_N,\tau_1,\tau_1,\ldots,\tau_j,\tau_j)\, d\tau_1 \cdots d\tau_j$$

and for $k \geqslant 1$,

$$h_{(N+2k)sym}(t_1,\ldots,t_N,\sigma_1,\sigma_1,\ldots,\sigma_k,\sigma_k) = k_{N+2k}(t_1,\ldots,t_N,\sigma_1,\sigma_1,\ldots,\sigma_k,\sigma_k)$$

$$+ \sum_{j=1}^{\infty} (-1)^j a(N+2k,j) \int_{-\infty}^{\infty} k_{N+2k+2j}(t_1,\ldots,t_N,\sigma_1,\sigma_1,$$

$$\ldots,\sigma_k,\sigma_k,\tau_1,\tau_1,\ldots,\tau_j,\tau_j)\, d\tau_1 \cdots d\tau_j$$

Using some elementary manipulations and variable relabelings gives

$$\int_{-\infty}^{\infty} h_{(N+2k)sym}(t_1,\ldots,t_N,\tau_1,\tau_1,\ldots,\tau_k,\tau_k)\, d\tau_1 \cdots d\tau_k$$

$$= \sum_{i=k}^{\infty} (-1)^{i-k} a(N+2k,i-k) \int_{-\infty}^{\infty} k_{N+2i}(t_1,\ldots,t_N,\tau_1,\tau_1,\ldots,\tau_i,\tau_i)\, d\tau_1 \cdots d\tau_i$$

Now, the right side of (132) can be written as

$$h_{Nsym}(t_1,\ldots,t_N) + \sum_{k=1}^{\infty} a(N,k) \int_{-\infty}^{\infty} h_{(N+2k)sym}(t_1,\ldots,t_N,\tau_1,\tau_1,\ldots,\tau_k,\tau_k)\, d\tau_1 \cdots d\tau_k$$

$$= k_N(t_1,\ldots,t_N) + \sum_{j=1}^{\infty} (-1)^j a(N,j) \int_{-\infty}^{\infty} k_{N+2j}(t_1,\ldots,t_N,\tau_1,\tau_1,\ldots,\tau_j,\tau_j)\, d\tau_1 \cdots d\tau_j$$

$$+ \sum_{k=1}^{\infty} a(N,k) \sum_{i=k}^{\infty} (-1)^{i-k} a(N+2k,i-k)$$

$$\int_{-\infty}^{\infty} k_{N+2i}(t_1,\ldots,t_N,\tau_1,\tau_1,\ldots,\tau_i,\tau_i)\, d\tau_1 \cdots d\tau_i$$

For the general term of the form

$$\int_{-\infty}^{\infty} k_{N+2q}(t_1,\ldots,t_N,\tau_1,\tau_1,\ldots,\tau_q,\tau_q)\, d\tau_1 \cdots d\tau_q$$

with $q \geqslant 1$, the coefficient is

$$(-1)^q a(N,q) + \sum_{k=1}^{q} (-1)^{q-k} a(N,k) a(N+2k,q-k)$$

Substituting the definition of $a(N,j)$ and using simple identities shows that this coefficient is 0. Thus the proof is complete.

Example 5.14 Consider the degree-3 polynomial system shown in Figure 5.2.

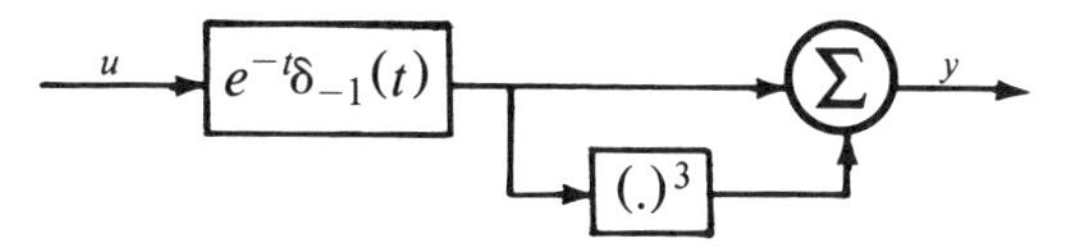

Figure 5.2. A degree-3 polynomial system.

The Wiener kernels are easily computed from the symmetric Volterra kernels

$$h_1(t) = e^{-t}\delta_{-1}(t)$$
$$h_{2sym}(t_1,t_2) = 0$$
$$h_{3sym}(t_1,t_2,t_3) = e^{-t_1}e^{-t_2}e^{-t_3}\delta_{-1}(t_1)\delta_{-1}(t_2)\delta_{-1}(t_3)$$

Using (132), there are only two nonzero Wiener kernels:

$$k_1(t) = (1 + \frac{3A}{2})e^{-t}\delta_{-1}(t)$$
$$k_3(t_1,t_2,t_3) = e^{-t_1}e^{-t_2}e^{-t_3}\delta_{-1}(t_1)\delta_{-1}(t_2)\delta_{-1}(t_3)$$

5.6 Remarks and References

Remark 5.1 An introductory discussion of the impulse response of a homogeneous system can be found in

M. Brilliant, "Theory of the Analysis of Nonlinear Systems," MIT RLE Technical Report No. 345, 1958 (AD 216-209).

One area in which impulse inputs are of general interest is in impulse sampler models for sampled data systems. This subject is discussed in

A. Bush, "Some Techniques for the Synthesis of Nonlinear Systems," MIT RLE Technical Report No. 441, 1966 (AD 634-122).

Sampled-data and discrete-time systems will be treated in Chapter 6.

Remark 5.2 The steady-state response of a homogeneous system to a sinusoidal input is briefly discussed in

D. George, "Continuous Nonlinear Systems," MIT RLE Technical Report No. 355, 1959 (AD 246-281).

and in

J. Barrett, "The Use of Functionals in the Analysis of Nonlinear Physical Systems," *Journal of Electronics and Control,* Vol. 15, pp.567-615, 1963.

The steady-state response of a nonlinear system to single- and multi-tone sinusoidal inputs, and the use of this response in electronic circuit analysis is the subject of

J. Bussgang, L. Ehrman, J. Graham, "Analysis of Nonlinear Systems with Multiple Inputs," *Proceedings of the IEEE,* Vol. 62, pp.1088-1119, 1974.

Many interesting examples and experimental results are included in this paper, and in other referenced reports by the authors. General formulas similar to those of Section 5.3 for the response to multi-tone inputs are derived in

E. Bedrosian, S. Rice, "The Output Properties of Volterra Systems (Nonlinear Systems with Memory) Driven by Harmonic and Gaussian Inputs," *Proceedings of the IEEE,* Vol. 59, pp.1688-1707, 1971.

For a recent book-length treatment at a more elementary level, see

D. Weiner, J. Spina, *Sinusoidal Analysis and Modeling of Weakly Nonlinear Circuits,* Van Nostrand Reinhold, New York, 1980.

Remark 5.3 The well known Manley-Rowe formulas are closely related to the topic of steady-state frequency response of a Volterra system. As originally developed in 1956, these formulas describe the constraints on power flow at various frequencies in a nonlinear capacitor. Since that time the formulas have been generalized to apply to a large class of nonlinear systems. See the following two books, the first of which is the more elementary.

R. Clay, *Nonlinear Networks and Systems,* John Wiley, New York, 1971.

P. Penfield, *Frequency-Power Formulas,* MIT Press, Cambridge, Massachusetts, 1960.

Remark 5.4 The response of a nonlinear system to random inputs was the topic in which N. Wiener first used the Volterra representation for nonlinear systems. Wiener's approach will be discussed further in Chapter 7 in the context of system identification. Material similar to that discussed in Section 5.4 is introduced in the report by George and the paper by Barrett cited above. The approach I have used follows

M. Rudko, D. Weiner, "Volterra Systems with Random Inputs: A Formalized Approach," *IEEE Transactions on Communications,* Vol. COM-26, pp.217-227, 1978 (Addendum: Vol. COM-27, pp. 636-638, 1979).

This paper presents a derivation of the general formula for the cross-spectral density given in (91), and a general reduction of the partial output power spectral density formula in (108).

A different approach can be found in the paper by Bedrosian and Rice mentioned in Remark 5.2. More general inputs such as a sum of sinusoidal and Gaussian signals also are considered by Bedrosian and Rice. Many related papers, and reprints of some papers cited above can be found in the collection

A. Haddad, ed., *Nonlinear Systems: Processing of Random Signals - Classical Analysis,* Benchmark Papers, Vol. 10, Dowden, Hutchinson, and Ross, Stroudsberg, Pennsylvania, 1975.

A derivation of the formula for the order-n autocorrelation of a Gaussian random process can be found in

M. Schetzen, *The Volterra and Wiener Theories of Nonlinear Systems,* John Wiley, New York, 1980.

J. Laning, R. Battin, *Random Processes in Automatic Control,* McGraw-Hill, New York, 1956.

Remark 5.5 A very important kind of Volterra system is a feedback system involving linear dynamic subsystems, and a static nonlinear element. When a Gaussian input process is applied to such a system, the description of the output statistics in terms of the given subsystems is a difficult problem. One approach, the so-called quasi-functional method, is discussed in

H. Smith, *Approximate Analysis of Randomly Excited Nonlinear Controls,* MIT Press, Cambridge, Massachusetts, 1966.

Remark 5.6 The idea of using orthogonal representations for the response of a nonlinear system to a random input has been investigated from a number of different, though closely related, viewpoints. The work of Wiener on nonlinear systems dates back to a report that is difficult to obtain.

N. Wiener, "Response of a Nonlinear Device to Noise," MIT Radiation Laboratory Report No. 165, 1942.

The most readily available account by Wiener is in a book of transcribed lectures:

N. Wiener, *Nonlinear Problems in Random Theory,* MIT Press, Cambridge, Massachusetts, 1958.

The basic mathematical paper

R. Cameron, W. Martin, "The Orthogonal Development of Nonlinear Functionals in Series of Fourier-Hermite Functionals," *Annals of Mathematics,* Vol. 48, pp. 385-392, 1947.

develops the representation of a nonlinear functional acting on a white Gaussian process by using the orthogonality properties of Hermite polynomials. A related representation, with emphasis on nonlinear system theory, is discussed in the paper by Barrett cited in Remark 5.2, and in

J. Barrett, "Hermite Functional Expansions and the Calculation of Output Autocorrelation and Spectrum for any Time-Invariant Nonlinear System with Noise Input," *Journal of Electronics and Control,* Vol. 16, pp. 107-113, 1964.

Informative reviews of many aspects of these early contributions can be found in

L. Zadeh, "On the Representation of Nonlinear Operators," *IRE Wescon Convention Record,* Part 2, pp. 105-113, 1957.

W. Root, "On System Measurement and Identification," in *System Theory,* Polytechnic Institute of Brooklyn Symposium Proceedings, Vol. 15, Polytechnic Press, New York, pp. 133-157, 1965.

R. Deutsch, *Nonlinear Transformations of Random Processes,* Prentice-Hall, Englewood Cliffs, New Jersey, 1962.

A more recent paper on the Wiener representation is

G. Palm, T. Poggio, "The Volterra Representation and the Wiener Expansion: Validity and Pitfalls," *SIAM Journal on Applied Mathematics,* Vol. 33, pp. 195-216, 1977.

This paper contains analyses of a number of delicate technical issues, particularly convergence properties. Finally, the Wiener representation is discussed in detail at an introductory level in

M. Schetzen, *The Volterra and Wiener Theories of Nonlinear Systems,* John Wiley, New York, 1980.

Remark 5.7 Orthogonal representations have been developed for nonlinear functionals of Poisson and other, more general, stochastic processes. See

H. Ogura, "Orthogonal Functionals of the Poisson Process," *IEEE Transactions on Information Theory,* Vol. IT-18, pp. 473-480, 1972.

A. Segall, T. Kailath, "Orthogonal Functionals of Independent-Increment Processes," *IEEE Transactions on Information Theory,* Vol. IT-22, pp. 287-298, 1976.

A general framework for orthogonal expansions is discussed in

S. Yasui, "Stochastic Functional Fourier Series, Volterra Series, and Nonlinear Systems Analysis," *IEEE Transactions on Automatic Control,* Vol. AC-24, pp. 230-241, 1979.

The reader should be forewarned that further pursuit of the topics of Section 5.5 leads quickly into the theory of stochastic integrals and other stratospheric mathematical tools. The reference list of any of the papers cited above can serve as a launch pad.

5.7 Problems

5.1. For a degree-n homogeneous system described in terms of the regular kernel

$$y(t) = \int_0^\infty h_{reg}(\sigma_1, \ldots, \sigma_n) u(t-\sigma_1-\cdots-\sigma_n) \cdots u(t-\sigma_n)\, d\sigma_1 \cdots d\sigma_n$$

find expressions for the response to the inputs $u(t) = \delta_0(t)$ and $u(t) = \delta_0(t) + \delta_0(t-T)$, $T > 0$.

5.2. For the system shown below, find an expression for the steady-state response to $u(t) = 2A\cos(\omega t)$ in terms of the subsystem transfer functions.

$$u \to \boxed{H_0(s)} \to \boxed{(.)^n} \to \boxed{H_1(s)} \to y$$

5.3. Show that the steady-state response of a degree-n homogeneous system to the input $u(t) = B + 2A\cos(\omega t)$ can be written in the form

$$y_{ss}(t) = \sum_{k=0}^{n} \binom{n}{k} B^{n-k} A^k \sum_{j=0}^{k} \binom{k}{j} H_{nsym}(\underbrace{0, \ldots, 0}_{n-k}; \underbrace{i\omega, \ldots, i\omega}_{j}; \underbrace{-i\omega, \ldots, -i\omega}_{k-j})\, e^{i(2j-k)\omega t}$$

Notice that all of the first n harmonics of the input frequency are included, not just those of the same parity as n.

5.4. Derive a formula for the steady-state response of a degree-n homogeneous system to a unit step function in terms of $H_{sym}(s_1, \ldots, s_n)$ and in terms of $H_{reg}(s_1, \ldots, s_n)$.

5.5. Find a necessary and sufficient condition that the frequencies in the two-tone steady-state response formula for a Volterra system be distinct for distinct values of M and N.

5.6. Suppose that the input to a degree-2 homogeneous system is a real, stationary, zero-mean, white Gaussian random process. Show that

$$R_{yy}(\tau) = \int_{-\infty}^{\infty}\int_{-\infty}^{\infty} h_{sym}(\sigma_1, \sigma_1) h_{sym}(\sigma_2, \sigma_2)\, d\sigma_1 d\sigma_2$$

$$+ 2\int_{-\infty}^{\infty}\int_{-\infty}^{\infty} h_{sym}(\tau+\sigma_1, \tau+\sigma_2) h_{sym}(\sigma_1, \sigma_2)\, d\sigma_1 d\sigma_2$$

5.7. Suppose that $u(t)$ is a stationary random input to a stationary, degree-n, homogeneous system. Show that if the input is applied at $t = -\infty$, then the output random process is stationary.

5.8. For a stationary, zero-mean, Gaussian random process $u(t)$ with autocorrelation $R_{uu}(\tau)$, give an expression for $R_{uu}^{(6)}(t_1, \ldots, t_6)$.

5.9. Express the result of Example 5.8 in terms of the symmetric system function.

5.10. Consider the modulation system diagramed below for the case where the message signal is $u(t) = A_m\cos(\omega_m t)$, $t \geqslant 0$.

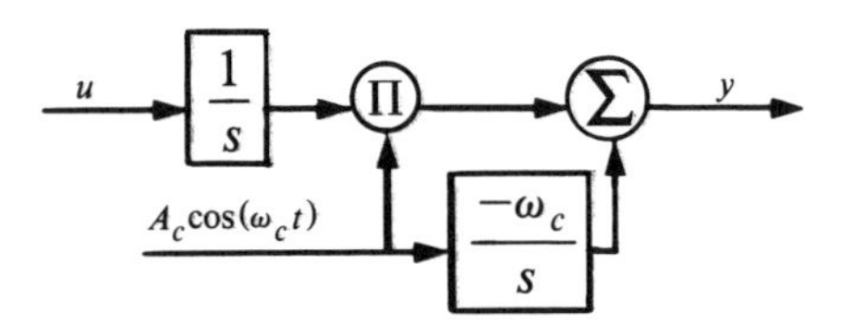

Show that

$$y(t) = A_c\cos(\omega_c t) + \frac{A_cA_m}{2\omega_m}\cos[(\omega_c+\omega_m)t] - \frac{A_cA_m}{2\omega_m}\cos[(\omega_c-\omega_m)t] - 1$$

Neglecting the constant term, when $A_m/A_c << 1$ this is called a narrow-band FM signal with sinusoidal modulation. Show that the FM modulation system in Example 3.1 also can be used to generate such a signal.

5.11. Show that (83) can be written in the form

$$S_{yu}(\omega) = \frac{1}{(2\pi)^n}\int_{-\infty}^{\infty} \hat{S}_{yu}(\omega_1-\alpha_2,\alpha_2-\alpha_3,\ldots,\alpha_n-\alpha_{n+1},\alpha_{n+1})\, d\alpha_2 \cdots d\alpha_{n+1}$$

and interpret this as an association-of-variables expression.

5.12. Suppose the input to a degree-n homogeneous system is real, stationary, zero-mean, Gaussian noise with autocorrelation $R_{uu}(\tau)$. Find an expression for the cross-correlation $R_{yu}(\tau)$ in terms of the symmetric kernel.

5.13. Suppose a two-tone input is applied to a degree-n homogeneous system. Show that in the steady-state response the number of frequency terms $e^{i[M\omega_1+N\omega_2]t}$, ignoring complex conjugates, is given by the number of integers M and N satisfying

$$\textit{parity}\ |M+N| = \textit{parity}\ n,\quad |M|+|N| \leqslant n$$

5.14. Use (132) to derive the following relationship between the *Wiener system function* $K_N(\omega_1,\ldots,\omega_N) = F[k_N(t_1,\ldots,t_N)]$ and the symmetric (Volterra) system functions $H_{nsym}(\omega_1,\ldots,\omega_n)$.

$$K_N(\omega_1,\ldots,\omega_N) = \sum_{j=0}^{\infty} \frac{(N+2j)!A^j}{N!j!2^j(2\pi)^j}$$

$$\int_{-\infty}^{\infty} H_{(N+2j)sym}(\omega_1,\ldots,\omega_N,\gamma_1,-\gamma_1,\ldots,\gamma_j,-\gamma_j)\, d\gamma_1 \cdots d\gamma_j$$

(Hint: Follow the familiar stategy of inverse transform, manipulation, transform.)

5.15. For the bilinear state equation

$$\dot{x}(t) = Ax(t) + Dx(t)u(t) + bu(t)$$

$$y(t) = cx(t)$$

show that if $D^2 = 0$ and $Db = 0$, then the Wiener kernels for the system are identical to the symmetric Volterra kernels.

5.16 Prove that $E[G_n[k_n,u(t+\tau)]G_n[k_n,u(t)]]$ in (128) can be written in terms of the Wiener system function (Problem 5.14) as

$$E[G_n[k_n,u(t+\tau)]G_n[k_n,u(t)]] = \frac{n!A^n}{(2\pi)^n} \int_{-\infty}^{\infty} |K_n(\omega_1,\ldots,\omega_n)|^2 e^{i(\omega_1+\cdots+\omega_n)\tau}\, d\omega_1 \cdots d\omega_n$$

5.17 Suppose a Volterra system is described by the regular kernels $h_{nreg}(t_1,\ldots,t_n)$, $n = 1,2,\ldots$. For a Gaussian white noise input with intensity P, show that the output expectation is

$$E[y(t)] = \sum_{n=1}^{\infty} P^n \int_0^{\infty} h_{2nreg}(0,\sigma_1,0,\sigma_2,\ldots,0,\sigma_n)\, d\sigma_1 \cdots d\sigma_n$$

5.18 Suppose the bilinear state equation (A,D,b,c,R^m) is driven by Gaussian white noise with intensity P. Use Problem 5.17 to show that

$$E[y(t)] = -Pc(A + PD^2)^{-1}Db$$

Discuss the conditions under which this result is meaningful.

CHAPTER 6

Discrete-Time Systems

Most of the nonlinear system theory that has been discussed so far for continuous-time systems can be developed for discrete-time systems. There are differences, of course, but these mostly are differences in technical detail or interpretation of the results. The situation is similar to the linear case where the continuous- and discrete-time theories look much the same.

In this chapter I will discuss briefly the salient features of Volterra series methods for discrete-time nonlinear systems. For simplicity only stationary systems will be considered. Special attention will be devoted to points where the discrete- and continuous-time theories differ, and much of the rather routine transcription of results will be left to the reader. In addition, two new classes of systems will be discussed: bilinear input/output systems, and two-dimensional linear systems. While the general classes of multilinear input/output systems and multidimensional linear systems are of interest in their own right, the simplest cases of each are introduced here to demonstrate the similarity in representations and analysis methods to the now familiar class of homogeneous systems.

6.1 Input/Output Representations in the Time Domain

Consider a discrete-time system representation of the form

$$y(k) = \sum_{i_1=0}^{\infty} \cdots \sum_{i_n=0}^{\infty} h(i_1,\ldots,i_n)u(k-i_1) \cdots u(k-i_n), \quad k = 0,1,2,\ldots \tag{1}$$

The input signal $u(k)$ and output signal $y(k)$ are real sequences that are assumed to be zero for $k < 0$. The kernel $h(i_1,\ldots,i_n)$ is real, and equal to zero if any argument is negative. It is a simple matter to verify that a system described by (1) is *stationary, causal,* and *degree-n homogeneous.* The upper limits on the summations can be lowered to k, but infinite upper limits are retained for notational simplicity.

Since for any k, $y(k)$ in (1) is given by a finite summation, there is no need even to mention technical hypotheses. In other words, issues like continuity and integrability in the continuous-time case do not arise in regard to the representation in (1). Also, notice that direct transmission terms are explicitly displayed in (1), and there is no need to consider impulsive kernels. For example, if

$$h(i_1,\ldots,i_n) = \begin{cases} 1, & i_1 = \cdots = i_n = 0 \\ 0, & otherwise \end{cases}$$

then the system can be written in the form

$$y(k) = u^n(k), \quad k = 0,1,2,\ldots$$

The familiar sum-over-permutations argument shows that the kernel in (1) can be replaced by the *symmetric kernel*

$$h_{sym}(i_1,\ldots,i_n) = \frac{1}{n!} \sum_{\pi(.)} h(i_{\pi(1)},\ldots,i_{\pi(n)}) \tag{2}$$

without loss of generality. (Recall that the summation is over all $n!$ permutations of $1,2,\ldots,n$.) From the symmetric kernel representation, a *triangular kernel* can be defined. However, some care is required because it cannot be argued that values of the kernel at particular arguments do not contribute to the sum, as was done for the integral of nonimpulsive kernels in the continuous-time case. That is, the values of the triangular kernel at boundary points of the triangular domain must be adjusted appropriately. One way to do this adjustment is to use the notation

$$h_{tri}(i_1,\ldots,i_n) = h_{sym}(i_1,\ldots,i_n)\hat{\delta}_{-1}(i_1-i_2,i_2-i_3,\ldots,i_{n-1}-i_n) \tag{3}$$

where the special *multivariable step function* is defined by

$$\hat{\delta}_{-1}(i_1,\ldots,i_{n-1}) = \begin{cases} 0, & \text{if } any\ i_j < 0 \\ 1, & i_1 = \cdots = i_{n-1} = 0 \\ \vdots & \\ \dfrac{n!}{m_1! \cdots m_j!}, & \begin{cases} i_1 = \cdots = i_{m_1-1} = 0, \ldots, \\ i_{j+1} = \cdots = i_{j+m_j-1} = 0 \end{cases} \\ \vdots & \\ n!, & i_1,\ldots,i_{n-1} > 0 \end{cases} \tag{4}$$

It is easy to verify that when $n = 2$ this setup yields consistent results in going from the symmetric kernel to the triangular kernel using (3), and then from the triangular kernel back to the symmetric kernel using (2). The higher-degree cases are less easy, but still straightforward. The uncircumflexed notation will be retained for the more traditional step function:

$$\delta_{-1}(k) = \begin{cases} 1, & k = 0,1,2,\ldots \\ 0, & k < 0 \end{cases}$$

The third special form is the *regular kernel* representation. Starting with the triangular kernel representation,

$$y(k) = \sum_{i_1=0}^{\infty} \cdots \sum_{i_n=0}^{\infty} h_{tri}(i_1,\ldots,i_n)u(k-i_1) \cdots u(k-i_n) \tag{5}$$

a simple change of variables argument gives

$$y(k) = \sum_{i_1=0}^{\infty} \cdots \sum_{i_n=0}^{\infty} h_{reg}(i_1,\ldots,i_n)u(k-i_1- \cdots -i_n) \tag{6}$$

$$u(k-i_2- \cdots -i_n) \cdots u(k-i_n)$$

where

$$\begin{aligned} h_{reg}(i_1,\ldots,i_n) &= h_{tri}(i_1+ \cdots +i_n,i_2+ \cdots +i_n,\ldots,i_n) \\ &= h_{sym}(i_1+ \cdots +i_n,i_2+ \cdots +i_n,\ldots,i_n)\hat{\delta}_{-1}(i_1,\ldots,i_{n-1}) \end{aligned} \tag{7}$$

Notice again that the upper limits of the summations in (5) and (6) can be replaced by finite quantities. But that makes the notation more complicated, so just as in the continuous-time case the infinities are used.

Although only stationary systems will be considered, general representations of the form

$$y(k) = \sum_{i_1=0}^{k} \cdots \sum_{i_n=0}^{k} h(k,i_1,\ldots,i_n)u(i_1) \cdots u(i_n) \tag{8}$$

will arise. It is natural to follow the continuous-time case and call a kernel $h(k,i_1,\ldots,i_n)$ *stationary* if

$$h(0,i_1-k,\ldots,i_n-k) = h(k,i_1,\ldots,i_n) \tag{9}$$

If this relationship holds, then setting

$$g(i_1,\ldots,i_n) = h(0,-i_1,\ldots,-i_n) \tag{10}$$

yields the representation

$$y(k) = \sum_{i_1=0}^{k} \cdots \sum_{i_n=0}^{k} g(k-i_1,\ldots,k-i_n)u(i_1) \cdots u(i_n) \tag{11}$$

which is equivalent to (8) since

$$g(k-i_1,\ldots,k-i_n) = h(0,i_1-k,\ldots,i_n-k) = h(k,i_1,\ldots,i_n) \tag{12}$$

A simple change of variables permits rewriting (11) in the form

$$y(k) = \sum_{j_1=0}^{k} \cdots \sum_{j_n=0}^{k} g(j_1,\ldots,j_n)u(k-j_1) \cdots u(k-j_n) \tag{13}$$

that is, in the form of (1).

With these basic representations in hand, the description of polynomial and Volterra systems is simply a matter of finite and infinite sums of homogeneous terms. Of course, the convergence issue becomes important for Volterra systems, but the basic approaches to convergence in the continuous-time case carry over directly. The topic of interconnections of discrete-time homogeneous, polynomial, or Volterra systems will not be discussed since the developments are easily transcribed from Section 1.4.

6.2 Input/Output Representations in the Transform Domain

For an n-variable function $f(i_1,\ldots,i_n)$ that is zero if any of the integers $i_1,\ldots,i_n$ is negative, that is, a *one-sided function,* the n-variable *z-transform* is defined by

$$F(z_1,\ldots,z_n) = Z[f(i_1,\ldots,i_n)]$$

$$= \sum_{i_1=0}^{\infty} \cdots \sum_{i_n=0}^{\infty} f(i_1,\ldots,i_n) z_1^{-i_1} \cdots z_n^{-i_n} \tag{14}$$

This can be viewed as a nonpositive power series in the complex variables $z_1,\ldots,z_n$, in which case convergence conditions must be included. However, for the functions that will be considered here (just as for the functions typically considered in discrete-time linear system theory) convergence regions always exist. Therefore, I will be very casual in this regard. Actually, (14) can be viewed as an algebraic object (formal series) in n indeterminates, in which case the question of convergence does not arise. While this, perhaps more sophisticated, viewpoint can be used to establish most of the results to be discussed, I will retain the more classical interpretation.

Example 6.1 Reminiscent of Example 2.1, consider the function

$$f(i_1,i_2) = i_1 - i_1\lambda^{-i_2}, \quad i_1,i_2 \geqslant 0$$

where λ is a constant. The z-transform of this function can be computed from the basic definition by writing

$$F(z_1,z_2) = \sum_{i_1=0}^{\infty}\sum_{i_2=0}^{\infty} (i_1-i_1\lambda^{-i_2}) z_1^{-i_1} z_2^{-i_2}$$

$$= \sum_{i_1=0}^{\infty}\sum_{i_2=0}^{\infty} i_1 z_1^{-i_1} z_2^{-i_2} - \sum_{i_1=0}^{\infty}\sum_{i_2=0}^{\infty} i_1\lambda^{-i_2} z_1^{-i_1} z_2^{-i_2}$$

$$= (\sum_{i_1=0}^{\infty} i_1 z_1^{-i_1})(\sum_{i_2=0}^{\infty} z_2^{-i_2}) - (\sum_{i_1=0}^{\infty} i_1 z_1^{-i_1})(\sum_{i_2=0}^{\infty} \lambda^{-i_2} z_2^{-i_2})$$

Summing each infinite series (or recalling single-variable z-transforms), it is clear that

$$F(z_1,z_2) = \frac{z_1}{(z_1-1)^2}\,\frac{z_2}{z_2-1} - \frac{z_1}{(z_1-1)^2}\,\frac{z_2}{z_2-\lambda}$$

$$= \frac{(1-\lambda)z_1z_2}{(z_1-1)^2(z_2-1)(z_2-\lambda)}$$

A careful look at the definition (14) and the calculations in Example 6.1 indicates immediately a couple of properties of the z-transform. These and other properties listed below are very similar in nature to properties of the Laplace transform, and the general proofs are easy. All functions are assumed to be one-sided, and the capital-letter notation is retained for the z-transform.

Theorem 6.1 The z-transform is linear:

$$Z[f(i_1,\ldots,i_n) + g(i_1,\ldots,i_n)] = F(z_1,\ldots,z_n) + G(z_1,\ldots,z_n)$$
$$Z[\alpha f(i_1,\ldots,i_n)] = \alpha F(z_1,\ldots,z_n), \quad \text{for } scalar\ \alpha \tag{15}$$

Theorem 6.2 If $f(i_1,\ldots,i_n)$ can be written as a product of two factors

$$f(i_1,\ldots,i_n) = h(i_1,\ldots,i_k)g(i_{k+1},\ldots,i_n) \tag{16}$$

then

$$F(z_1,\ldots,z_n) = H(z_1,\ldots,z_k)G(z_{k+1},\ldots,z_n) \tag{17}$$

Theorem 6.3 If $f(i_1,\ldots,i_n)$ is given by the single-variable convolution

$$f(i_1,\ldots,i_n) = \sum_{j=0}^{\infty} h(j)g(i_1-j,\ldots,i_n-j) \tag{18}$$

then

$$F(z_1,\ldots,z_n) = H(z_1 \cdots z_n)G(z_1,\ldots,z_n) \tag{19}$$

Theorem 6.4 If $f(i_1,\ldots,i_n)$ is given by the n-fold convolution

$$f(i_1,\ldots,i_n) = \sum_{j_1=0}^{\infty} \cdots \sum_{j_n=0}^{\infty} h(j_1,\ldots,j_n)g(i_1-j_1,\ldots,i_n-j_n) \tag{20}$$

then

$$F(z_1,\ldots,z_n) = H(z_1,\ldots,z_n)G(z_1,\ldots,z_n) \tag{21}$$

Theorem 6.5 If $I_1,\ldots,I_n$ are nonnegative integers, then

$$Z[f(i_1-I_1,\ldots,i_n-I_n)] = z_1^{-I_1} \cdots z_n^{-I_n} F(z_1,\ldots,z_n) \tag{22}$$

The basic formula for the inverse z-transform is a multivariable contour integration:

$$f(i_1,\ldots,i_n) = \frac{1}{(2\pi i)^n} \int_{\Gamma_n} \cdots \int_{\Gamma_1} F(z_1,\ldots,z_n) z_1^{i_1-1} \cdots z_n^{i_n-1}\, dz_1 \cdots dz_n \tag{23}$$

where each Γ_j is an appropriate contour in the z_j complex plane. For reasons that should be obvious, this formula is difficult to use. An alternative approach is to obtain the values of $f(i_1, \ldots, i_n)$ as the coefficients in the series expansion of $F(z_1, \ldots, z_n)$ into nonpositive powers of $z_1, \ldots, z_n$. If $F(z_1, \ldots, z_n)$ is a rational function, this series expansion can be simply a matter of division of the numerator polynomial by the denominator polynomial. But I should point out that some care must be exercised, because not every rational function corresponds to a z-transform. The distinction is that a z-transform must correspond to a nonpositive power series. Requiring a rational function to be proper or strictly proper is not a remedy.

Example 6.2 The rational function

$$F(z_1,z_2) = \frac{1}{z_1+z_2}$$

is not a z-transform since division gives

$$\frac{1}{z_1+z_2} = z_1^{-1} - z_1^{-2}z_2 + z_1^{-3}z_2^2 - \cdots$$

or

$$\frac{1}{z_1+z_2} = z_2^{-1} - z_1z_2^{-2} - z_1^2z_2^{-3} - \cdots$$

neither of which can be written as a nonpositive power series. On the other hand,

$$F(z_1,z_2) = \frac{z_1z_2}{z_1z_2-1}$$

is a z-transform since it corresponds to the nonpositive power series

$$F(z_1,z_2) = 1 + z_1^{-1}z_2^{-1} + z_1^{-2}z_2^{-2} + \cdots$$

The corresponding function can be written as

$$f(i_1,i_2) = \begin{cases} 1, & i_1 = i_2 = 0, 1, 2, \ldots \\ 0, & otherwise \end{cases}$$

The z-transform representation is used for degree-n homogeneous systems in just the same way that the Laplace transform is used in the continuous-time case. A transfer function for a degree-n homogeneous discrete-time system is defined as the z-transform of a kernel for the system. For example, the symmetric transfer function is

$$H_{sym}(z_1, \ldots, z_n) = Z[h_{sym}(i_1, \ldots, i_n)] \tag{24}$$

Unfortunately, though probably not unexpectedly, to represent the input/output relation (1) directly in terms of $U(z)$, $Y(z)$, and $H_{sym}(z_1, \ldots, z_n)$ seems to be impossible. The usual device is to write (1) as the pair of equations:

$$y_n(k_1,\ldots,k_n) = \sum_{i_1=0}^{\infty} \cdots \sum_{i_n=0}^{\infty} h_{sym}(i_1,\ldots,i_n)u(k_1-i_1)\cdots u(k_n-i_n)$$

$$y(k) = y_n(k_1,\ldots,k_n)\,\Big|_{k_1=\cdots=k_n=k} \tag{25}$$

Then Theorem 6.4 permits rewriting the first equation in the form

$$Y_n(z_1,\ldots,z_n) = H_{sym}(z_1,\ldots,z_n)U(z_1)\cdots U(z_n) \tag{26}$$

while the second equation is an association of variables that involves contour integrations of the form

$$Y(z) = \frac{1}{(2\pi i)^{n-1}} \int_{\Gamma_1} \cdots \int_{\Gamma_{n-1}} \frac{Y_n(z_1,z_2/z_1,\ldots,z/z_{n-1})}{z_1\cdots z_{n-1}}\,dz_1\cdots dz_{n-1} \tag{27}$$

The representation of (1) in terms of the triangular transfer function takes the same form.

In the case of the regular transfer function, the formulas in (25) and (26) do not directly apply. However, by suitably restricting the class of input signals, a much more explicit formula can be derived. This result is similar to Theorem 2.10, although I will present it in detail with a proof that is much different from that in the continuous-time case, and that requires no hypotheses on the form of the regular transfer function.

The first step is to establish a basic expression for the z-transform of the input/output expression (6). Write the regular transfer function

$$H_{reg}(z_1,\ldots,z_n) = Z[h_{reg}(i_1,\ldots,i_n)] \tag{28}$$

in the form

$$H_{reg}(z_1,\ldots,z_n) = \sum_{i_1=0}^{\infty} \cdots \sum_{i_{n-1}=0}^{\infty} H_{i_1\cdots i_{n-1}}(z_n)z_1^{-i_1}\cdots z_{n-1}^{-i_{n-1}} \tag{29}$$

where each $H_{i_1\cdots i_{n-1}}(z_n)$ is defined according to

$$H_{i_1\cdots i_{n-1}}(z_n) = \sum_{i_n=0}^{\infty} h_{reg}(i_1,\ldots,i_n)z_n^{-i_n},\quad i_1,\ldots,i_{n-1} = 0,1,2,\ldots \tag{30}$$

Lemma 6.1 The z-transform of the output of a degree-n homogeneous, discrete-time system can be written in the form

$$Y(z) = \sum_{i_1=0}^{\infty} \cdots \sum_{i_{n-1}=0}^{\infty} H_{i_1\cdots i_{n-1}}(z) \sum_{k=i_1+\cdots+i_{n-1}}^{\infty} u(k-i_1-\cdots-i_{n-1})$$

$$u(k-i_2-\cdots-i_{n-1})\cdots u(k-i_{n-1})u(k)z^{-k} \tag{31}$$

Proof Taking the z-transform of $y(k)$ as given in (6) yields

$$Y(z) = \sum_{k=0}^{\infty} \sum_{i_1=0}^{\infty} \cdots \sum_{i_n=0}^{\infty} h_{reg}(i_1,\ldots,i_n)u(k-i_1-\cdots-i_n)$$

$$u(k-i_2-\cdots-i_n)\cdots u(k-i_n)z^{-k}$$

Replacing the summation index k by $j = k-i_n$ gives

$$Y(z) = \sum_{i_1=0}^{\infty} \cdots \sum_{i_{n-1}=0}^{\infty} [\sum_{i_n=0}^{\infty} h_{reg}(i_1,\ldots,i_n)z^{-i_n}]$$

$$\sum_{j=-i_n}^{\infty} u(j-i_1-\cdots-i_{n-1})u(j-i_2-\cdots-i_{n-1})\cdots u(j-i_{n-1})u(j)z^{-j}$$

Now the result is clear from (30) and the assumption that $u(k) = 0$ for $k < 0$.

This lemma provides an alternative to the association-of-variables method for performing input/output calculations. Furthermore, a more direct expression for $Y(z)$ can be obtained for a ubiquitous class of inputs.

Theorem 6.6 Suppose a degree-n homogeneous, discrete-time system is described by the regular transfer function $H_{reg}(z_1,\ldots,z_n)$, and the input is of the form

$$U(z) = \sum_{j=1}^{m} \frac{a_j z}{z-\lambda_j}, \quad \lambda_j \neq 0,\ j = 1,\ldots,m \tag{32}$$

Then

$$Y(z) = \sum_{j_1=1}^{\infty} \cdots \sum_{j_{n-1}=1}^{\infty} a_{j_1} \cdots a_{j_{m-1}}$$

$$H_{reg}(\frac{z}{\lambda_{j_1}\cdots\lambda_{j_{n-1}}}, \frac{z}{\lambda_{j_2}\cdots\lambda_{j_{n-1}}}, \ldots, \frac{z}{\lambda_{j_{n-1}}}, z)\,U(\frac{z}{\lambda_{j_1}\cdots\lambda_{j_{n-1}}}) \tag{33}$$

Proof The z-transform in (32) clearly corresponds to the input signal

$$u(k) = \sum_{j=1}^{m} a_j\lambda_j^k, \quad k = 0,1,\ldots$$

Substituting into (31) gives

$$Y(z) = \sum_{i_1=0}^{\infty} \cdots \sum_{i_{n-1=0}}^{\infty} H_{i_1 \cdots i_{n-1}}(z) \sum_{k=i_1+\cdots+i_{n-1}} [\sum_{j=1}^{m} a_j\lambda_j^{k-i_1-\cdots-i_{n-1}}]$$

$$[\sum_{j_1=1}^{m} a_{j_1}\lambda_{j_1}^{k-i_2-\cdots-i_{n-1}}]\cdots[\sum_{j_{n-1}=1}^{m} a_{j_{n-1}}\lambda_{j_{n-1}}^{k}]z^{-k}$$

$$= \sum_{j_1=1}^{m} \cdots \sum_{j_{n-1}=1}^{m} a_{j_1}\cdots a_{j_{n-1}} \sum_{i_1=0}^{\infty} \cdots \sum_{i_{n-1}=0}^{\infty} H_{i_1 \cdots i_{n-1}}(z)$$

$$\sum_{j=1}^{m} a_j \sum_{k=i_1+\cdots+i_{n-1}}^{\infty} \lambda_j^{k-i_1-\cdots-i_{n-1}} \lambda_{j_1}^{k-i_2-\cdots-i_{n-1}} \cdots \lambda_{j_{n-1}}^{k} z^{-k}$$

Now replace the index k by $r = k-i_1-\cdots-i_{n-1}$ to obtain

$$Y(z) = \sum_{j_1=1}^{m} \cdots \sum_{j_{n-1}=1}^{m} a_{j_1} \cdots a_{j_{n-1}} \sum_{i_1=0}^{\infty} \cdots \sum_{i_{n-1}=0}^{\infty} H_{i_1 \cdots i_{n-1}}(z) \left(\frac{z}{\lambda_{j_1} \cdots \lambda_{j_{n-1}}}\right)^{-i_1}$$

$$\left(\frac{z}{\lambda_{j_2} \cdots \lambda_{j_{n-1}}}\right)^{-i_2} \cdots \left(\frac{z}{\lambda_{j_{n-1}}}\right)^{-i_{n-1}} \sum_{j=1}^{m} a_j \sum_{r=0}^{\infty} \lambda_j^r \left(\frac{z}{\lambda_{j_1} \cdots \lambda_{j_{n-1}}}\right)^{-r}$$

$$= \sum_{j_1=1}^{m} \cdots \sum_{j_{n-1}=1}^{m} a_{j_1} \cdots a_{j_{n-1}} H_{reg}\left(\frac{z}{\lambda_{j_1} \cdots \lambda_{j_{n-1}}}, \ldots, \frac{z}{\lambda_{j_{n-1}}}, z\right) U\left(\frac{z}{\lambda_{j_1} \cdots \lambda_{j_{n-1}}}\right)$$

Theorem 6.6 is general enough to cover a wide range of situations, and, although somewhat messy in appearance, the calculation of system responses is relatively straightforward. For example, if $H_{reg}(z_1,\ldots,z_n)$ is proper rational, then $Y(z)$ also will be proper rational, and partial fraction expansion can be used to compute $y(k)$.

Before leaving the topic of transform representations, I should point out a few simple relationships between the various transfer functions. Using (7), and a simple change of variables,

$$H_{reg}(z_1,\ldots,z_n) = \sum_{i_1=0}^{\infty} \cdots \sum_{i_n=0}^{\infty} h_{tri}(i_1+\cdots+i_n, i_2+\cdots+i_n, \ldots, i_n) z_1^{-i_1} \cdots z_n^{-i_n}$$

$$= \sum_{j_1=0}^{\infty} \cdots \sum_{j_n=0}^{\infty} h_{tri}(j_1,\ldots,j_n) z_1^{-j_1} \left(\frac{z_2}{z_1}\right)^{-j_2} \cdots \left(\frac{z_n}{z_{n-1}}\right)^{-j_n}$$

$$= H_{tri}(z_1, z_2/z_1, \ldots, z_n/z_{n-1})$$

This relationship is easily inverted to obtain

$$H_{tri}(z_1,\ldots,z_n) = H_{reg}(z_1, z_1 z_2, \ldots, z_1 \cdots z_n)$$

It is much messier to consider the symmetric transfer function. A basic relationship implied by (2) is

$$H_{sym}(z_1,\ldots,z_n) = \frac{1}{n!} \sum_{\pi(.)} H_{tri}(z_{\pi(1)},\ldots,z_{\pi(n)})$$

Then (35) gives

$$H_{sym}(z_1,\ldots,z_n) = \frac{1}{n!} \sum_{\pi(.)} H_{reg}(z_{\pi(1)}, z_{\pi(1)} z_{\pi(2)}, \ldots, z_1 \cdots z_n)$$

To compute H_{reg} or H_{tri} from H_{sym}, it seems that the best way to proceed is to find the symmetric kernel, use (3) or (7) to obtain the regular or triangular kernel, and then compute the z-transform. This is an unpleasant prospect at best, but there are some tricks that can be used in simple cases.

Example 6.3 For the $n = 2$ case,

$$H_{sym}(z_1,z_2) = \frac{1}{2} H_{reg}(z_1,z_1z_2) + \frac{1}{2} H_{reg}(z_2,z_1z_2)$$

$$= \frac{1}{2} \sum_{i_1=0}^{\infty} \sum_{i_2=0}^{\infty} h_{reg}(i_1,i_2) z_1^{-i_1} (z_1z_2)^{-i_2}$$

$$+ \frac{1}{2} \sum_{i_1=0}^{\infty} \sum_{i_2=0}^{\infty} h_{reg}(i_1,i_2) z_2^{-i_1} (z_1z_2)^{-i_2}$$

Thus, a simple change of variables gives

$$2H_{sym}(z_1,z_2/z_1) = \sum_{i_1=0}^{\infty} \sum_{i_2=0}^{\infty} h_{reg}(i_1,i_2) z_1^{-i_1} z_2^{-i_2}$$

$$+ \sum_{i_1=0}^{\infty} \sum_{i_2=0}^{\infty} h_{reg}(i_1,i_2) z_1^{i_1} z_2^{-(i_1+i_2)}$$

Clearly, the first term on the right side is $H_{reg}(z_1,z_2)$, while the second term contains only positive powers of z_1 plus 1/2 of each z_1^0 term in $2H_{sym}(z_1,z_2/z_1)$. Thus, $H_{reg}(z_1,z_2)$ can be obtained by dividing out $2H_{sym}(z_1,z_2/z_1)$, deleting all terms involving positive powers of z_1, and multiplying each z_1^0 term by 1/2. For the particular case

$$H_{sym}(z_1,z_2) = \frac{z_1z_2}{z_1z_2-1}$$

changing variables and dividing gives

$$2H_{sym}(z_1,z_2/z_1) = \frac{2z_2}{z_2-1} = 2(1 + z_2^{-1} + z_2^{-2} + \cdots)$$

Then, since the complete series is composed of z_1^0 terms,

$$H_{reg}(z_1,z_2) = 1 + z_2^{-1} + z_2^{-2} + \cdots = \frac{z_2}{z_2-1}$$

6.3 Obtaining Input/Output Representations from State Equations

All of the methods discussed in Chapter 3 can be adapted to the discrete-time case with relatively little change. Rather than fill several pages by doing this, I will concentrate on a judicious combination of the variational equation method and the Carleman linearization method, and consider a general class of state equations at the outset. As mentioned before, a nice feature of the discrete-time case is that the issue of impulsive kernels does not arise. That is, direct transmission terms are naturally included in the discrete-time input/output representation. However, the reader will notice that these terms do complicate considerably the general forms for the kernels.

State equations of the form

$$x(k+1) = f[x(k),u(k)], \quad k = 0,1,\ldots$$
$$y(k) = h[x(k),u(k)] \tag{34}$$

will be treated, where $x(k)$ is $n \times 1$ and $u(k)$ and $y(k)$ are scalars. It is assumed that the initial state is $x(0) = 0$, that $f(0,0) = 0$, and that $h(0,0) = 0$. This is done for simplicity, though if $x(0) = x_0 = \neq 0$, and $f(x_0,0) = x_0$, then x_0 is an equilibrium state and a simple variable change can be used to obtain the zero-initial-state formulation. (If x_0 is not an equilibrium state, then more subtle machinations are required to recast the problem into the form considered here.)

The final assumption on (34) is that the functions $f(x,u)$ and $h(x,u)$ are such that they can be represented using a Taylor's formula about $x = 0$, $u = 0$ of order sufficient to permit calculating the polynomial input/output representation to the degree desired. Then the given state equation can be replaced by an approximating state equation of the form

$$x(k+1) = \sum_{i=0}^{N} \sum_{j=0}^{N} F_{ij} x^{(i)}(k) u^j(k), \quad F_{00} = 0$$
$$y(k) = \sum_{i=0}^{N} \sum_{j=0}^{N} H_{ij} x^{(i)}(k) u^j(k), \quad H_{00} = 0 \tag{35}$$

where $x^{(i)}(0) = 0$, $i = 1,\ldots,N$, and the standard Kronecker product notation is used. Just as in the continuous-time case, the crucial fact is that the kernels through degree N corresponding to (35) will be identical to the kernels through degree N corresponding to (34). (The reader well versed in Chapter 3 will notice that the upper limits on the sums in (35) need not be taken too seriously. There are a number of terms in (35) which will not contribute to the degree-N polynomial representation.)

The next step is to develop difference equations for $x^{(2)}(k)$, $x^{(3)}(k)$, and so forth, corresponding to the difference equation for $x(k)$ in (35). This is a simple matter in principle, though the form of the equation is different from the continuous-time case because no product rule is involved in expressing $x^{(j)}(k+1)$ in terms of $x^{(j-1)}(k+1)$. For example, the difference equation for $x^{(2)}(k)$ is given by

$$x^{(2)}(k+1) = x(k+1) \otimes x(k+1)$$
$$= [\sum_{i=0}^{N}\sum_{j=0}^{N} F_{ij} x^{(i)}(k) u^j(k)] \otimes [\sum_{i=0}^{N}\sum_{j=0}^{N} F_{ij} x^{(i)}(k) u^j(k)] \tag{36}$$

Using implicit summation, this will result in a difference equation of the form

$$x^{(2)}(k+1) = \sum_{i,j \geq 0} [\sum_{\substack{k+q=i \\ m+n=j}} F_{km} \otimes F_{qn}] x^{(i)}(k) u^j(k) \tag{37}$$

where the initial condition is $x^{(2)}(0) = 0$. This equation has the same form as the difference equation for $x(k)$ in (35), and it should be clear that the equations for $x^{(3)}(k)$, $x^{(4)}(k)$, . . . will also. Now, set

$$x^{\otimes}(k) = \begin{bmatrix} x^{(1)}(k) \\ \vdots \\ x^{(N)}(k) \end{bmatrix} \tag{38}$$

This leads to an approximating equation through degree N in the so-called *state-affine* form

$$x^{\otimes}(k+1) = \sum_{i=0}^{N-1} A_i x^{\otimes}(k)u^i(k) + \sum_{i=1}^{N} b_i u^i(k), \quad x^{\otimes}(0) = 0$$
$$y(k) = \sum_{i=0}^{N-1} c_i x^{\otimes}(k)u^i(k) + \sum_{i=1}^{N} d_i u^i(k) \tag{39}$$

where the upper limits in the summations are chosen to include the terms needed to compute kernels of degree $\leqslant N$. Of course, the dimension of this state equation is quite high, but for a general derivation this is less of a problem than the plethora of terms. Notice that a bilinear discrete-time state equation is a pleasingly simple case.

To solve the state-affine difference equation in (39), I will use the variational equation method, and drop the now superfluous Kronecker symbol. The procedure is to assume an input signal of the form $\alpha u(k)$, α an arbitrary real number, and a solution of the form

$$x(k) = \alpha x_1(k) + \alpha^2 x_2(k) + \alpha^3 x_3(k) + \cdots \tag{40}$$

Substituting into the state equation and equating the coefficients of like powers of α yields the variational equations

$$x_1(k+1) = A_0 x_1(k) + b_1 u(k), \quad x_1(0) = 0$$
$$x_2(k+1) = A_0 x_2(k) + A_1 x_1(k)u(k) + b_2 u^2(k), \quad x_2(0) = 0$$
$$x_3(k+1) = A_0 x_3(k) + A_1 x_2(k)u(k) + A_2 x_1(k)u^2(k) + b_3 u^3(k), \quad x_3(0) = 0$$
$$\vdots$$
$$x_N(k+1) = \sum_{i=0}^{N-1} A_i x_{N-i}(k)u^i(k) + b_N u^N(k), \quad x_N(0) = 0 \tag{41}$$

These equations can be solved easily, and writing the solutions recursively gives (for $k > 0$)

$$x_1(k) = \sum_{i=0}^{k-1} A_0^{k-1-i} b_1 u(i)$$
$$x_2(k) = \sum_{i=0}^{k-1} A_0^{k-1-i}[A_1 x_1(i)u(i) + b_2 u^2(i)]$$

$$x_3(k) = \sum_{i=0}^{k-1} A_0^{k-1-i}[A_1x_2(i)u(i) + A_2x_1(i)u^2(i) + b_3u^3(i)]$$

$$\vdots$$

$$x_N(k) = \sum_{i=0}^{k-1} A_0^{k-1-i}[\sum_{j=1}^{N-1} A_jx_{N-j}(i)u^j(i) + b_Nu^N(i)] \tag{42}$$

Unraveling this recursive set yields rather complicated solution formulas for the variational equations. The first three expressions are listed below.

$$x_1(k) = \sum_{i_1=0}^{k-1} A_0^{k-1-i_1}b_1u(i_1)$$

$$x_2(k) = \sum_{i_1=0}^{k-1}\sum_{i_2=0}^{i_1-1} A_0^{k-1-i_1}A_1A_0^{i_1-1-i_2}b_1u(i_1)u(i_2) + A_0^{k-1-i_1}b_2u^2(i_1)$$

$$x_3(k) = \sum_{i_1=0}^{k-1}\sum_{i_2=0}^{i_1-1}\sum_{i_3=0}^{i_2-1} A_0^{k-1-i_1}A_1A_0^{i_1-1-i_2}A_1A_0^{i_2-1-i_3}b_1u(i_1)u(i_2)u(i_3)$$

$$+ A_0^{k-1-i_1}A_1A_0^{i_1-1-i_2}b_2u(i_1)u^2(i_2) + A_0^{k-1-i_1}A_2A_0^{i_1-1-i_2}b_1u^2(i_1)u(i_2)$$

$$+ A_0^{k-1-i_1}b_3u^3(i_1) \tag{43}$$

Before proceeding to a general result, it is convenient to convert the first two solution expressions in (43) to the regular form. Of course, a vector kernel for $x_1(k)$ is easy to describe in regular form. Write

$$x_1(k) = \sum_{i_1=1}^{k} A_0^{i_1-1}b_1u(k-i_1) = \sum_{i_1=0}^{k} g(i_1)u(k-i_1) \tag{44}$$

where, using a step function to indicate $g(0) = 0$,

$$g(i_1) = A_0^{i_1-1}b_1\delta_{-1}(i_1-1) \tag{45}$$

Now, $x_2(k)$ can be written in the triangular, vector-kernel expression

$$x_2(k) = \sum_{i_1=0}^{k}\sum_{i_2=0}^{i_1} w_{tri}(k,i_1,i_2)u(i_1)u(i_2)$$

where

$$w_{tri}(k,i_1,i_2) = \begin{cases} A_0^{k-1-i_1}A_1A_0^{i_1-1-i_2}b_1, & k > i_1 > i_2 \geqslant 0 \\ A_0^{k-1-i_1}b_2, & k > i_1 = i_2 \geqslant 0 \\ 0, & otherwise \end{cases}$$

Or, using unit step and unit pulse functions,

$$w_{tri}(k,i_1,i_2) = A_0^{k-1-i_1}A_1A_0^{i_1-1-i_2}b_1\delta_{-1}(k-1-i_1)\delta_{-1}(i_1-1-i_2)$$
$$+ A_0^{k-1-i_1}b_2\delta_{-1}(k-1-i_1)\delta_0(i_1-i_2), \quad i_1,i_2 \geqslant 0 \tag{46}$$

To check stationarity, note that

$$w_{tri}(0,i_1-k,i_2-k) = w_{tri}(k,i_1,i_2) \tag{47}$$

so a triangular kernel in stationary form is

$$g_{tri}(i_1,i_2) = w_{tri}(0,-i_1,-i_2)$$
$$= A_0^{i_1-1}A_1A_0^{i_2-1-i_1}b_1\delta_{-1}(i_1-1)\delta_{-1}(i_2-1-i_1)$$
$$+ A_0^{i_1-1}b_2\delta_{-1}(i_1-1)\delta_0(i_2-i_1), \quad i_1,i_2 \geqslant 0 \tag{48}$$

Over the first triangular domain, the triangular kernel is

$$g_{tri}(i_1,i_2) = A_0^{i_2-1}A_1A_0^{i_1-1-i_2}b_1\delta_{-1}(i_2-1)\delta_{-1}(i_1-1-i_2)$$
$$+ A_0^{i_2-1}b_2\delta_{-1}(i_2-1)\delta_0(i_1-i_2) \tag{49}$$

so that the regular kernel is given by

$$g_{reg}(i_1,i_2) = g_{tri}(i_1+i_2,i_2)$$
$$= A_0^{i_2-1}A_1A_0^{i_1-1}b_1\delta_{-1}(i_2-1)\delta_{-1}(i_1-1) + A_0^{i_2-1}b_2\delta_{-1}(i_2-1)\delta_0(i_1) \tag{50}$$

for $i_1,i_2 \geqslant 0$.

Taking into account the output equation in (39), it is clear that the degree-2 term, $y_2(k)$, in the output is given by

$$y_2(k) = c_0x_2(k) + c_1x_1(k)u(k) + d_2u^2(k)$$
$$= \sum_{i_1=0}^{k}\sum_{i_2=0}^{k} c_0g_{reg}(i_1,i_2)u(k-i_1-i_2)u(k-i_2)$$
$$+ \sum_{i_1=0}^{k} c_1g(i_1)u(k-i_1)u(k) + d_2u^2(k)$$
$$= \sum_{i_1=0}^{k}\sum_{i_2=0}^{k} h_{reg}(i_1,i_2)u(k-i_1-i_2)u(k-i_2) \tag{51}$$

where, from (45) and (50),

$$h_{reg}(i_1,i_2) = \begin{cases} d_2, & i_1 = i_2 = 0 \\ c_1A_0^{i_1-1}b_1, & i_1 > 0,\ i_2 = 0 \\ c_0A_0^{i_2-1}b_2, & i_1 = 0,\ i_2 > 0 \\ c_0A_0^{i_2-1}A_1A_0^{i_1-1}b_1, & i_1,i_2 > 0 \end{cases} \tag{52}$$

To perform this calculation in general is a very messy exercise in the manipulation of summations and indices. Therefore, I will omit the details and simply present the result. The degree-n regular kernel corresponding to the state-affine state equation (39) is given by

$$h_{reg}(i_1,\ldots,i_n) = \begin{cases} d_n, & i_1 = \cdots = i_n = 0 \\ c_{n-1}A_0^{i_1-1}b_1, & i_1 > 0,\ i_2 = \cdots = i_n = 0 \\ \vdots & \\ c_qA_0^{i_{n-q}-1}A_rA_0^{i_{n-q-r}-1}\cdots A_0^{i_j-1}b_j, & \begin{cases} i_j,\ldots,i_{n-q-r},i_{n-q} > 0 \\ all\ others = 0 \end{cases} \\ \vdots & \\ c_0A_0^{i_n-1}A_1A_0^{i_{n-1}-1}\cdots A_1A_0^{i_1-1}b_1, & i_1,\ldots,i_n > 0 \end{cases} \tag{53}$$

Notice that the sum of the subscripts in each term of this expression is n, and that the subscript on the coefficient preceding each $A_0^{i_k-1}$ determines the index k. There will be a total of 2^n terms in a general degree-n kernel.

Example 6.4 Discrete-time state-affine systems arise directly in the description of bilinear continuous-time systems with sampled input signals. For simplicity only the degree-2 homogeneous case will be discussed, and the impulse model will be used for the sampled signal. That is, the system is described by

$$y(t) = \int_0^\infty \int_0^\infty h_{reg}(\sigma_1,\sigma_2)u(t-\sigma_1-\sigma_2)u(t-\sigma_2)\,d\sigma_2 d\sigma_1$$

where

$$h_{reg}(t_1,t_2) = ce^{At_2}De^{At_1}b,\quad t_1,t_2 \geqslant 0$$

and the input signal is

$$u(t) = \sum_{k=0}^{\infty} u(kT)\delta_0(t-kT)$$

where T is the sampling period. Then the output at the mth sampling instant is given by

$$y(mT) = \int_0^\infty \int_0^\infty h_{reg}(\sigma_1,\sigma_2) \sum_{k_1=0}^{\infty} u(k_1T)\delta_0(mT-\sigma_1-\sigma_2-k_1T)$$

$$\sum_{k_2=0}^{\infty} u(k_2T)\delta_0(mT-\sigma_2-k_2T)\,d\sigma_2 d\sigma_1$$

$$= \sum_{k_1=0}^{\infty}\sum_{k_2=0}^{\infty} h_{reg}(k_2T-k_1T,mT-k_2T)u(k_1T)u(k_2T)$$

Restoring this expression to regular form requires changes of variables of summation. First replace k_2 by $j_2 = m - k_2$, and replace k_1 by $j_1 = m - j_2 - k_1$. Then using the fact that the input signal and the regular kernel both are zero for negative arguments gives

$$y(mT) = \sum_{j_1=0}^{\infty} \sum_{j_2=0}^{\infty} h_{reg}(j_1T,j_2T)u(mT-j_1T-j_2T)u(mT-j_2T)$$

Thus it is clear that the regular kernel for this discrete-time representation is

$$h(j_1T,j_2T) = c(e^{AT})^{j_2}D(e^{AT})^{j_1}b \ , \quad j_1,j_2 \geq 0$$

Now with the definitions

$$A_0 = e^{AT} \ , \quad A_1 = D \ , \quad b_1 = e^{AT}b \ , \quad b_2 = Db$$
$$c_0 = ce^{AT} \ , \quad c_1 = cD \ , \quad d_2 = cDb$$

this kernel conforms to the state-affine kernel specified in (53).

As a final comment, observe that it is straightforward to compute the degree-n regular transfer function from (53). Indeed,

$$\begin{aligned} H_{reg}(z_1,\ldots,z_n) &= \sum_{i_1=0}^{\infty} \cdots \sum_{i_n=0}^{\infty} h_{reg}(i_1,\ldots,i_n)z_1^{-i_1} \cdots z_n^{-i_n} \\ &= d_n + c_{n-1}(z_1I - A_0)^{-1}b_1 + \cdots \\ &+ c_q(z_{n-q}I - A_0)^{-1}A_r(z_{n-q-r}I - A_0)^{-1} \cdots (z_jI - A_0)^{-1}b_j + \cdots \\ &+ c_0(z_nI - A_0)^{-1}A_1(z_{n-1}I - A_0)^{-1}A_1 \cdots A_1(z_1I - A_0)^{-1}b_1 \end{aligned} \quad (54)$$

though this expression, like (53), is not very explicit, and a certain amount of digging is needed to produce all 2^n terms.

Example 6.5 The degree-3 regular transfer function for the state-affine state equation (39) is

$$\begin{aligned} H_{reg}(z_1,z_2,z_3) &= d_3 + c_2(z_1I - A_0)^{-1}b_1 + c_1(z_2I - A_0)^{-1}b_2 \\ &+ c_0(z_3I - A_0)^{-1}b_3 + c_1(z_2I - A_0)^{-1}A_1(z_1I - A_0)^{-1}b_1 \\ &+ c_0(z_3I - A_0)^{-1}A_2(z_1I - A_0)^{-1}b_1 + c_0(z_3I - A_0)^{-1}A_1(z_2I - A_0)^{-1}b_2 \\ &+ c_0(z_3I - A_0)^{-1}A_1(z_2I - A_0)^{-1}A_1(z_1I - A_0)^{-1}b_1 \end{aligned}$$

If the state equation actually is bilinear, then the only surviver among these terms is the last one.

6.4 State-Affine Realization Theory

The realization problem for discrete-time systems will be discussed mainly in terms of the regular-transfer-function input/output representation, and state-affine state equations (realizations). Thus, the realization theory considered here is somewhat more general than that in Chapter 4. In fact, the bilinear realization theory for discrete-time systems will appear as a relatively uncomplicated special case. (The bilinear theory also can be obtained from Chapter 4 by modifications only slightly more difficult than wholesale replacement of *s*'s by *z*'s.) I will concentrate on homogeneous and polynomial systems here, and leave Volterra systems to the original research literature.

Recall that a transfer function is called *rational* if it can be written as a ratio of polynomials:

$$H(z_1,\ldots,z_n) = \frac{P(z_1,\ldots,z_n)}{Q(z_1,\ldots,z_n)} \tag{55}$$

A rational transfer function is called *proper (strictly proper)* if degree $P(z_1,\ldots,z_n)$ in each variable is no greater than (less than) degree $Q(z_1,\ldots,z_n)$ in the corresponding variable. A rational transfer function is called *recognizable* if $Q(z_1,\ldots,z_n) = Q_1(z_1)\cdots Q_n(z_n)$, where each $Q_j(z_j)$ is a single variable polynomial. As in Chapter 4, it is assumed that the numerator and denominator polynomials are relatively prime to rule out trivial issues.

A *state-affine realization* of a degree-n homogeneous or polynomial system takes the form

$$x(k+1) = \sum_{i=0}^{n-1} A_i x(k)u^i(k) + \sum_{i=1}^{n} b_i u^i(k)$$
$$y(k) = \sum_{i=0}^{n-1} c_i x(k)u^i(k) + \sum_{i=1}^{n} d_i u^i(k) \tag{56}$$

where the (finite) dimension of the state vector $x(k)$ is called the *dimension of the realization.* Notice that the upper limits on the summations in (56) have been set in accordance with the degree of the system.

Using this formulation a basic result on realizability can be stated as follows.

Theorem 6.7 A degree-n homogeneous discrete-time system is state-affine realizable if and only if the regular transfer function of the system is a proper, recognizable function.

Proof If the system has a state-affine realization, then the regular transfer function can be written as in (54). Writing each $(z_kI - A_0)^{-1}$ in the classical-adjoint-over-determinant form, and placing all the terms over a common denominator shows that $H_{reg}(z_1,\ldots,z_n)$ is a proper, recognizable function.

If $H_{reg}(z_1,\ldots,z_n)$ is a proper, recognizable function, then it can be written in the form

$$H_{reg}(z_1,\ldots,z_n) = \frac{P(z_1,\ldots,z_n)}{Q_1(z_1)\cdots Q_n(z_n)} \tag{57}$$

where

$$P(z_1,\ldots,z_n) = \sum_{j_1=0}^{m_1}\cdots\sum_{j_n=0}^{m_2} p_{j_1\cdots j_n} z_1^{j_1}\cdots z_n^{j_n}$$

$$Q_i(z_i) = z_i^{m_i} + \sum_{j_i=0}^{m_i-1} q_{i,j_i} z_i^{j_i},\quad i = 1,\ldots,n \tag{58}$$

Just as in the continuous-time case, the numerator polynomial can be written in a matrix factored form

$$P(z_1,\ldots,z_n) = Z_n\cdots Z_2Z_1P$$

where Z_j contains entries chosen from $0, 1, z_j,\ldots,z_j^{m_j}$, $j = 1,\ldots,n$, and P is a vector of coefficients. Thus the regular transfer function can be written in the factored form

$$H_{reg}(z_1,\ldots,z_n) = \frac{Z_n}{Q_n(z_n)}\cdots\frac{Z_2}{Q_2(z_2)}\frac{Z_1P}{Q_1(z_1)}$$

$$= G_n(z_n)\cdots G_2(z_2)G_1(z_1)$$

Each $G_j(z_j)$ is a matrix with proper rational entries, and thus linear-realization techniques can be used to write

$$G_j(z_j) = \hat{C}_j(z_jI - \hat{A}_j)^{-1}\hat{B}_j + \hat{D}_j,\quad j = 1,\ldots,n$$

Now consider the state-affine realization (56) specified as follows. Let A_0 be block diagonal, and A_j be zero except possibly on the jth block subdiagonal:

$$A_0 = \begin{bmatrix} \hat{A}_1 & 0 & \cdots & 0 \\ 0 & \hat{A}_2 & \cdots & 0 \\ \vdots & \vdots & \vdots & \vdots \\ 0 & 0 & \cdots & \hat{A}_n \end{bmatrix}$$

$$A_1 = \begin{bmatrix} 0 & \cdots & 0 & 0 \\ \hat{B}_2\hat{C}_1 & \cdots & 0 & 0 \\ \vdots & \vdots & \vdots & \vdots \\ 0 & \cdots & \hat{B}_n\hat{C}_{n-1} & 0 \end{bmatrix}$$

$$A_j = \begin{bmatrix} 0 & \cdots & 0 & 0 & \cdots & 0 \\ \vdots & \vdots & \vdots & \vdots & & \vdots \\ \hat{B}_{j+1}\hat{D}_j\cdots\hat{D}_2\hat{C}_1 & \cdots & 0 & 0 & \cdots & 0 \\ \vdots & \vdots & \vdots & \vdots & & \vdots \\ 0 & \cdots & \hat{B}_n\hat{D}_{n-1}\cdots\hat{D}_{n-j+1}\hat{C}_{n-j} & 0 & \cdots & 0 \end{bmatrix}$$

Let each b_j have zero entries except possibly for the jth block entry according to

$$b_1 = \begin{bmatrix} \hat{B}_1 \\ 0 \\ \vdots \\ 0 \end{bmatrix} , \quad b_j = \begin{bmatrix} 0 \\ \vdots \\ \hat{B}_j\hat{D}_{j-1} \cdots \hat{D}_1 \\ \vdots \\ 0 \end{bmatrix} , \quad j = 2,\ldots,n$$

Let each c_j have zero entries except possibly for the $(n-j)^{th}$ block entry according to

$$c_0 = [0 \ \cdots \ 0 \ \hat{C}_n]$$

$$c_j = [0 \ \cdots \ (\hat{D}_n \cdots \hat{D}_{n-j+1}\hat{C}_{n-j}) \ \cdots \ 0] \ , \ j = 1,\ldots,n-1$$

and, finally, let $d_1 = \cdots = d_{n-1} = 0$, $d_n = \hat{D}_n\hat{D}_{n-1} \cdots \hat{D}_1$. The regular transfer functions corresponding to this state-affine realization can be computed from (54). Because of the special block structure, it turns out that all the transfer functions of degree $\neq n$ are zero, and the degree-n transfer function is given by

$$H_{reg}(z_1,\ldots,z_n) = (\hat{D}_n + \hat{C}_n(z_nI-\hat{A}_n)^{-1}\hat{B}_n) \ \cdots \ (\hat{D}_1 + \hat{C}_1(z_1I-\hat{A}_1)^{-1}\hat{B}_1)$$
$$= G_n(z_n) \ \cdots \ G_1(z_1)$$

This calculation, which is left as an uninteresting exercise, completes the proof.

Corollary 6.1 A degree-n homogeneous, discrete-time system is bilinear realizable if and only if the regular transfer function of the system is a strictly proper, recognizable function.

Before proceeding to the construction of minimal state-affine realizations, perhaps a slight digression on realizability is in order. So far I have presented all the realization results in terms of the regular transfer function. These results are easily transcribed to the triangular transfer function representation since a simple change of variables relates the two transfer functions. However, it is much more difficult to discuss realizability in terms of the symmetric transfer function. One way to approach this topic is to use relationships between the regular and symmetric transfer functions. This topic was discussed briefly in Section 6.2 for the degree-2 case, and that discussion will be followed up here.

Example 6.6 From the relationship

$$H_{sym}(z_1,z_2) = \frac{1}{2}H_{reg}(z_1,z_1z_2) + \frac{1}{2}H_{reg}(z_2,z_1z_2)$$

it is clear that the symmetric transfer function for a state-affine-realizable degree-2 homogeneous system must have the form

$$H_{sym}(z_1,z_2) = \frac{P(z_1,z_2)}{Q_1(z_1)Q_1(z_2)Q_2(z_1z_2)}$$

where $Q_1(z_1)$ and $Q_2(z_2)$ are single-variable polynomials, and $P(z_1,z_2)$ is a 2-variable polynomial. But the numerator cannot be arbitrary; there also are constraints on the form of $P(z_1,\ldots,z_n)$. These constraints are rather subtle to work out, so I will be content to continue Example 6.3 by indirect methods. The symmetric transfer function

$$H_{sym}(z_1,z_2) = \frac{z_1z_2}{z_1z_2-1}$$

is state-affine realizable since the corresponding regular transfer function is the proper, recognizable rational function

$$H_{reg}(z_1,z_2) = \frac{z_2}{z_2-1}$$

The construction of a state-affine realization as given in the proof of Theorem 6.7 is exceedingly simple in this case, giving

$$\hat{C}_1 = \hat{A}_1 = \hat{B}_1 = 0, \;\; \hat{D}_1 = 1$$
$$\hat{C}_2 = \hat{A}_2 = \hat{B}_2 = \hat{D}_2 = 1$$

Thus a state-affine realization is

$$x(k+1) = \begin{bmatrix} 0 & 0 \\ 0 & 1 \end{bmatrix} x(k) + \begin{bmatrix} 0 \\ 1 \end{bmatrix} u^2(k)$$
$$y(k) = [0 \quad 1]x(k) + u^2(k)$$

Note that this case is so simple that by defining a new input $\hat{u}(t) = u^2(t)$ the realization is linear.

For the polynomial-system case, the input/output representation that is natural to consider is the sequence of regular transfer functions of the homogeneous subsystems. Then the basic realizability result (and its proof) is simply a restatement of Theorem 4.9 in Section 4.3.

Theorem 6.8 A polynomial, discrete-time system is state-affine realizable if and only if the regular transfer function of each homogeneous subsystem is a proper, recognizable function.

To construct minimal-dimension state-affine realizations for polynomial systems, a shift operator approach much like that in Chapter 4 will be used. Of course, there are more kinds of terms to be dealt with in the state-affine case, and nonpositive power series rather than negative power series are involved due to the definition of the z-transform. However, the basic ideas are just the same.

For a given finite-length sequence of regular transfer functions

$$\hat{H}(z_1,\ldots,z_N) = (\, H(z_1),\ H_{reg}(z_1,z_2),\ldots,\ H_{reg}(z_1,\ldots,z_N),0,\ldots) \tag{59}$$

where

$$H_{reg}(z_1,\ldots,z_k) = \sum_{i_1=0}^{\infty} \cdots \sum_{i_k=0}^{\infty} h_{reg}(i_1,\ldots,i_k)z_1^{-i_1} \cdots z_k^{-i_k} \tag{60}$$

the minimal realization problem can be stated as follows. Find matrices $A_0, \ldots, A_{N-1}$ of dimension $m \times m$, $b_1, \ldots, b_N$ of dimension $m \times 1$, $c_0, \ldots, c_{N-1}$ of dimension $1 \times m$, and scalars $d_1, \ldots, d_N$ such that (53) is satisfied for $n = 1, \ldots, N$, the right side of (53) is 0 for $n > N$, and such that m is as small as possible. These matrices specify a state-affine realization of the form (56) of dimension m, and the shorthand notation (A_j,b_j,c_j,d_j,R^m) will be used for such a realization.

Given any nonpositive power series

$$V_k(z_1,\ldots,z_k) = \sum_{i_1=0}^{\infty} \cdots \sum_{i_k=0}^{\infty} v_{i_1 \cdots i_k} z_1^{-i_1} \cdots z_k^{-i_k} \tag{61}$$

define a *shift operator* via

$$SV_k(z_1,\ldots,z_k) = \sum_{i_1=0}^{\infty} \cdots \sum_{i_k=0}^{\infty} v_{i_1+1,i_2 \cdots i_k} z_1^{-i_1} \cdots z_k^{-i_k} \tag{62}$$

It is easy to verify that the shift is a linear operator that can be interpreted as

$$SV_k(z_1,\ldots,z_k) = z_1[V_k(z_1,\ldots,z_k) - V_k(\infty,z_2,\ldots,z_k)] \tag{63}$$

and that $SV_k(z_1,\ldots,z_k)$ is also a nonpositive power series. An *index operator* is defined by

$$TV_k(z_1,\ldots,z_k) = \begin{cases} 0, \ k = 1 \\ \displaystyle\sum_{i_1=0}^{\infty} \cdots \sum_{i_{k-1}=0}^{\infty} v_{0i_1 \cdots i_{k-1}} z_1^{-i_1} \cdots z_{k-1}^{-i_{k-1}}, \ k > 1 \end{cases} \tag{64}$$

that is

$$TV_k(z_1,\ldots,z_k) = V_k(\infty,z_1,\ldots,z_{k-1}), \ \ k > 1 \tag{65}$$

Again, it is not hard to see that T is a linear operator and that $TV_k(z_1,\ldots,z_k)$ is a nonpositive power series. The same symbols S and T will be used regardless of the number of variables in the series to which the operators are applied. Then these definitions can be extended to finite-length sequences of nonpositive power series in the obvious way:

$$S(\ V(z_1),\ V_2(z_1,z_2),\ V_3(z_1,z_2,z_3),\ \ldots) = (\ SV(z_1),\ SV_2(z_1,z_2),\ SV_3(z_1,z_2,z_3),\ \ldots) \tag{66}$$

$$T(\ V(z_1),\ V_2(z_1,z_2),\ V(z_1,z_2,z_3),\ \ldots) = (\ TV_2(z_1,z_2),\ TV_3(z_1,z_2,z_3),\ \ldots) \tag{67}$$

Now suppose a degree-N polynomial system is specified as in (59). Then a collection of linear spaces of finite-length sequences of nonpositive power series is defined as follows.

$$U_1 = span \{ \hat{H}(z_1,\ldots,z_N),\ S\hat{H}(z_1,\ldots,z_N),\ S^2\hat{H}(z_1,\ldots,z_N),\ \ldots\}$$
$$U_2 = span \{ TU_1,\ STU_1,\ S^2TU_1,\ \ldots\}$$
$$\vdots$$
$$U_N = span \{ TU_{N-1},\ STU_{N-1},\ S^2TU_{N-1},\ \ldots\}$$
$$U = span \{ U_1,\ U_2,\ \ldots,\ U_N\} \tag{68}$$

Then S and T can be viewed as linear operators with U as the domain and range.

Define a set of *initialization operators* $L_j{:}R \to U$ according to

$$L_1 r = S\hat{H}(z_1,\ldots,z_N)r$$
$$L_2 r = ST\hat{H}(z_1,\ldots,z_N)r$$
$$\vdots$$
$$L_N r = ST^{N-1}\hat{H}(z_1,\ldots,z_N)r \tag{69}$$

and define a set of *evaluation operators* $E_j{:}U \to R$ as follows. If

$$\hat{V}(z_1,\ldots,z_N) = (\ V_1(z_1),\ V_2(z_1,z_2),\ \ldots,\ V_N(z_1,\ldots,z_N),\ 0,\ \ldots)$$

is an element of U, then

$$E_0\hat{V}(z_1,\ldots,z_N) = V_1(\infty)$$
$$E_1\hat{V}(z_1,\ldots,z_N) = E_0T\hat{V}(z_1,\ldots,z_N) = V_2(\infty,\infty)$$
$$\vdots$$
$$E_{N-1}\hat{V}(z_1,\ldots,z_N) = E_0T^{N-1}\hat{V}(z_1,\ldots,z_N) = V_N(\infty,\ldots,\infty) \tag{70}$$

Finally, let $d_j{:}R \to R$ be specified by

$$d_j = H_{reg}(z_1,\ldots,z_j)\,\big|_{z_1=\cdots=z_j=\infty}\ ,\quad j = 1,\ldots,N \tag{71}$$

To show that these all are linear operators on their respective domains is very easy.

Now it can be shown that if U is finite dimensional, then (ST^j,L_j,E_j,d_j,U) is an abstract, finite-dimensional, state-affine realization of the given polynomial system. Once this is done, finding a concrete realization involves replacing U by R^m and finding the matrix representations $A_j = ST^j$, $b_j = L_j$, $c_j = E_j$, and interpreting the scalar operators d_j as constants. The proof that this process yields a minimal-dimension state-affine realization will be omitted since it is complicated. In fact, I will omit most of the demonstration that (ST^j,L_j,E_j,d_j,U) is a realization. To indicate how the calculation goes, consider the case of a degree-N polynomial system with $N > 3$,

$$\hat{H}(z_1,\ldots,z_N) = (\ H(z_1),\ H_{reg}(z_1,z_2),\ H_{reg}(z_1,z_2,z_3),\ \ldots)$$
$$= (\ \sum_{i_1=0}^{\infty} h(i_1)z_1^{-i_1},\ \sum_{i_1=0}^{\infty}\sum_{i_2=0}^{\infty} h_{reg}(i_1,i_2)z_1^{-i_1}z_2^{-i_2},$$

$$\sum_{i_1=0}^{\infty} \sum_{i_2=0}^{\infty} \sum_{i_3=0}^{\infty} h_{reg}(i_1,i_2,i_3)z_1^{-i_1}z_2^{-i_2}z_3^{-i_3}, \ldots)$$

Then a selection of degree-3 terms in (53) can be verified as follows. The constant term is

$$d_3 = H_{reg}(\infty,\infty,\infty) = h_{reg}(0,0,0)$$

The term $c_2 A_0^{j_1-1} b_1$ in (53) corresponds to

$$E_2 S^{j_1-1} L_1 = E_0 T^2 S^{j_1} \hat{H}(z_1,\ldots,z_N)$$

$$= E_0 T^2 (\sum_{i_1=0}^{\infty} h(i_1+j_1) z_1^{-i_1}, \sum_{i_1=0}^{\infty} \sum_{i_2=0}^{\infty} h_{reg}(i_1+j_1,i_2) z_1^{-i_1} z_2^{-i_2},$$

$$\sum_{i_1=0}^{\infty} \sum_{i_2=0}^{\infty} \sum_{i_3=0}^{\infty} h_{reg}(i_1+j_1,i_2,i_3)z_1^{-i_1}z_2^{-i_2}z_3^{-i_3}, \ldots)$$

$$= E_0 (\sum_{i_1=0}^{\infty} h_{reg}(j_1,0,i_1) z_1^{-i_1}, \ldots) = h_{reg}(j_1,0,0)$$

As a final illustration, $c_0 A_0^{j_3-1} A_2 A_0^{j_1-1} b_1$ corresponds to

$$E_0 S^{j_3-1}(ST^2) S^{j_1-1} L_1 = E_0 S^{j_3} T^2 S^{j_1} \hat{H}(z_1,\ldots,z_N)$$

$$= E_0 S^{j_3} (\sum_{i_1=0}^{\infty} h_{reg}(j_1,0,i_1) z_1^{-i_1}, \ldots)$$

$$= E_0 (\sum_{i_1=0}^{\infty} h_{reg}(j_1,0,i_1+j_3) z_1^{-i_1}, \ldots)$$

$$= h_{reg}(j_1,0,j_3)$$

All this shifting, indexing, and evaluating rapidly becomes a lot of fun and I urge the reader to do a few more terms. However, the investment in notation and calculation needed to verify the realization in general is probably unprofitable.

Example 6.7 Just to fix the nature of the calculations, consider a simple polynomial system described by

$$\hat{H}(z_1,z_2) = (\frac{z_1}{z_1-1} , \frac{z_2}{z_2-1} , 0,\ldots)$$

(Here I will not work with the power-series form of the regular transfer functions, for simplicity.) Application of the shift operator gives

$$S\hat{H}(z_1,z_2) = \left(\frac{z_1}{z_1-1}, 0, \ldots\right)$$

$$S^2\hat{H}(z_1,z_2) = S\hat{H}(z_1,z_2)$$

so that

$$U_1 = span \left\{\left(\frac{z_1}{z_1-1}, \frac{z_2}{z_2-1}, , 0, \ldots\right), \left(\frac{z_1}{z_1-1}, 0, \ldots\right)\right\}$$

Application of the index operator gives

$$T\hat{H}(z_1,z_2) = \left(\frac{z_1}{z_1-1}, 0, \ldots\right)$$

and easy calculations show that

$$ST\hat{H}(z_1,z_2) = T\hat{H}(z_1,z_2)$$

$$TS\hat{H}(z_1,z_2) = 0$$

Thus $U_2 \subset U_1$, and the linear space U can be taken to be U_1. Since U has dimension 2, it can be replaced by R^2 with the basis elements chosen according to

$$\begin{bmatrix}1\\0\end{bmatrix} = \left(\frac{z_1}{z_1-1}, \frac{z_2}{z_2-1}, 0, \ldots\right)$$

$$\begin{bmatrix}0\\1\end{bmatrix} = \left(\frac{z_1}{z_1-1}, 0, \ldots\right)$$

In terms of this basis, the shift and index operators are given by the matrices

$$S = \begin{bmatrix}0 & 0\\1 & 1\end{bmatrix}, \quad T = \begin{bmatrix}0 & 0\\1 & 0\end{bmatrix}$$

Thus,

$$A_0 = S = \begin{bmatrix}0 & 0\\1 & 1\end{bmatrix}, \quad A_1 = ST = \begin{bmatrix}0 & 0\\1 & 0\end{bmatrix}$$

The initialization operators are represented by

$$L_1 = S\hat{H}(z_1,z_2) = \begin{bmatrix}0\\1\end{bmatrix}, \quad L_2 = ST\hat{H}(z_1,z_2) = \begin{bmatrix}0\\1\end{bmatrix}$$

so that

$$b_1 = b_2 = \begin{bmatrix}0\\1\end{bmatrix}$$

The evaluation operators give

$$E_0\hat{H}(z_1,z_2) = 1 \ , \ E_0S\hat{H}(z_1,z_2) = 1$$
$$E_1\hat{H}(z_1,z_2) = 1 \ , \ E_1S\hat{H}(z_1,z_2) = 0$$

from which the corresponding matrix representations are

$$c_0 = [1 \quad 1] \ , \quad c_1 = [1 \quad 0]$$

Finally, it is clear that the constant terms are given by $d_1 = d_2 = 1$. Thus, a minimal state-affine realization of the given system is

$$x(k+1) = \begin{bmatrix} 0 & 0 \\ 1 & 1 \end{bmatrix} x(k) + \begin{bmatrix} 0 & 0 \\ 1 & 0 \end{bmatrix} x(k)u(k) + \begin{bmatrix} 0 \\ 1 \end{bmatrix} u(k) + \begin{bmatrix} 0 \\ 1 \end{bmatrix} u^2(k)$$
$$y(k) = [1 \quad 1]\, x(k) + [1 \quad 0] x(k)u(k) + u(k) + u^2(k)$$

The extension of this approach to the Volterra system case should be evident in broad outline. The kinds of difficulties that arise are indicated in Section 4.4, and the general theory is discussed in detail in the research literature cited in Section 6.8.

6.5 Response Characteristics of Discrete-Time Systems

The response of both homogeneous and polynomial discrete-time systems to various classes of input signals can be analyzed using much the same approach as in Chapter 5. To substantiate this claim, I will outline how some of the analysis goes for unit pulse and sinusoidal inputs. For random input signals, some of the results paralleling the continuous-time case will be derived from a less informal viewpoint than was adopted in Section 5.5.

Consider first the response of a degree-n homogeneous system to inputs composed of sums of delayed unit pulses, where the *unit pulse* is defined by

$$\delta_0(k) = \begin{cases} 1, & k = 0 \\ 0, & otherwise \end{cases} \tag{72}$$

In terms of the symmetric kernel representation, the calculations in Section 5.1 carry over directly, and so they will not be repeated here. However, I will go through some simple calculations in terms of the regular kernel representation in order to point out one perhaps surprising feature.

For the homogeneous system

$$y(k) = \sum_{i_1=0}^{\infty} \cdots \sum_{i_n=0}^{\infty} h_{nreg}(i_1, \ldots, i_n) u(k-i_1- \cdots -i_n) \cdots u(k-i_n) \tag{73}$$

with the input $u(k) = \delta_0(k)$, simple inspection yields the response

$$y(k) = h_{nreg}(0, \ldots, 0,k), \quad k = 0,1, \ldots \tag{74}$$

A more interesting situation occurs when the input is composed of two unit pulses,

$$u(k) = \delta_0(k) + a\delta_0(k-K) \tag{75}$$

where a is a real number, and K is a positive integer. The calculation of the corresponding response as outlined below is simple, though a bit lengthy. The response formula (73) gives

$$y(k) = \sum_{i_1=0}^{\infty} \cdots \sum_{i_n=0}^{\infty} h_{nreg}(i_1,\ldots,i_n)[\delta_0(k-i_1-\cdots-i_n)+a\delta_0(k-K-i_1-\cdots-i_n)]$$

$$\cdots [\delta_0(k-i_{n-1}-i_n)+a\delta_0(k-K-i_{n-1}-i_n)][\delta_0(k-i_n)+a\delta_0(k-K-i_n)] \tag{76}$$

Inspection of the last bracketed term on the right side shows that the summand will be nonzero only for two values of i_n, namely, $i_n = k$, and $i_n = k - K$. Thus,

$$y(k) = \sum_{i_1=0}^{\infty} \cdots \sum_{i_{n-1}=0}^{\infty} h_{nreg}(i_1,\ldots,i_{n-1},k)[\delta_0(-i_1-\cdots-i_{n-1})$$

$$+ a\delta_0(-K-i_1-\cdots-i_{n-1})] \cdots [\delta_0(-i_{n-1}) + a\delta_0(-K-i_{n-1})]$$

$$+ \sum_{i_1=0}^{\infty} \cdots \sum_{i_{n-1}=0}^{\infty} ah_{nreg}(i_1,\ldots,i_{n-1},k-K)[\delta_0(K-i_1-\cdots-i_{n-1})$$

$$+ a\delta_0(-i_1-\cdots-i_{n-1})] \cdots [\delta_0(K-i_{n-1}) + a\delta_0(-i_{n-1})] \tag{77}$$

In the first term on the right side of (77), a little thought shows that the only nonzero summand occurs when $i_1 = \cdots = i_{n-1} = 0$. The second term is somewhat less easy to penetrate, so I will carry the calculation one step further. The summand will be nonzero only for the values $i_{n-1} = K$, and $i_{n-1} = 0$. Thus,

$$y(k) = h_{nreg}(0,\ldots,0,k)$$

$$+ \sum_{i_1=0}^{\infty} \cdots \sum_{i_{n-2}=0}^{\infty} ah_{nreg}(i_1,\ldots,i_{n-2},K,k-K)[\delta_0(-i_1-\cdots-i_{n-2})$$

$$+ a\delta_0(-i_1-\cdots-i_{n-2}-K)] \cdots [\delta_0(-i_{n-2}) + a\delta_0(-i_{n-2}-K)]$$

$$+ \sum_{i_1=0}^{\infty} \cdots \sum_{i_{n-2}=0}^{\infty} a^2h_{nreg}(i_1,\ldots,i_{n-2},0,k-K)[\delta_0(K-i_1-\cdots-i_{n-2})$$

$$+ a\delta_0(-i_1-\cdots-i_{n-2})] \cdots [\delta_0(K-i_{n-2}) + a\delta_0(-i_{n-2})] \tag{78}$$

Again, from the first summation, only the summand with $i_1 = \cdots = i_{n-2} = 0$ is nonzero. The second summation should be reduced further, but a pattern rapidly emerges that yields the response formula

$$y(k) = h_{nreg}(0,\ldots,0,k) + ah_{nreg}(0,\ldots,0,K,k-K)$$
$$+ a^2 h_{nreg}(0,\ldots,0,K,0,k-K) + a^3 h_{nreg}(0,\ldots,0,K,0,0,k-K)$$
$$+ \cdots + a^n h_{nreg}(0,\ldots,0,k-K) \tag{79}$$

(Rather than insert unit step functions in this expression, it is left understood that the regular kernel is zero if any argument is negative.)

The interesting thing about the response formula (79) is that if the system described by $h_{nreg}(i_1,\ldots,i_n)$ is bilinear realizable, and $n > 2$, then $y(k)$ is identically zero. This follows from the fact that the regular kernel corresponding to a homogeneous bilinear state equation is zero if any argument is zero. The general statement is that a degree-n homogeneous bilinear state equation has zero response to an input that contains at most $n-1$ nonzero values. The general proof in the style of the two-pulse case requires a very messy calculation. A much shorter proof is suggested in Problem 6.4. At any rate, this special property of discrete-time bilinear state equations indicates their somewhat restricted input/output behavior. In contrast, state-affine state equations are quite general, as should be clear from Section 6.3.

Frequency-response properties of the type discussed in Chapter 5 carry over more or less directly to the discrete-time case. To illustrate this I will briefly consider the steady-state response of a discrete-time, homogeneous system to the input signal

$$u(k) = 2A\cos(\omega k) = Ae^{i\omega k} + Ae^{-i\omega k} \tag{80}$$

The output can be written in terms of the symmetric kernel as

$$y(k) = \sum_{i_1=0}^{k} \cdots \sum_{i_n=0}^{k} h_{nsym}(i_1,\ldots,i_n) \prod_{j=1}^{n} [Ae^{i\omega(k-i_j)} + Ae^{-i\omega(k-i_j)}] \tag{81}$$

Letting $\lambda_1 = i\omega$ and $\lambda_2 = -i\omega$, expanding the n-fold product, and rearranging the summations gives

$$y(k) = A^n \sum_{k_1=1}^{2} \cdots \sum_{k_n=1}^{2} [\sum_{i_1=0}^{k} \cdots \sum_{i_n=0}^{k} h_{nsym}(i_1,\ldots,i_n)\exp(-\sum_{j=1}^{n} \lambda_{k_j} i_j)]\exp(\sum_{j=1}^{n} \lambda_{k_j} k)$$

Assuming convergence of the bracketed summations as k becomes large, $y(k)$ becomes arbitrarily close to the steady-state response defined by

$$y_{ss}(k) = A^n \sum_{k_1=1}^{2} \cdots \sum_{k_n=1}^{2} H_{nsym}(e^{\lambda_{k_1}},\ldots,e^{\lambda_{k_n}}) \exp(\sum_{j=1}^{n} \lambda_{k_j} k) \tag{82}$$

There are many terms in (82) with identical exponents, and these can be collected together using the symmetry of the transfer function. Let

$$G_{m,n-m}(e^{\lambda_1},e^{\lambda_2}) = \binom{n}{m} H_{nsym}(\underbrace{e^{\lambda_1},\ldots,e^{\lambda_1}}_{m};\ \underbrace{e^{\lambda_2},\ldots,e^{\lambda_2}}_{n-m}) \tag{83}$$

for $m = 0, 1, \ldots, n$. Then, replacing λ_1 by $i\omega$ and λ_2 by $-i\omega$,

$$y_{ss}(k) = A^n[G_{n,0}(e^{i\omega},e^{-i\omega})e^{in\omega k} + G_{0,n}(e^{i\omega},e^{-i\omega})e^{-in\omega k}]$$
$$+ A^n[G_{n-1,1}(e^{i\omega},e^{-i\omega})e^{i(n-2)\omega k} + G_{1,n-1}(e^{i\omega},e^{-i\omega})e^{-i(n-2)\omega k}]$$
$$+ \cdots + \begin{cases} A^n G_{n/2,n/2}(e^{i\omega},e^{-i\omega}), & n \text{ even} \\ A^n[G_{\frac{n+1}{2},\frac{n-1}{2}}(e^{i\omega},e^{-i\omega})e^{i\omega k} + G_{\frac{n-1}{2},\frac{n+1}{2}}(e^{i\omega},e^{-i\omega})e^{-i\omega k}], & n \text{ odd} \end{cases} \tag{84}$$

Using standard identities and the fact that

$$G_{m,n-m}(e^{i\omega},e^{-i\omega}) = G_{n-m,m}(e^{-i\omega},e^{i\omega})$$

gives the steady-state response expression

$$y_{ss}(k) = 2A^n |G_{n,0}(e^{i\omega},e^{-i\omega})| \cos[n\omega k + \angle G_{n,0}(e^{i\omega},e^{-i\omega})]$$
$$+ 2A^n |G_{n-1,1}(e^{i\omega},e^{-i\omega})| \cos[(n-2)\omega k + \angle G_{n-1,1}(e^{i\omega},e^{-i\omega})] + \cdots$$
$$+ \begin{cases} A^n G_{n/2,n/2}(e^{i\omega},e^{-i\omega}), & n \text{ even} \\ 2A^n |G_{\frac{n+1}{2},\frac{n-1}{2}}(e^{i\omega},e^{-i\omega})| \cos[\omega k + \angle G_{\frac{n+1}{2},\frac{n-1}{2}}(e^{i\omega},e^{-i\omega})], & n \text{ odd} \end{cases} \tag{85}$$

This calculation should be enough to indicate that the results of Sections 5.2 and 5.3 can be developed for the discrete-time case with ease.

Nonlinear systems with random inputs, the last major topic of Chapter 5, can be developed for discrete-time systems in a simple, informal manner that parallels the continuous-time case. Rather than do this, I will discuss orthogonal representations for discrete-time nonlinear systems with random inputs from a more general and more rigorous viewpoint. (While these more general ideas can be carried back to continuous-time systems, it is much easier to approach rigor without mortis in the discrete-time case.)

The development of orthogonal representations for nonlinear systems with random input signals will be based on the notion of orthogonalizing a random process. A discrete-time random process will be written in the form

$$u = \{u(k)\ ;\ k = \ldots, -1, 0, 1, 2, \ldots\} \tag{86}$$

and it will be assumed throughout that u is real, (strict-sense) stationary, and such that $|E[u^n(k)]| < \infty$ for all nonnegative integers n. Furthermore, it will be assumed throughout that the random process u is *independent of order n* for all nonnegative integers n. That is, for distinct indices $i_1, \ldots, i_n$, and any polynomials $p_1(x), \ldots, p_n(x)$,

$$E[p_1(u(i_1)) \cdots p_n(u(i_n))] = E[p_1(u(i_1))] \cdots E[p_n(u(i_n))] \tag{87}$$

This is a restrictive assumption, but it plays a crucial role in the development. It can be shown that a white Gaussian random process satisfies this assumption, and so the setting here includes the discrete-time version of the case discussed in Section 5.5.

Definition 6.1 The random process u is called *polynomial orthogonalizable* if there exist real, symmetric polynomial functions

$$\Phi_n(i_1,\ldots,i_n,u) = \Phi_n(u(i_1),\ldots,u(i_n)), \quad n = 0,1,2,\ldots \tag{88}$$

such that for all integers $i_1,\ldots,i_n,j_1,\ldots,j_m$,

$$E[\Phi_n(i_1,\ldots,i_n,u)\Phi_m(j_1,\ldots,j_m,u)]$$
$$= \begin{cases} E[\Phi_n(i_1,\ldots,i_n,u)\Phi_n(j_1,\ldots,j_n,u)], & n = m \\ 0, \; n \neq m \end{cases} \tag{89}$$

Such a set will be called a *polynomial orthogonal representation* for u.

An approach to finding polynomial orthogonal representations for a random process can be given as follows. The notation that often will be used for Φ_n involves collecting together repeated arguments and showing the number of occurrences. From symmetry, it is clear that this reordering is immaterial.

Lemma 6.2 Suppose that $\psi_n(x)$, $n = 0,1,2,\ldots$, is a set of single-variable polynomials with $\psi_0(x) = 1$ and such that for the random process u,

$$E[\psi_n(u(k))\psi_m(u(k))] = \begin{cases} E[\psi_n^2(u(k))] < \infty, & n = m \\ 0, \; n \neq m \end{cases} \tag{90}$$

Then the random process u is polynomial orthogonalizable, and a polynomial orthogonal representation is given by

$$\Phi_n(\underbrace{i_1,\ldots,i_1}_{n_1};\ldots;\underbrace{i_p,\ldots,i_p}_{n_p},u) = \psi_{n_1}(u(i_1))\cdots\psi_{n_p}(u(i_p)) \tag{91}$$

where $i_1,\ldots,i_p$ are distinct integers, and $n_1+\cdots+n_p = n$.

Proof It is clear that each Φ_n defined in (91) is a symmetric polynomial function. Furthermore, with some abuse of notation that arises in collecting together repeated arguments in the style of (91), $E[\Phi_n(i_1,\ldots,i_n,u)\Phi_m(j_1,\ldots,j_m,u)]$ can be written in the form

$$E[\Phi_n(\underbrace{i_1,\ldots,i_1}_{n_1};\ldots;\underbrace{i_p,\ldots,i_p}_{n_p},u)\Phi_m(\underbrace{j_1,\ldots,j_1}_{m_1};\ldots;\underbrace{j_q,\ldots,j_q}_{m_q},u)]$$
$$= E[\psi_{n_1}(u(i_1))\cdots\psi_{n_p}(u(i_p))\psi_{m_1}(u(j_1))\cdots\psi_{m_q}(u(j_q))] \tag{92}$$

where $i_1,\ldots,i_p$ are distinct with $n_1+\cdots+n_p = n$, and $j_1,\ldots,j_q$ are distinct with $m_1+\cdots+m_q = m$. If $m \neq n$ either there is a distinct integer, say i_1, in the set $\{i_1,\ldots,i_p,j_1,\ldots,j_q\}$, or there are two identical integers, say $i_1 = j_1$, such that $n_1 \neq m_1$. Using the independence assumption to write (92) as a product of expected values, in the former case one of the factors will be $E[\psi_{n_1}(u(i_1))]$, which is zero since $\psi_{n_1}(x)$ is orthogonal to $\psi_0(x) = 1$. In the latter case one of the factors will be $E[\psi_{n_1}(u(i_1))\psi_{m_1}(u(i_1))]$, which also is zero. Thus (89) has been verified, although it is convenient to further note here that when $n = m$, (92) gives zero unless $\{j_1, \ldots, j_n\}$ is a permutation of $\{i_1, \ldots, i_n\}$. If the permutation condition holds, then $E[\Phi_n(i_1, \ldots, i_n, u)\Phi_n(j_1, \ldots, j_n, u)]$ is given by

$$E[\Phi_n^2(\underbrace{i_1, \ldots, i_1}_{n_1}; \ldots; \underbrace{i_p, \ldots, i_p}_{n_p}, u)] = E[\psi_{n_1}^2(u(i_1))] \cdots E[\psi_{n_p}^2(u(i_p))] \tag{93}$$

where $i_1, \ldots, i_p$ are distinct, and $n_1 + \cdots + n_p = n$.

The example that will be carried throughout this topic corresponds to the Wiener orthogonal representation discussed for the continuous-time case in Section 5.5.

Example 6.8 Suppose the random process u is zero-mean, Gaussian, and white, with intensity $E[u^2(k)] = A$. Then it can be verified that u satisfies the order n independence condition. To construct a polynomial orthogonal representation, take ψ_n, $n = 0, 1, 2, \ldots$ to be the Hermite polynomials given by

$$\psi_n(x) = \sum_{r=0}^{[n/2]} \frac{(-1)^r n! A^r}{r! 2^r (n-2r)!} x^{n-2r} \tag{94}$$

where $[n/2]$ is the largest integer $\leqslant n/2$. The first few Hermite polynomials are

$$\psi_0(x) = 1, \quad \psi_1(x) = x, \quad \psi_2(x) = x^2 - A$$

It is left to the references to verify (90) in this case, and to obtain the identity

$$E[\psi_n^2(u(k))] = n! A^n \tag{95}$$

However, I should point out that arguments reminiscent of the proofs of Theorems 5.1 and 5.2 can be used in place of an appeal to the literature. At any rate, the Hermite polynomials lead to a polynomial orthogonal representation for zero-mean, white Gaussian random processes via the definition in (91).

The following mathematical framework will be convenient in developing the representation for nonlinear systems with random inputs. Let $F[u(k)]$ be a real-valued functional of the sample function $u(k)$ of the random process u. Assume that $E[F^2[u(k)]] < \infty$, and denote by $L_2(u)$ the Hilbert space of such functionals F and G with inner product

$$<F,G> = E[F[u(k)]G[u(k)]] \tag{96}$$

Suppose $\Phi_0, \Phi_1, \ldots$ is a polynomial orthogonal representation for u constructed as in Lemma 6.2. Then for each n and $i_1, \ldots, i_n$, $\Phi_n(i_1, \ldots, i_n, u)$ is an element of $L_2(u)$. If $f_n(i_1, \ldots, i_n)$ is a real-valued function that satisfies

$$\sum_{i_1=-\infty}^{\infty} \cdots \sum_{i_n=-\infty}^{\infty} f_n^2(i_1, \ldots, i_n) < \infty \tag{97}$$

then

$$\sum_{i_1=-\infty}^{\infty} \cdots \sum_{i_n=-\infty}^{\infty} f_n(i_1, \ldots, i_n)\Phi_n(i_1, \ldots, i_n, u) \tag{98}$$

is an element of $L_2(u)$. (The demonstration of this fact will be omitted. In the sequel, functions f_n that are nonzero for only finitely many arguments will be considered, and in this case the claim is clear.) It is left to the reader to show that, when considering expressions of the form (98), the symmetry of $\Phi_n(i_1, \ldots, i_n, u)$ implies that without loss of generality $f_n(i_1, \ldots, i_n)$ can be assumed to be symmetric.

Now consider a stationary, causal system $y(k) = H[u(k)]$, where the input is a sample function from the real, stationary, independent of order n, random process u. To represent the system as an element of $L_2(u)$, assume that k is fixed, $E[y^2(k)] < \infty$, and for simplicity that the system is *finite memory.* That is, there exists a positive integer M such that $y(k)$ depends only on the values $u(k)$, $u(k-1), \ldots, u(k-M)$. Such a system will be denoted by the functional notation

$$y(k) = H[u(k-j),\ j = 0, 1, \ldots, M] \tag{99}$$

All the machinery is now available to develop a representation for a system of the form (99), which explicitly involves a polynomial orthogonal representation for u. In particular, consider a representation of the form

$$\begin{aligned} y_N(k) &= H_N[u(k-j),\ j = 0, 1, \ldots, M] \\ &= \sum_{n=0}^{N} \sum_{i_1=0}^{M} \cdots \sum_{i_n=0}^{M} k_n(i_1, \ldots, i_n)\Phi_n(k-i_1, \ldots, k-i_n, u) \end{aligned} \tag{100}$$

where each $k_n(i_1, \ldots, i_n)$ is symmetric. Clearly H_N belongs to $L_2(u)$, and the system representation is stationary, finite memory, and causal. The objective is to choose the coefficient functions $k_n(i_1, \ldots, i_n)$ so that (100) approximates (99) in the mean-square sense. That is, choose k_0, $k_1(i_1), \ldots, k_N(i_1, \ldots, i_N)$ to minimize the error

$$\begin{aligned} \|y(k) - y_N(k)\|^2 &= \langle y(k)-y_N(k),\ y(k)-y_N(k) \rangle \\ &= E[(y(k) - y_N(k))^2] \end{aligned} \tag{101}$$

Definition 6.2 The symmetric functions $k_n(i_1, \ldots, i_n)$ that minimize (101) are called *Fourier kernels* (relative to the Φ_n), and the resulting functional (100) is called a *functional Fourier series representation* of the system.

Theorem 6.9 Suppose a polynomial orthogonal representation for u is constructed as in Lemma 6.2 using the polynomials $\psi_0(x), \psi_1(x), \ldots$. Then the nth Fourier kernel is given by

$$k_n(\underbrace{i_1, \ldots, i_1}_{n_1}; \ldots; \underbrace{i_p, \ldots, i_p}_{n_p}) = \frac{n_1! \cdots n_p!}{n!E[\psi_{n_1}^2(u(k))] \cdots E[\psi_{n_p}^2(u(k))]} E[y(k)\psi_{n_1}(u(k-i_1)) \cdots \psi_{n_p}(u(k-i_p))] \tag{102}$$

where $i_1, \ldots, i_p$ are distinct, $n_1 + \cdots + n_p = n$, and $n = 0, 1, \ldots, N$.

Proof Using an abbreviated notation for (100) with all arguments discarded, the error criterion (101) can be written in the form

$$\|y - y_N\|^2 = \langle y - \sum_{n=0}^{N} k_n X\Phi_n,\ y - \sum_{n=0}^{N} k_n X\Phi_n \rangle$$

$$= \langle y,y \rangle - 2\langle y, \sum_{n=0}^{N} k_n X\Phi_n \rangle + \langle \sum_{n=0}^{N} k_n X\Phi_n,\ \sum_{n=0}^{N} k_n X\Phi_n \rangle$$

Using the easily verified result

$$\langle k_n X\Phi_n, k_m X\Phi_m \rangle = 0, \quad n \neq m$$

and writing the inner products as expectations gives

$$\|y - y_N\|^2 = E[y^2] - 2\sum_{n=0}^{N} E[y(k_n X\Phi_n)] + \sum_{n=0}^{N} E[(k_n X\Phi_n)^2]$$

Now, expand the notation of the terms on the right side. First

$$E[y(k_n X\Phi_n)] = E[y(k)\sum_{i_1=0}^{M} \cdots \sum_{i_n=0}^{M} k_n(i_1,\ldots,i_n)\Phi_n(k-i_1,\ldots,k-i_n,u)]$$

$$= \sum_{i_1=0}^{M} \cdots \sum_{i_n=0}^{M} k_n(i_1,\ldots,i_n) E[y(k)\Phi_n(k-i_1,\ldots,k-i_n,u)]$$

Using symmetry properties and the construction for Φ_n in Lemma 6.2, a general term in $E[y(k_n X\Phi_n)]$ can be isolated as follows. Suppose $i_1,\ldots,i_p$, $1 \leqslant p \leqslant n$, are distinct nonnegative integers, and $n_1,\ldots,n_p$ are positive integers with $n_1+\cdots+n_p = n$. Then all those terms containing n_j occurrences of the argument i_j, $j = 1,\ldots,p$, are identical, and the collection of these terms can be written as

$$\frac{n!}{n_1!\cdots n_p!} k_n(\underbrace{i_1,\ldots,i_1}_{n_1};\ldots;\underbrace{i_p,\ldots,i_p}_{n_p})\ E[y(k)\psi_{n_1}(u(k-i_1))\cdots\psi_{n_p}(u(k-i_p))]$$

In a similar manner,

$$E[(k_n X\Phi_n)^2] = E[\sum_{i_1=0}^{M} \cdots \sum_{i_n=0}^{M} k_n(i_1,\ldots,i_n)\Phi_n(k-i_1,\ldots,k-i_n,u)$$

$$\sum_{j_1=0}^{M} \cdots \sum_{j_n=0}^{M} k_n(j_1,\ldots,j_n)\Phi_n(k-j_1,\ldots,k-j_n,u)]$$

$$= \sum_{i_1=0}^{M} \cdots \sum_{i_n=0}^{M} \sum_{j_1=0}^{M} \cdots \sum_{j_n=0}^{M} k_n(i_1,\ldots,i_n) k_n(j_1,\ldots,j_n)$$

$$E[\Phi_n(k-i_1,\ldots,k-i_n)\Phi_n(k-j_1,\ldots,k-j_n,u)]$$

As discussed earlier, the expected value on the right side is zero unless $j_1,\ldots,j_n$ is a permutation of $i_1,\ldots,i_n$. Using this fact, and symmetry properties, again a general term can be isolated. Suppose $i_1,\ldots,i_p$ and $n_1,\ldots,n_p$ are just as above. Then the collection of terms containing n_j occurences of the argument i_j, $j = 1,\ldots,p$, can be written as

$$[\frac{n!}{n_1!\cdots n_p!}]^2 k_n^2(\underbrace{i_1,\ldots,i_1}_{n_1};\ldots;\underbrace{i_p,\ldots,i_p}_{n_p})\; E[\psi_{n_1}^2(u(k-i_1))]\;\cdots\; E[\psi_{n_p}^2(u(k-i_p))]$$

Now, the error criterion can be expressed as a sum of terms of the general types given above, with the sum ranging over all distinct nonnegative integers $i_1,\ldots,i_p$, $p = 1,\ldots,n$, and over all distributions of these integers as given by the positive integers $n_1,\ldots,n_p$, with $n_1+\cdots+n_p = n$. That is, in a vague summation notation,

$$\begin{aligned}\|y(k) - y_N(k)\|^2 = E[y^2(k)] & \\ + \sum \frac{(-2)n!}{n_1!\cdots n_p!}\; & k_n(\underbrace{i_1,\ldots,i_1}_{n_1};\ldots;\underbrace{i_p,\ldots,i_p}_{n_p}) \\ & E[y(k)\psi_{n_1}(u(k-i_1))\;\cdots\;\psi_{n_p}(u(k-i_p))] \\ + (\frac{n!}{n_1!\cdots n_p!})^2 & k_n^2(\underbrace{i_1,\ldots,i_1}_{n_1};\ldots;\underbrace{i_p,\ldots,i_p}_{n_p}) \\ & E[\psi_{n_1}^2(u(k-i_1))]\;\cdots\; E[\psi_{n_p}^2(u(k-i_p))] \end{aligned} \tag{103}$$

Minimization of the quadratic criterion is straightforward, and the result is easily seen to be the Fourier kernel specified in (102).

Example 6.9 Consider again the case where u is zero-mean, white, and Gaussian with intensity A. Using the results of Example 6.8, it is straightforward to calculate the first few terms in the functional Fourier series representation explicitly in terms of u rather than the orthogonal representation of u. In terms of the abbreviated notation for (100),

$$\begin{aligned} k_0X\Phi_0 &= k_0 \\ k_1X\Phi_1 &= \sum_{i_1=0}^{M} k_1(i_1)\psi_1(u(k-i_1)) = \sum_{i_1=0}^{M} k_1(i_1)u(k-i_1) \\ k_2X\Phi_2 &= \sum_{i_1=0}^{M}\sum_{i_2=0}^{M} k_2(i_1,i_2)\Phi_2(k-i_1,k-i_2,u) \\ &= \sum_{i_1=0}^{M}\sum_{i_2=0}^{M} k_2(i_1,i_2)u(k-i_1)u(k-i_2) - A\sum_{i=0}^{M} k_2(i,i) \end{aligned} \tag{104}$$

and so on. These terms should begin to look familiar from the continuous-time case in Section 5.5. Indeed, a messy general argument shows that

$$k_n X\Phi_n = \sum_{r=0}^{[n/2]} \frac{(-1)^r n! A^r}{r! 2^r (n-2r)!} \sum_{i_1=0}^{M} \cdots \sum_{i_{n-2r}=0}^{M} \sum_{j_1=0}^{M} \cdots \sum_{j_r=0}^{M}$$

$$k_n(i_1, \ldots, i_{n-2r}, j_1, j_1, \ldots, j_r, j_r) u(k-i_1) \cdots u(k-i_{n-2r}) \tag{105}$$

Thus, the Wiener orthogonal representation can be viewed as a particular case (of the general functional Fourier series) that displays explicitly the system input rather than an orthogonal representation of the input.

A natural question to ask about the functional Fourier series representation concerns the convergence properties as more terms are added. While this question will not be analyzed in detail, assuming certain completeness properties of the orthogonal polynomials $\psi_n(x)$, $n = 0, 1, \ldots$, it can be shown that

$$\lim_{N \to \infty} \| y(k) - y_N(k) \|^2 = 0 \tag{106}$$

That is, the given system can be approximated to desired accuracy by a finite functional Fourier series of the form

$$y(k) = \sum_{n=0}^{N} k_n X\Phi_n$$

$$= \sum_{n=0}^{N} \sum_{i_1=0}^{M} \cdots \sum_{i_n=0}^{M} k_n(i_1, \ldots, i_n) \Phi_n(k-i_1, \ldots, k-i_n, u) \tag{107}$$

Finally, the orthogonality property of the functional Fourier series offers a simple expression for the autocorrelation function of the system output, the derivation of which is left as an exercise.

6.6 Bilinear Input/Output Systems

The theory of nonlinear systems with input/output behavior that can be described by an n-linear operator is closely related to the theory of homogeneous nonlinear systems. To illustrate, I will consider the special case of stationary systems that can be represented by a bilinear (2-linear) operator. (The terminology is dangerous. Systems described by bilinear operators should not be confused with systems described by bilinear state equations.) It is a straightforward matter to carry the discussion back to the continuous-time case, though some differences do appear. It is less straightforward to generalize the theory to the n-linear case simply because of the morass of algebra, and the resulting need to develop a more abstract and subtle notation.

Consider a stationary, causal, discrete-time system that has two scalar inputs, $u_1(k)$ and $u_2(k)$, and a scalar output $y(k)$. Such a system can be represented in operator notation by

$$y = F[u_1, u_2] \tag{108}$$

where F is an operator with appropriately defined, real, linear function spaces for the domain and range. The system is called a *bilinear input/output system* if F is a bilinear operator; that is, if F is linear in each argument. More precisely, F is a bilinear operator if

$$F[\alpha_1 u_1+\hat{\alpha}_1\hat{u}_1, \alpha_2 u_2+\hat{\alpha}_2\hat{u}_2] = \alpha_1\alpha_2 F[u_1,u_2] + \alpha_1\hat{\alpha}_2 F[u_1,\hat{u}_2] + \hat{\alpha}_1\alpha_2 F[\hat{u}_1,u_2] + \hat{\alpha}_1\hat{\alpha}_2 F[\hat{u}_1,\hat{u}_2] \quad (109)$$

for all real α_1, $\hat{\alpha}_1$, α_2, $\hat{\alpha}_2$ and all input signals $u_1(k)$, $\hat{u}_1(k)$, $u_2(k)$, and $\hat{u}_2(k)$.

The main part of the discussion here will be concerned with developing a more explicit form for the input/output representation. A simple way to accomplish this is to consider bilinear input/output systems that are realizable by a general class of state equations. Then special properties of these state equations can be used to obtain properties of the corresponding input/output representation. I will pursue this approach for bilinear input/output systems that are realizable by state equations of the form

$$x(k+1) = f[x(k),u_1(k),u_2(k)], \quad k = 0,1,2,\ldots$$
$$y(k) = h(x(k)), \quad x(0) = 0 \quad (110)$$

where $x(k)$ is the n-dimensional state vector, and f and h are assumed to be analytic functions satisfying $f(0,0,0) = 0$ and $h(0) = 0$. The choice of the equilibrium initial state at zero and the analyticity requirements can be relaxed in various ways without changing the essential features of the results. Also, more general output equations, namely those of the form

$$y(k) = h[x(k),u_1(k),u_2(k)]$$

can be handled by the methods to be used, although the formulas and block diagrams become more complicated.

The next step is to use power series expansions of f and h in (110) to rewrite the state equation description, and then to consider input signals of the form indicated in (109). By writing $x(k)$ and $y(k)$ as expansions in terms of α_1, $\hat{\alpha}_1$, α_2, and $\hat{\alpha}_2$, and imposing the bilinear input/output condition (109), various terms in the resulting variational equations can be eliminated to obtain a simpler state equation. I will simplify this procedure somewhat by imposing the input/output condition (implied by (109))

$$F[\alpha_1 u_1, \alpha_2 u_2] = \alpha_1\alpha_2 F[u_1,u_2] \quad (111)$$

Removing terms in the state equation that are incompatible with (111) will yield a simple structural form for realizations of bilinear input/output systems. (It will become more or less apparent that further simplification is not obtained by imposing the more complicated condition (109), although a proof of this fact will not be given.)

Using the familiar Kronecker product notation, the state equation (110) can be written in the form

$$x(k+1) = A_1x(k) + A_2x(k) \otimes x(k) + D_1x(k)u_1(k) + D_2x(k)u_2(k)$$
$$+ b_1u_1(k) + b_2u_2(k) + b_3u_1(k)u_2(k) + \cdots$$
$$y(k) = c_1x(k) + c_2x(k) \otimes x(k) + \cdots \tag{112}$$

where only the terms that enter into the subsequent development are displayed. For input signals $\alpha_1u_1(k)$ and $\alpha_2u_2(k)$, assume

$$x(k) = \alpha_1x_1(k) + \alpha_2x_2(k) + \alpha_1\alpha_2x_3(k) + \cdots \tag{113}$$

Again, only those terms in α_1 and α_2 are displayed that will yield an output $y(k)$ consistent with (111). Substituting (113) into the state equation and equating coefficients of like terms in α_1, α_2, and $\alpha_1\alpha_2$, yields the following state equation description for the bilinear input/output system represented by (110):

$$x_1(k+1) = A_1x_1(k) + b_1u_1(k), \quad x_1(0) = 0$$
$$x_2(k+1) = A_1x_2(k) + b_2u_2(k), \quad x_2(0) = 0$$
$$x_3(k+1) = A_1x_3(k) + A_2[x_1(k) \otimes x_2(k) + x_2(k) \otimes x_1(k)] + D_1x_2(k)u_1(k)$$
$$+ D_2x_1(k)u_2(k) + b_3u_1(k)u_2(k), \quad x_3(0) = 0$$
$$y(k) = c_1x_3(k) + c_2[x_1(k) \otimes x_2(k) + x_2(k) \otimes x_1(k)] \tag{114}$$

This set of equations can be put into a simpler form, although at considerable expense in dimension, by applying the Carleman linearization idea to the equation for $x_3(k)$. Let

$$\hat{x}_3(k) = x_1(k) \otimes x_2(k) + x_2(k) \otimes x_1(k) \tag{115}$$

Then a straightforward computation shows that $\hat{x}_3(k)$ satisfies

$$\hat{x}_3(k+1) = A_1 \otimes A_1\hat{x}_3(k) + [A_1 \otimes b_1 + b_1 \otimes A_1]x_2(k)u_1(k) + [A_1 \otimes b_2$$
$$+ b_2 \otimes A_1]x_1(k)u_2(k) + [b_1 \otimes b_2 + b_2 \otimes b_1]u_1(k)u_2(k) \tag{116}$$

where $\hat{x}_3(0) = 0$. Now let

$$z_3(k) = \begin{bmatrix} x_3(k) \\ \hat{x}_3(k) \end{bmatrix} \tag{117}$$

and combine the equations for $x_3(k)$ and $\hat{x}_3(k)$ to obtain

$$z_3(k+1) = \begin{bmatrix} A_1 & A_2 \\ 0 & A_1 \otimes A_2 \end{bmatrix} z_3(k) + \begin{bmatrix} D_1 \\ A_1 \otimes b_1 + b_1 \otimes A_1 \end{bmatrix} x_2(k)u_1(k)$$
$$+ \begin{bmatrix} D_2 \\ A_1 \otimes b_2 + b_2 \otimes A_1 \end{bmatrix} x_1(k)u_2(k)$$
$$+ \begin{bmatrix} b_3 \\ b_1 \otimes b_2 + b_2 \otimes b_1 \end{bmatrix} u_1(k)u_2(k) \tag{118}$$

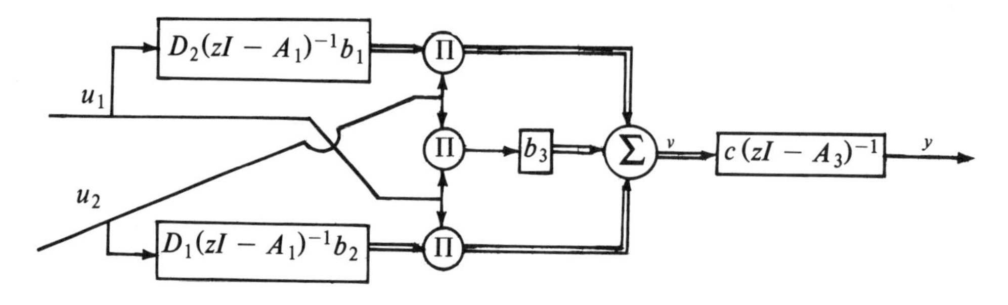

Figure 6.1 Interconnection realization of a bilinear input/output system.

Of course, the output equation can be written in the form

$$y(k) = [c_1 \quad c_2]z_3(k) \tag{119}$$

Summarizing in a simpler notation, the bilinear input/output system (110) also can be described by a state equation of the form

$$x_1(k+1) = A_1x_1(k) + b_1u_1(k),\quad x_1(0) = 0$$

$$x_2(k+1) = A_1x_2(k) + b_2u_2(k),\quad x_2(0) = 0$$

$$x_3(k+1) = A_3x_3(k) + D_1x_2(k)u_1(k) + D_2x_1(k)u_2(k) + b_3u_1(k)u_2(k),\quad x_3(k) = 0$$

$$y(k) = cx_3(k) \tag{120}$$

The simple structural form of (120) is indicated by the interconnection diagram shown in Figure 6.1, where vector quantities are denoted by double arrows. Of course, this interconnection realization is usually far from minimal since the dimension of the state equation (120) is $3n + n^2$. Some reduction in dimension could be obtained by using the reduced Kronecker product, but the result still would be far from minimal.

A concrete form for the input/output representation of a bilinear input/output system described by (110) can be derived from the interconnection structure shown in Figure 6.1. The derivation involves the familiar procedure of tracing the various signals through the diagram until the output signal is reached. Clearly, for $k > 0$,

$$D_2x_1(k)u_2(k) = \sum_{j_2=0}^{k-1} D_2A_1^{k-j_2-1}b_1u_1(j_2)u_2(k)$$

$$D_1x_2(k)u_1(k) = \sum_{j_2=0}^{k-1} D_1A_1^{k-j_2-1}b_2u_1(k)u_2(j_2)$$

$$v(k) = D_2x_1(k)u_2(k) + D_1x_2(k)u_1(k) + b_3u_1(k)u_2(k)$$

$$y(k) = \sum_{j_1=0}^{k-1} cA_3^{k-j_1-1} v(j_1) \tag{121}$$

Putting these equations together gives the input/output formula

$$\begin{aligned} y(k) = &\sum_{j_1=0}^{k-1} \sum_{j_2=0}^{j_1-1} cA_3^{k-j_1-1} D_2 A_1^{j_1-j_2-1} b_1 u_1(j_2) u_2(j_1) \\ &+ \sum_{j_1=0}^{k-1} \sum_{j_2=0}^{j_1-1} cA_3^{k-j_1-1} D_1 A_1^{j_1-j_2-1} b_2 u_1(j_1) u_2(j_2) \\ &+ \sum_{j_1=0}^{k-1} cA_3^{k-j_1-1} b_3 u_1(j_1) u_2(j_1) \end{aligned} \tag{122}$$

Thus, a bilinear input/output system described by (110) can be represented in the form

$$y(k) = \sum_{j_1=0}^{k-1} \sum_{j_2=0}^{k-1} h(k-j_1,k-j_2) u_1(j_1) u_2(j_2) \tag{123}$$

where, after some rearrangement of (122), the kernel is given by

$$h(i_1,i_2) = \begin{cases} cA_3^{i_1-1} D_1 A_1^{i_2-i_1-1} b_2, & 0 < i_1 < i_2 \\ cA_3^{i_2-1} D_2 A_1^{i_1-i_2-1} b_1, & 0 < i_2 < i_1 \\ cA_3^{i_1-1} b_3, & 0 < i_1 = i_2 \\ 0, & \mathit{otherwise} \end{cases} \tag{124}$$

In addition to representing a bilinear input/output system, it is apparent that (123) corresponds to a causal and stationary system.

Using the 2-variable z-transform, a *transfer function* representation can be defined for bilinear input/output systems:

$$H(z_1,z_2) = \sum_{i_1=0}^{\infty} \sum_{i_2=0}^{\infty} h(i_1,i_2) z_1^{-i_1} z_2^{-i_2} \tag{125}$$

The special structure for the kernel as displayed in (124) also implies a special structure for the transfer function. Substituting (124) into (125) gives

$$\begin{aligned} H(z_1,z_2) = &\sum_{i_1=1}^{\infty} \sum_{i_2=i_1+1}^{\infty} cA_3^{i_1-1} D_1 A_1^{i_2-i_1-1} b_2 z_1^{-i_1} z_2^{-i_2} \\ &+ \sum_{i_2=1}^{\infty} \sum_{i_1=i_2+1}^{\infty} cA_3^{i_2-1} D_2 A_1^{i_1-i_2-1} b_1 z_1^{-i_1} z_2^{-i_2} \\ &+ \sum_{i_1=i_2=1}^{\infty} cA_3^{i_1-1} b_3 z_1^{-i_1} z_2^{-i_2} \end{aligned} \tag{126}$$

To illustrate the remainder of the calculation, I will work out the first term on the right side of (126) in detail. Replacing the summation index i_2 by $j_2 = i_2 - i_1 - 1$ and using the identity

$$\sum_{i=0}^{\infty} A^i z^{-i} = z(zI - A)^{-1} \tag{127}$$

allows the first term to be rewritten as follows:

$$\sum_{i_1=1}^{\infty} \sum_{j_2=0}^{\infty} cA_3^{i_1-1} D_1 A_1^{j_2} b_2 z_1^{-i_1} z_2^{-(j_2+i_1+1)} = \sum_{i_1=1}^{\infty} cA_3^{i_1-1} D_1 (z_2 I - A_1)^{-1} b_2 (z_1 z_2)^{-i_1}$$

$$= c(z_1 z_2 I - A_3)^{-1} D_1 (z_2 I - A_1)^{-1} b_2$$

Performing this kind of calculation on the remaining two terms in (126) yields

$$H(z_1,z_2) = c(z_1 z_2 I - A_3)^{-1} D_1 (z_2 I - A_1)^{-1} b_2$$
$$+ c(z_1 z_2 I - A_3)^{-1} D_2 (z_1 I - A_1)^{-1} b_1 + c(z_1 z_2 I - A_3)^{-1} b_3 \tag{128}$$

Thus a general form has been obtained for the transfer function of a bilinear input/output system that can be described by a state equation of the form (110).

I will leave further discussion of the theory of bilinear input/output systems to the literature cited in Section 6.8. It should be clear at this point that such systems can be studied using methods similar to those developed for Volterra/Wiener representations. Input/output calculations in the transform domain involve the association-of-variables technique, and for certain types of input signals explicit response formulas can be derived. The structural form of the transfer function (or kernel) can be used to describe elementary conditions for realizability in terms of an interconnection structure. Finally, the reader surely has noticed that by setting $u_1(k) = u_2(k) = u(k)$, the bilinear input/output system reduces to a homogeneous system of degree 2. All of this indicates the symbiotic relationship between research in multilinear input/output systems and in homogeneous systems.

6.7 Two-Dimensional Linear Systems

The theory of multidimensional linear systems involves representations that resemble the Volterra/Wiener representation for nonlinear systems. Two-dimensional, stationary, discrete-time linear systems constitute the most widely studied case, and I will discuss the basics of this theory in order to exhibit the connections to nonlinear system theory. Motivation for the study of two-dimensional discrete-time systems comes principally from the processing (or filtering) of two-dimensional signals, notably in image or array processing, and geophysics.

The basic input/output representation for a two-dimensional, stationary, discrete-time linear system can be written in the form

$$y(k_1,k_2) = \sum_{i_1=0}^{k_1} \sum_{i_2=0}^{k_2} h(k_1-i_1,k_2-i_2)u(i_1,i_2), \quad k_1,k_2 = 0,1,2,\ldots \tag{129}$$

The input $u(k_1,k_2)$ and output $y(k_1,k_2)$ are real two-dimensional (or, doubly indexed) signals that are defined for integer arguments, but that are assumed to be zero if either argument is negative. Linearity is easily verified: in the obvious notation, the response to $\alpha u_1(k_1,k_2)+\beta u_2(k_1,k_2)$ is $\alpha y_1(k_1,k_2)+\beta y_2(k_1,k_2)$ for any scalars α and β. Stationarity corresponds to a delay invariance property, which, in the context of (129), can be stated as follows. If $u_2(k_1,k_2) = u_1(k_1-K_1,k_2-K_2)$, then $y_2(k_1,k_2) = y_1(k_1-K_1,k_2-K_2)$ for all nonnegative integer pairs K_1, K_2. Notice that the concept of causality is not mentioned, though something vaguely like that is built into the representation. The reader might enjoy consulting his muse on the types of operations on an array of data $u(k_1,k_2)$ that might be described by (129).

Using the 2-variable z-transform, and the convolution property in Theorem 6.4, gives the input/output representation

$$Y(z_1,z_2) = H(z_1,z_2)U(z_1,z_2) \tag{130}$$

where $H(z_1,z_2) = Z[h(k_1,k_2)]$ is called the *transfer function* of the system. It should be immediately obvious from earlier chapters how to use (130) to investigate the response properties of the system for various classes of input signals.

Example 6.10 The simplest (nonzero) input signal is the unit pulse input, which is defined in the two-dimensional setting by

$$u_0(k_1,k_2) = \begin{cases} 1, & k_1 = k_2 = 0 \\ 0, & otherwise \end{cases}$$

From (129), the response clearly is

$$y(k_1,k_2) = h(k_1,k_2), \quad k_1,k_2 = 0,1,2,\ldots$$

or, since $U_0(z_1,z_2) = 1$,

$$Y(z_1,z_2) = H(z_1,z_2)$$

Of course, in a digital filtering context it is the steady-state frequency response properties of the system that are of prime importance. It is rather easy to work out these properties, and therefore that task is left to Section 6.9.

There are several types of state equation representations that can be adopted for the study of two-dimensional linear systems. I will work with the general form

$$x(k_1+1,k_2+1) = A_1x(k_1+1,k_2) + A_2x(k_1,k_2+1) + B_1u(k_1+1,k_2) + B_2u(k_1,k_2+1)$$
$$y(k_1,k_2) = cx(k_1,k_2), \quad k_1,k_2 = 0,1,2,\ldots \tag{131}$$

where $x(k_1,k_2)$ is an n x 1 vector. Iterating this equation for the first few values of k_1 and k_2 shows that the initial conditions, more appropriately called boundary conditions, required for solution are the values $x(k_1,0)$, $k_1 = 0,1,\ldots$, and

$x(0,k_2)$, $k_2 = 0,1,\ldots$. This multiplicity of boundary conditions indicates that $x(k_1,k_2)$ is not a state vector for the system in any precise sense of the term. That is, knowledge of the value of $x(k_1,k_2)$ and the input signal does not suffice to determine the value of $x(k_1+K_1,k_2+K_2)$. Stated in the context of array processing, a single value of $x(k_1,k_2)$ does not specify the "state" of the array. Rather, the equation for $x(k_1,k_2)$ gives the recursion necessary to specify the array in a pointwise fashion. Thus I will call $x(k_1,k_2)$ a *local state vector* for the two-dimensional system, and call n the *local dimension* of the system (there goes the terminology, again). Further consideration of the nature of a "global state" will be left to the literature since my interest here is more mechanical than philosophical.

There are intuitive ways to arrive at a choice for the form of the local state equation for two-dimensional systems. This intuition is based on viewing the system as an array processor, and imagining various methods by which the values in the array might be generated. I will go through one of these just to give some motivation for the choice in (131).

Example 6.11 Suppose the values $y(k_1,k_2)$ in a certain array can be generated by a combination of a horizontal recursion and a vertical recursion. Let $x_h(k_1,k_2)$ be the local horizontal state, and $x_v(k_1,k_2)$ be the local vertical state, and suppose the local states propagate according to

$$\begin{aligned} x_h(k_1+1,k_2) &= A_1x_h(k_1,k_2) + A_2x_v(k_1,k_2) + B_1u(k_1,k_2) \\ x_v(k_1,k_2+1) &= A_3x_h(k_1,k_2) + A_4x_v(k_1,k_2) + B_2u(k_1,k_2) \\ y(k_1,k_2) &= c_1x_h(k_1,k_2) + c_2x_v(k_1,k_2), \quad k_1,k_2 = 0,1,\ldots \end{aligned} \tag{132}$$

Of course, the input signal $u(k_1,k_2)$ must be specified, and it is clear that the boundary conditions that must be specified are the values of $x_h(0,k_2)$, and $x_v(k_1,0)$ (the left-hand and bottom edges of the array). These local state equations can be put into the form (131) by defining $x(k_1,k_2)$ as

$$x(k_1,k_2) = \begin{bmatrix} x_h(k_1,k_2) \\ x_v(k_1,k_2) \end{bmatrix}$$

Then a straightforward calculation gives

$$\begin{aligned} x(k_1+1,k_2+1) = &\begin{bmatrix} 0 & 0 \\ A_3 & A_4 \end{bmatrix} x(k_1+1,k_2) + \begin{bmatrix} A_1 & A_2 \\ 0 & 0 \end{bmatrix} x(k_1,k_2+1) \\ &+ \begin{bmatrix} B_1 \\ 0 \end{bmatrix} u(k_1,k_2+1) + \begin{bmatrix} 0 \\ B_2 \end{bmatrix} u(k_1+1,k_2) \end{aligned}$$

$$y(k_1,k_2) = [c_1 \quad c_2]x(k_1,k_2)$$

which shows that (132) can be viewed as a special case of (131).

The transfer function corresponding to the local state equation in (131) is easy to compute using the result of Problem 6.3. For zero boundary conditions, the state equation can be written in the transform-domain form

$$z_1z_2X(z_1,z_2) = A_1z_1X(z_1,z_2) + A_2z_2X(z_1,z_2) + B_1z_1U(z_1,z_2) + B_2z_2U(z_1,z_2) \tag{133}$$

Solving gives

$$X(z_1,z_2) = (z_1z_2I - A_1z_1 - A_2z_2)^{-1}(B_1z_1 + B_2z_2)U(z_1,z_2) \tag{134}$$

so that the input/output relationship takes the form

$$Y(z_1,z_2) = c(z_1z_2I - A_1z_1 - A_2z_2)^{-1}(B_1z_1 + B_2z_2)U(z_1,z_2) \tag{135}$$

Thus the transfer function corresponding to (131) can be written in the form

$$H(z_1,z_2) = c(z_1z_2I - A_1z_1 - A_2z_2)^{-1}(B_1z_1 + B_2z_2) \tag{136}$$

From the transform-domain solution of the local state equation, an "array-domain" solution can be derived as follows. Using an identity of the form (127) permits writing the matrix inverse in (134) in the form

$$\begin{aligned}(z_1z_2I - A_1z_1 - A_2z_2)^{-1} &= \sum_{i=0}^{\infty} (A_1z_1 + A_2z_2)^i(z_1z_2)^{-(i+1)} \\ &= \sum_{i_1=0}^{\infty}\sum_{i_2=0}^{\infty} A^{i_1,i_2} z_1^{-i_1}z_2^{-i_2}\end{aligned} \tag{137}$$

where A^{i_1,i_2} might appropriately be called the *two-dimensional transition matrix.* Equating coefficients of like terms in (137) shows that the first few values of A^{i_1,i_2} are

$$\begin{aligned}&A^{0,i} = A^{i,0} = 0, \quad i = 0,1,2,\ldots \\ &A^{1,3} = A_1^2 \\ &A^{2,1} = A_2, \quad A^{3,1} = A_2^2, \quad A^{2,2} = A_1A_2 + A_2A_1\end{aligned} \tag{138}$$

Now the convolution property, Theorem 6.4 in Section 6.2, in conjunction with (137) and (134) can be used to obtain an expression for the local state. First note that

$$\begin{aligned}X(z_1,z_2) &= \Big(\sum_{i_1=0}^{\infty}\sum_{i_2=0}^{\infty} A^{i_1,i_2}z_1^{-i_1}z_2^{-i_2}\Big)(B_1z_1 + B_2z_2)\sum_{j_1=0}^{\infty}\sum_{j_2=0}^{\infty} u(j_1,j_2)z_1^{-j_1}z_2^{-j_2} \\ &= \sum_{i_1=0}^{\infty}\sum_{i_2=0}^{\infty}\sum_{j_1=0}^{\infty}\sum_{j_2=0}^{\infty} z_1A^{i_1,i_2}B_1u(j_1,j_2)z_1^{-(i_1+j_1)}z_2^{-(i_2+j_2)} \\ &\quad + \sum_{i_1=0}^{\infty}\sum_{i_2=0}^{\infty}\sum_{j_1=0}^{\infty}\sum_{j_2=0}^{\infty} z_2A^{i_1,i_2}B_2u(j_1,j_2)z_1^{-(i_1+j_1)}z_2^{-(i_2+j_2)}\end{aligned} \tag{139}$$

Replacing j_1 by $k_1 = j_1+i_1$ and j_2 by $k_2 = j_2+i_2$, and making use of the "one-sidedness" of the input signal gives

$$X(z_1,z_2) = \sum_{k_1=0}^{\infty}\sum_{k_2=0}^{\infty}\sum_{i_1=0}^{k_1}\sum_{i_2=0}^{k_2} z_1A^{i_1,i_2}B_1u(k_1-i_1,k_2-i_2)z_1^{-k_1}z_2^{-k_2}$$

$$+ \sum_{k_1=0}^{\infty} \sum_{k_2=0}^{\infty} \sum_{i_1=0}^{k_1} \sum_{i_2=0}^{k_2} z_2 A^{i_1,i_2} B_2 u(k_1-i_1,k_2-i_2) z_1^{-k_1} z_2^{-k_2} \tag{140}$$

It follows that the solution of the local state equation (131) is given by

$$x(k_1,k_2) = \sum_{i_1=0}^{k_1} \sum_{i_2=0}^{k_2} (A^{i_1+1,i_2} B_1 + A^{i_1,i_2+1} B_2) u(k_1-i_1,k_2-i_2) \tag{141}$$

There are a number of structural features of local state equations of the form (131) that are similar to familiar properties in one-dimensional linear system theory. To illustrate, I will briefly discuss reachability and observability concepts for local state equations.

Definition 6.3 A state x_1 of the local state equation (131) is called a *reachable state* (from zero boundary conditions) if there exists an input signal such that for some $K_1,K_2 < \infty$, $x(K_1,K_2) = x_1$. The local state equation is called *reachable* if every state is reachable.

From (141) it is clear that a state x_1 is a reachable state if and only if

$$x_1 \in span \{ (A^{i_1+1,i_2} B_1 + A^{i_1,i_2+1} B_2) \mid i_1,i_2 = 0,1,\ldots\} \tag{142}$$

Before restating this condition in the traditional form of a rank condition for reachability of the state equation, it is necessary to establish the following result (which is related to a two-dimensional version of the Cayley-Hamilton Theorem).

Lemma 6.3 For the two-dimensional state transition matrix defined in (137),

$$span \{ A^{i_1,i_2} \mid i_1,i_2 = 0,1,\ldots\} = span \{ A^{i_1,i_2} \mid i_1,i_2 = 0,1,\ldots,n\} \tag{143}$$

Proof Expressing the matrix inverse in (137) in the classical adjoint-over-determinant form gives

$$adj\ (z_1z_2I - A_1z_1 - A_2z_2) = \det\ (z_1z_2I - A_1z_1 - A_2z_2) \sum_{i_1=0}^{\infty} \sum_{i_2=0}^{\infty} A^{i_1,i_2} z_1^{-i_1} z_2^{-i_2}$$

On the left side of this expression there are no terms with nonpositive powers of z_1 or z_2. On the right side, $\det\ (z_1z_2I - A_1z_1 - A_2z_2)$ is a polynomial of degree n in z_1 and degree n in z_2, while the nonzero terms in the double summation can occur only for $i_1,i_2 \geqslant 1$. Thus equating coefficients of like terms of the form $z_1^{-i_1} z_2^{-i_2}$, $i_1,i_2 \geqslant 0$, shows that when $i_1 > n$ or $i_2 > n$ there is a nontrivial linear combination of the matrices A^{i_1,i_2} that is zero. Clearly this conclusion implies (143).

Theorem 6.10 The local state equation (131) is reachable if and only if the matrix

$$[B_1 | B_2 | A^{1,1}B_1 | A^{1,1}B_2 | A^{2,1}B_1 + A^{1,2}B_2 | \cdots | A^{n+1,n}B_1 + A^{n,n+1}B_2] \tag{144}$$

has (full) rank n.

Proof Although sparsely denoted, the matrix in (144) contains as columns all $n \times 1$ vectors of the form $(A^{i_1+1,i_2}B_1 + A^{i_1,i_2+1}B_2)$ with $i_1 \leqslant n$, $i_2 \leqslant n$. Thus, the result is an easy consequence of Lemma 6.3 and the condition for state reachability in (142).

The appropriate definition of observability for the local state equation (131) is based on the nonexistence of boundary conditions that at the output are indistinguishable from the zero boundary conditions.

Definition 6.4 The local state equation (131) is called *observable* if there is no set of nonzero boundary conditions such that with identically zero input, the output is identically zero.

The development of conditions to characterize observability can be based on an analysis of the response of (131) to zero inputs and nonzero boundary conditions. Such an analysis followed by an application of Lemma 6.3 leads to Theorem 6.11, the proof of which is left to the reader.

Theorem 6.11 The local state equation (131) is observable if and only if the matrix

$$\begin{bmatrix} c \\ cA^{1,2} \\ cA^{2,1} \\ \vdots \\ cA^{n,n} \end{bmatrix} \tag{145}$$

has (full) rank n.

At the time of this writing, a realization theory for two-dimensional linear systems in terms of local state equations of the form (131) has not been completely worked out. It is clear from (136) that proper rationality of a given transfer function $H(z_1,z_2)$ is a necessary condition for realizability. Further inspection reveals that another necessary condition is that both the numerator and denominator polynomials of $H(z_1,z_2)$ must be zero when $z_1 = z_2 = 0$. In other words, these polynomials must not have nonzero constant terms. These necessary conditions also are sufficient, and a proof can be given by constructing a realization for a general transfer function that satisfies the conditions. To write out such a realization would be tiresome, so I will indicate vaguely what one looks like with an example and leave the general form to the literature.

Example 6.12 Consider the two-dimensional linear system described by

$$H(z_1,z_2) = \frac{b_{10}z_1 + b_{01}z_2}{z_1z_2 + a_{10}z_1 + a_{01}z_2}$$

A simple calculation shows that a realization for this system is

$$x(k_1+1,k_2+1) = \begin{bmatrix} 0 & 0 \\ -a_{01} & -a_{10} \end{bmatrix} x(k_1+1,k_2) + \begin{bmatrix} -a_{01} & -a_{10} \\ 0 & 0 \end{bmatrix} x(k_1,k_2+1)$$

$$+ \begin{bmatrix} 0 \\ 1 \end{bmatrix} u(k_1+1,k_2) + \begin{bmatrix} 1 \\ 0 \end{bmatrix} u(k_1,k_2+1)$$

$$y(k_1,k_2) = [b_{01} \quad b_{10}]x(k_1,k_2)$$

where all the initial conditions are zero.

Of course, the construction of minimal-dimension realizations for two-dimensional linear systems is of great interest, and much remains to be done in this area. In the one-dimensional case, the concepts of reachability and observability are useful tools in developing a theory of minimal realizations. However, the following example shows that in the two-dimensional case the situation is more complicated, and that perhaps the reachability and observability definitions discussed earlier are not the best choices.

Example 6.13 For the transfer function.

$$H(z_1,z_2) = \frac{z_1 - z_2}{z_1z_2 + z_1 + z_2}$$

the realization given in Example 6.12 becomes

$$x(k_1+1,k_2+1) = \begin{bmatrix} 0 & 0 \\ -1 & -1 \end{bmatrix} x(k_1+1,k_2) + \begin{bmatrix} -1 & -1 \\ 0 & 0 \end{bmatrix} x(k_1,k_2+1)$$

$$+ \begin{bmatrix} 0 \\ 1 \end{bmatrix} u(k_1+1,k_2) + \begin{bmatrix} 1 \\ 0 \end{bmatrix} u(k_1,k_2+1)$$

$$y(k_1,k_2) = [1 \quad -1]x(k_1,k_2)$$

A quick calculation shows that this local state equation is both reachable and observable. But it is not minimal since another realization is given by

$$x(k_1+1,k_2+1) = -x(k_1+1,k_2) - x(k_1,k_2+1) + u(k_1+1,k_2) - u(k_1,k_2+1)$$

$$y(k_1,k_2) = x(k_1,k_2)$$

Finally, it is easy to show that a bilinear input/output system (or for that matter, a degree-2 homogeneous system) can be modeled using a two-dimensional linear system. This involves nothing more than comparing the transform-domain input/output equations for the two classes of systems. Such a comparison shows that a bilinear input/output system with transfer function $H(z_1,z_2)$ can be viewed as follows. From the input signals $u_1(k)$ and $u_2(k)$, form an array $u(k_1,k_2) = u_1(k_1)u_2(k_2)$. Process this array with the two-dimensional linear system with transfer function $H(z_1,z_2)$ to obtain the array $y(k_1,k_2)$. Then set $y(k) = y(k,k)$, that is, let $y(k)$ be the diagonal of the array. Schematically this implementation is shown in Figure 6.2.

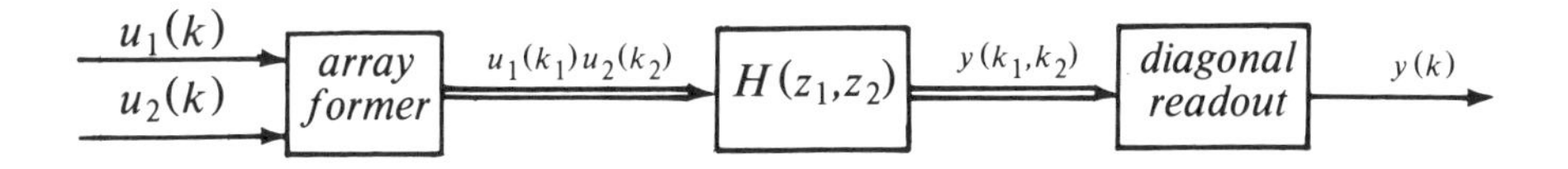

Figure 6.2. An implementation of a bilinear input/output system using a two-dimensional linear system.

6.8 Remarks and References

Remark 6.1 Early works dealing with Volterra series representations for discrete-time systems include

P. Alper, "A Consideration of the Discrete Volterra Series," *IEEE Transactions on Automatic Control,* Vol. AC-10, pp. 322-327, 1965.

A. Bush, "Some Techniques for the Synthesis of Nonlinear Systems," MIT RLE Technical Report No. 441, 1966 (AD 634-122).

H. Barker, S. Ambati, "Nonlinear Sampled-Data System Analysis by Multidimensional Z-Transforms," *Proceedings of the IEE,* Vol. 119, pp. 1407-1413, 1972.

All of these papers discuss the basic time-domain and transform-domain representations for discrete-time systems.

Remark 6.2 The development of state-affine realization theory in Section 6.4 is drawn from

S. Clancy, W. Rugh, "On the Realization Problem for Stationary Homogeneous Discrete-Time Systems," *Automatica,* Vol. 14, pp. 357-366, 1978.

S. Clancy, W. Rugh, "The Regular Transfer Function and Bilinear and State-Affine Realizations for Stationary, Homogeneous, Discrete-Time Systems," *Proceedings of the 1978 Conference on Information Sciences and Systems,* Electrical Engineering Department, The Johns Hopkins University, Baltimore, pp. 167-172, 1978.

A. Frazho, "Shift Operators and State-Affine Realization Theory," *Proceedings of the 19th IEEE Conference on Decision and Control,* Albuquerque, New Mexico, pp. 904-909, 1980.

The first of these papers contains a further discussion of the division-deletion method given in Example 6.3 for computing H_{reg} from H_{sym}. Another approach to the state-affine realization question is given in

E. Sontag, "Realization Theory of Discrete-Time Nonlinear Systems: I. The Bounded Case," *IEEE Transactions on Circuits and Systems,* Vol. CAS-26, pp. 342-356, 1979.

It should be noted that the papers by Frazho and Sontag cover much more general systems than those discussed in Section 6.4. Also, the noncommutative series approach (Remark 4.3) can be used in the discrete-time case. See

M. Fliess, "Un Codage Non Commutatif pour Certains Systemes Echantillonnes Non Lineaires," *Information and Control,* Vol. 38, pp. 264-287, 1978.

Remark 6.3 Calculations of the output mean, auto- and cross-correlations, and spectral densities for a discrete-time system with white noise input are given in

G. Cariolaro, G. Di Masi, "Second-Order Analysis of the Output of a Discrete-Time Volterra System Driven by White Noise," *IEEE Transactions on Information Theory,* Vol. IT-26, pp. 175-184, 1980.

The treatment of orthogonal representations in Section 6.5 is based on

S. Yasui, "Stochastic Functional Fourier Series, Volterra Series, and Nonlinear System Analysis," *IEEE Transactions on Automatic Control,* Vol. AC-24, pp. 230-242, 1979.

This paper treats a number of additional topics, a few of which will be discussed in Chapter 7. A detailed discussion of the properties of Hermite polynomials with regard to nonlinear system theory can be found in

M. Schetzen, *The Volterra and Wiener Theories of Nonlinear Systems,* John Wiley, New York, 1980.

Remark 6.4 The theory of multilinear input/output systems, in particular bilinear input/output systems, was spurred by

R. Kalman, "Pattern Recognition Properties of Multilinear Machines," IFAC International Symposium on Technical and Biological Problems of Control, Yeravan, USSR, 1968 (AD 731-304).

The basic interconnection structure representation for such a system was presented in

M. Arbib, "A Characterization of Multilinear Systems," *IEEE Transactions on Automatic Control,* Vol. AC-14, pp. 699-702, 1969.

Both of these papers use modern algebraic representations and the Nerode equivalence concept in essential ways. More recent works that continue the development from abstract viewpoints include

E. Fornasini, G. Marchesini, "Algebraic Realization Theory of Bilinear Discrete-Time Input/Output Maps," *Journal of The Franklin Institute,* Vol. 301, pp. 143-161, 1976.

B. Anderson, M. Arbib, E. Manes, "Foundations of System Theory: Multidecomposable Systems," *Journal of The Franklin Institute,* Vol. 301, pp. 497-508, 1976.

The methods I have used to introduce the theory of bilinear input/output systems follow more closely those in

E. Gilbert, "Bilinear and 2-Power Input-Output Maps: Finite Dimensional Realizations and the Role of Functional Series," *IEEE Transactions on Automatic Control,* Vol. AC-23, pp. 418-425, 1978.

Further developments regarding the structure shown in Figure 6.1 can be found in

J. Pearlman, "Canonical Forms for Bilinear Input/Output Maps," *IEEE Transactions on Automatic Control,* Vol. AC-23, pp. 595-602, 1978.

and a discussion of the difficulties involved in the minimal realization problem for multilinear input/output systems is given in

J. Pearlman, "Realizability of Multilinear Input/Output Maps," *International Journal of Control,* Vol. 32, pp. 271-283, 1980.

Remark 6.5 There has been a rapid growth of interest in the theory of multidimensional linear systems since the early 1970s. Several aspects of this theory are discussed in the special issue on multidimensional systems of the *Proceedings of the IEEE,* Vol. 65, 1977. The particular local state equation representation I have considered was introduced in

E. Fornasini, G. Marchesini, "Doubly Indexed Dynamical Systems: State-Space Models and Structural Properties," *Mathematical Systems Theory,* Vol. 12, pp. 59-72, 1978.

The general form for the realization of a given transfer function given there can be adapted to the setting in Section 6.7. The local state equation discussed in Example 6.9 is introduced in

R. Roesser, "A Discrete State Space Model for Linear Image Processing," *IEEE Transactions on Automatic Control,* Vol. AC-20, pp. 1-10, 1975.

The relation between bilinear input/output systems and linear two-dimensional systems given in Section 6.7 is exploited to study stability and realization properties of bilinear input/output systems in

E. Kamen, "On the Relationship between Bilinear Maps and Linear Two-Dimensional Maps," *Nonlinear Analysis,* Vol. 3, pp. 467-481, 1979.

6.9 Problems

6.1. Suppose two homogeneous, discrete-time, nonlinear systems are connected in cascade. Derive an expression for a kernel for the overall system in terms of the subsystem kernels.

6.2. If $Z[f(k)] = F(z)$, find the 2-variable z transform $Z[f(k_1+k_2)]$.

6.3. If

$$Z[f(k_1,\ldots,k_n)] = F(z_1,\ldots,z_n)$$

show that

$$Z[f(k_1+1,k_2,\ldots,k_n)] = z_1F(z_1,\ldots,z_n) - f(0,k_2,\ldots,k_n)$$

6.4. Using a block-form bilinear realization, show that the response of a degree-n (greater than 1), homogeneous, discrete-time system to an input of the form

$$u(k) = \delta_0(k) + a_1\delta_0(k-K_1) + \cdots + a_{n-2}\delta_0(k-K_{n-2})$$

is identically zero.

6.5. Compute the symmetric transfer function of the state-affine realization in Example 6.6 using a discrete-time version of the growing exponential method.

6.6. Write the state-affine state equation (56) in the form

$$x(k+1) = A_0x(k) + A[u(k)]x(k) + b[u(k)]$$

$$y(k) = c[u(k)]x(k) + d[u(k)]$$

with the obvious definitions of the functions $A(u)$, $b(u)$, $c(u)$, and $d(u)$. Then show the input/output relationship for the system can be written in the form

$$y(k) = \sum_{n=1}^{N} \sum_{i_1=0}^{k} \cdots \sum_{i_n=0}^{k} c[u(k)]A_0^{i_n-1}A[u(k-i_n)]A_0^{i_{n-1}-1}A[u(k-i_{n-1}-i_n)]$$

$$\cdots A_0^{i_2-1}A[u(k-i_2-\cdots-i_n)]A_0^{i_1-1}b[u(k-i_1-\cdots-i_n)]$$

6.7. Suppose a discrete-time, homogeneous system is described by the proper, recognizable regular transfer function $H_{reg}(z_1,\ldots,z_n)$. Derive a formula for the steady-state response of the system to the input $u(k) = 2A\cos(\omega k)$.

6.8. Suppose $u(k)$ is a real random signal defined on $k = 0,1,\ldots,K$. Discuss the possibilities of using the system diagramed below as an autocorrelation computer.

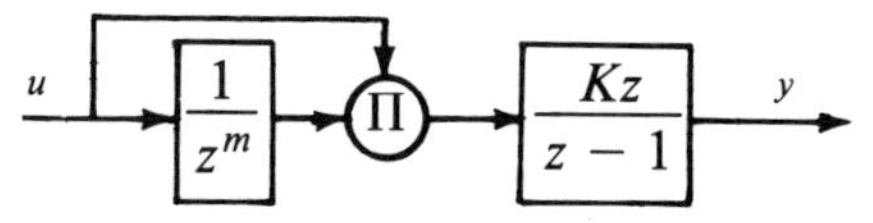

6.9. Complete the details in the proof of Theorem 6.9.

6.10. For a system described by (107), derive an expression for the output autocorrelation $R_{yy}(j)$.

6.11. Verify the realization in Example 6.12.

CHAPTER 7

Identification

The term identification will be used in a very broad sense to mean the obtaining of information about the kernels or transfer functions in a Volterra/Wiener representation of an unknown system from input/output experiments. This information usually will be in the form of values of the kernels or transfer functions for particular numerical values of the arguments. However, I also will discuss some simple parameter identification problems that arise when a particular structure is assumed for the unknown system, or when an expansion of the kernels in terms of known functions is assumed. It will become clear in the course of the discussion that much remains to be done.

Stationary polynomial systems will be considered, and usually the input/output experiments will involve the application of input signals of one of the types considered in Chapter 5 or Section 6.5. As a matter of convenience, sometimes the discussion will be in terms of continuous-time systems, and sometimes in terms of discrete-time systems.

7.1 Introduction

The determination of kernel values for an unknown system from a general input/output experiment is a linear problem. This is most easily demonstrated for the case of a discrete-time, polynomial system where, for technical simplicity, it is assumed that the system has *finite memory M,* and that the degree-0 term is zero. Assuming one-sided input signals, such a system can be described by the triangular kernel representation

$$y(k) = \sum_{n=1}^{N} \sum_{i_1=0}^{M} \sum_{i_2=0}^{i_1} \cdots \sum_{i_n=0}^{i_{n-1}} h_{ntri}(i_1,\ldots,i_n)u(k-i_1)\cdots u(k-i_n) \qquad (1)$$

Now suppose that for the input-signal values $u(0),\ldots,u(K)$, the corresponding output-signal values $y(0),\ldots,y(K)$ are known. Then it is straightforward from (1) to write a linear matrix equation in terms of the unknown kernel values:

$$Y = HU \tag{2}$$

where

$$Y = [y(0) \ \cdots \ y(K)]$$

$$H = [h_1(0)\ h_1(1) \cdots h_1(M)\ h_{2tri}(0,0)\ h_{2tri}(1,0)\ h_{2tri}(1,1) \cdots\ h_{Ntri}(M,\ldots,M)]$$

$$U = \begin{bmatrix} u(0) & u(1) & u(2) & \cdots \\ 0 & u(0) & u(1) & \cdots \\ 0 & 0 & u(0) & \cdots \\ \vdots & \vdots & 0 & \cdots \\ u^2(0) & u^2(1) & \vdots & \\ 0 & u(0)u(1) & & \\ 0 & u^2(0) & & \\ 0 & 0 & & \\ \vdots & \vdots & & \\ u^N(0) & & & \\ 0 & & & \\ \vdots & & & \end{bmatrix} \tag{3}$$

Now it is clear that if K is such that U is a square matrix, and if U is invertible, then the kernel values are given by $H = YU^{-1}$. If K is larger or smaller than this value, or if U is not invertible, then least-squares techniques such as pseudo-inversion can be used to obtain approximations to the kernel values.

While this development indicates the nature of the kernel-determination problem, it should be clear that the dimensions involved are very large in most cases of interest. For example, there are on the order of $(M+1)^n$ values of a degree n kernel with memory M. As a result, the solution of the linear equation $Y = HU$ can be quite difficult. These considerations lead naturally to the introduction of approximation techniques involving expansions of the kernels in terms of known functions.

Suppose it is assumed that each of the triangular kernels in (1) can be represented as a linear combination of products of known functions $\phi_0(k),\ \phi_1(k),\ldots,\phi_J(k)$. In particular, it is assumed that

$$h_{ntri}(k_1,\ldots,k_n) = \sum_{j_1=0}^{J} \sum_{j_2=0}^{J} \cdots \sum_{j_n=0}^{J} \alpha_{j_1 \cdots j_n} \phi_{j_1}(k_1) \cdots \phi_{j_n}(k_n) \tag{4}$$

Then (1) can be rewritten in the form

$$y(k) = \sum_{n=1}^{N} \sum_{j_1=0}^{J} \cdots \sum_{j_n=0}^{J} \alpha_{j_1 \cdots j_n} \sum_{i_1=0}^{M} \cdots \sum_{i_n=0}^{i_{n-1}} \phi_{j_1}(i_1) \tag{5}$$

$$\cdots \phi_{j_n}(i_n) u(k-i_1) \cdots u(k-i_n)$$

or, in a simpler notation,

$$y(k) = \sum_{n=1}^{N} \sum_{j_1=0}^{J} \cdots \sum_{j_n=0}^{J} \alpha_{j_1 \cdots j_n} \Phi_{j_1 \cdots j_n}(k) \tag{6}$$

with the obvious definition of $\Phi_{j_1 \cdots j_n}(k)$. For a known input signal $u(0),u(1),\ldots,u(K)$ and known corresponding response $y(0),y(1),\ldots,y(K)$, the $\Phi_{j_1 \cdots j_n}(k)$ are known, and (6) yields a set of linear equations for the unknown coefficients $\alpha_{j_1 \cdots j_n}$. If J is small, then the dimension of the system of equations is much smaller than the dimension of (2). That is, there can be many fewer expansion coefficients than kernel values.

Further investigation of the details of these general approaches will be left to the reader. For the remainder of the chapter, I will be concerned with identification methods based on particular types of input signals.

7.2 Identification Using Impulse Inputs

Continuous-time linear system identification based on the impulse response is widely discussed, even on occasion used, and so it seems necessary to discuss the corresponding situation for the nonlinear case. The reader should be forewarned, however, that the theoretical discussion has only limited potential for application. Suppose a degree-n homogeneous system is described by

$$y(t) = \int_0^t h_{nsym}(t-\sigma_1,\ldots,t-\sigma_n)u(\sigma_1) \cdots u(\sigma_n)\, d\sigma_1 \cdots d\sigma_n \tag{7}$$

Then from Section 5.1, the response to $u_0(t) = \delta_0(t)$ is $y_0(t) = h_{nsym}(t,\ldots,t)$. The response to

$$u_{p-1}(t) = \delta_0(t) + \delta_0(t-T_1) + \cdots + \delta_0(t-T_{p-1}) \tag{8}$$

for $p = 2,3,\ldots,n$, where $T_1,\ldots,T_{p-1}$ are distinct positive numbers, is

$$y_p(t) = \sum_m \frac{n!}{m_1! \cdots m_p!} h_{nsym}(\underbrace{t,\ldots,t}_{m_1};\ldots;\underbrace{t-T_{p-1},\ldots,t-T_{p-1}}_{m_p}) \tag{9}$$

where $\sum_m$ is a p-fold sum over all integers $m_1,\ldots,m_p$ such that $0 \leqslant m_i \leqslant n$, and $m_1+\cdots+m_p = n$. Based on these response formulas, an identification strategy for homogeneous systems is easy to explain for the degree-2 case.

For a degree-2 system, the responses to $u_0(t)$ and $u_1(t)$ are, respectively,

$$\begin{aligned} y_0(t) &= h_{2sym}(t,t) \\ y_1(t) &= h_{2sym}(t,t) + 2h_{2sym}(t,t-T_1) + h_{2sym}(t-T_1,t-T_1) \end{aligned} \tag{10}$$

Thus, the values of the symmetric kernel at equal arguments are given directly by values of $y_0(t)$. To determine the value of the symmetric kernel at any two distinct arguments, say t_1 and t_2, with $t_1 > t_2$, simply take $T_1 = t_1 - t_2$ for then (10) easily gives

$$h_{2sym}(t_1,t_2) = \frac{1}{2}[y_1(t_1) - y_0(t_1) - y_0(t_2)] \tag{11}$$

This kind of analysis can be generalized to degree-n homogeneous systems. That is, values of the system kernel at particular arguments can be found by properly combining the set of system responses to a set of input signals of the form $u_0(t),\ldots,u_{n-1}(t)$, as given in (8). The precise details of the general calculation are messy, and so I will leave them to the motivated reader and the literature.

The kind of calculation just covered can also be used in the polynomial system case. Again the degree-2 case will illustrate the development. Consider a system described by

$$y(t) = \int_0^t h_1(t-\sigma)u(\sigma)\,d\sigma + \int_0^t h_{2sym}(t-\sigma_1,t-\sigma_2)u(\sigma_1)u(\sigma_2)\,d\sigma_1 d\sigma_2 \tag{12}$$

The responses to $u_0(t)$ and $u_1(t)$ from (8) are listed below:

$$\begin{aligned} y_0(t) &= h_1(t) + h_{2sym}(t,t) \\ y_1(t) &= h_1(t) + h_1(t-T_1) + h_{2sym}(t,t) + 2h_{2sym}(t,t-T_1) \\ &\quad + h_{2sym}(t-T_1,t-T_1) \end{aligned} \tag{13}$$

Now, to show how to determine the value of the degree-2 kernel $h_{2sym}(t_1,t_2)$ for specified $t_1 > t_2$, I can proceed just as in the degree-2 homogeneous case. Setting $T_1 = t_1 - t_2$, an easy calculation gives

$$h_{2sym}(t_1,t_2) = \frac{1}{2}[y_1(t_1) - y_0(t_1) - y_0(t_2)] \tag{14}$$

But what about the degree-1 kernel? It is clear from (13) that values of this kernel must be separated from values of the degree-2 kernel at equal arguments. The issue of interpolation arises here, and one approach is to notice that $2u_0(t)$ yields the response

$$y_2(t) = 2h_1(t) + 4h_{2sym}(t,t) \tag{15}$$

Then $y_0(t)$ and $y_2(t)$ yield a set of equations that can be written in vector form

$$\begin{bmatrix} y_0(t) \\ y_2(t) \end{bmatrix} = \begin{bmatrix} 1 & 1 \\ 2 & 4 \end{bmatrix} \begin{bmatrix} h_1(t) \\ h_{2sym}(t,t) \end{bmatrix} \tag{16}$$

Solving yields

$$\begin{aligned} h_1(t) &= 2y_0(t) - \frac{1}{2}y_2(t) \\ h_{2sym}(t,t) &= -y_0(t) + \frac{1}{2}y_2(t) \end{aligned} \tag{17}$$

Thus, these types of kernel values can be obtained at any value of $t \geqslant 0$.

For higher-degree polynomial systems, this analysis can be continued. But the details become increasingly fussy, and the interpolation idea involving impulses of various weights becomes increasingly barren from a feasibility viewpoint. Thus I drop the subject here, although similar ideas will arise in conjunction with less drastic input signals.

The question of how these symmetric-kernel evaluations might be used depends very much on the situation at hand. In most of the applications to date, sufficient values have been obtained to make plots of the kernel, and these have been analyzed to determine characteristics of the system. Little can be said in general since the analysis depends so much on the physical system being modeled.

From a general viewpoint, the ability to determine a mathematical model of a system from kernel values depends critically on the assumptions that are made about the (unknown) system. For example, a functional form for the kernels might be assumed, in which case the determination of the parameters in the functional form is another step in the system identification process. This kind of assumption can be conveniently implemented by assuming an interconnection structure for the unknown system, or by assuming the system can be described by a particular type of state-equation realization. Since little can be said about the general case at present, I will be content with a simple example which, incidently, indicates that the symmetric kernel is not always the most convenient choice of representation.

Example 7.1 Suppose it is known that a system can be described by a differential equation of the form

$$\ddot{y}(t) + a_1\dot{y}(t) + a_0 y(t) = b_0 u(t) + d_0 y(t)u(t)$$

or, equivalently, the bilinear state equation (A,D,b,c,R^2):

$$\dot{x}(t) = \begin{bmatrix} 0 & 1 \\ -a_0 & -a_1 \end{bmatrix} x(t) + \begin{bmatrix} 0 & 0 \\ d_0 & 0 \end{bmatrix} x(t)u(t) + \begin{bmatrix} 0 \\ b_0 \end{bmatrix} u(t)$$

$$y(t) = [1 \quad 0]x(t)$$

To avoid trivial cases, assume $b_0, d_0 \neq 0$. The results of Problem 5.1, in conjunction with the general form

$$h_{nreg}(t_1,\ldots,t_n) = ce^{At_n}De^{At_{n-1}}D \cdots e^{At_1}b, \quad n = 1,2,\ldots$$

give the unit-impulse response of the system in the form

$$y_0(t) = \sum_{n=1}^{\infty} h_{nreg}(0,\ldots,0,t) = h_1(t) = ce^{At}b$$

where the facts that $D^2 = 0$ and $Db = 0$ have been used. Now, it can be assumed from linear system theory that c, A, and B, equivalently, a_0, a_1, and b_0 can be calculated from this unit-impulse response. To determine D, that is, d_0, the response of the system to $\delta_0(t) + \delta_0(t-T)$, $T > 0$ will be used. This response can be written in the form

$$y_1(t) = \sum_{n=1}^{\infty} [h_{nreg}(0,\ldots,0,t) + h_{nreg}(0,\ldots,0,T,t-T)$$

$$+ h_{nreg}(0,\ldots,0,T,0,t-T) + \cdots + h_{nreg}(0,\ldots,0,t-T)]$$

$$= ce^{At}b + ce^{A(t-T)}b\delta_{-1}(t-T) + ce^{A(t-T)}De^{AT}b\delta_{-1}(t-T)$$

It is left as an easy exercise to show that since c, A, and b are known, d_0 can be computed from the value of $y_1(t)$ for any $t > T$.

7.3 Identification Based on Steady-State Frequency Response

The steady-state response of homogeneous and polynomial systems to sinusoidal inputs provides the basis for another approach to the identification problem. The ideas are similar to well known linear-system frequency response methods for finding values of the transfer function. Specifically, suppose a stable linear system is described by the transfer function $H(s)$. Then, following the review in Section 5.2, the (complex) value of $H(i\omega)$ for fixed, real ω can be determined by measuring the magnitude and phase of the steady-state response to $u(t) = 2A\cos(\omega t)$. Actually, two evaluations are determined since $H(-i\omega)$ is given by the complex conjugate of the measured complex number $H(i\omega)$.

Again I will begin the discussion of nonlinear systems by considering a degree-2 homogeneous system described in terms of the symmetric transfer function. From Section 5.2, the steady-state response to $u(t) = 2A\cos(\omega t)$ is

$$y_{ss}(t) = 2A^2H_{2sym}(i\omega,-i\omega) + 2A^2|H_{2sym}(i\omega,i\omega)|\cos[2\omega t + \angle H_{2sym}(i\omega,i\omega)] \quad (18)$$

Thus, the values of $H_{2sym}(i\omega,-i\omega)$ and $H_{2sym}(i\omega,i\omega)$ can be determined. But this does not provide enough information in general to uniquely determine the system transfer function.

Example 7.2 Consider the degree-2 systems shown in Figure 7.1. Either by computing the symmetric transfer functions and substituting into (18), or by tracing the input

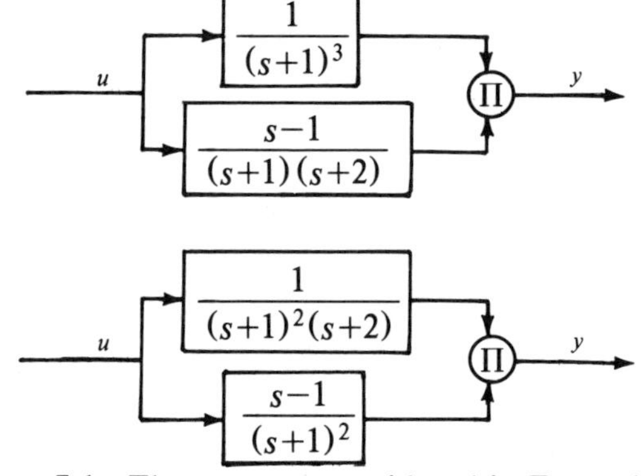

Figure 7.1. The systems considered in Example 7.2.

$2A\cos(\omega t)$ through the systems, it can be shown that the steady-state responses to single-tone inputs are identical. Also it can be verified that the responses to other types of inputs are not identical, although this should be clear. The calculations are as boring as the result is unfortunate, and thus I omit the details.

One way to circumvent this situation is to use a more complicated input signal. For example, consider the response of a degree-2 system to a two-tone input:

$$u(t) = 2A_1\cos(\omega_1 t) + 2A_2\cos(\omega_2 t) \tag{19}$$

From Example 5.5, the steady-state response in terms of the symmetric transfer function is given by

$$\begin{aligned} y_{ss}(t) &= 2A_1^2 H_{2sym}(i\omega_1,-i\omega_1) + 2A_2^2 H_{2sym}(i\omega_2,-i\omega_2) \\ &+ 4A_1A_2|H_{2sym}(-i\omega_1,i\omega_2)|\cos[(\omega_2-\omega_1)t + \angle H_{2sym}(-i\omega_1,i\omega_2)] \\ &+ 4A_1A_2|H_{2sym}(i\omega_1,i\omega_2)|\cos[(\omega_1+\omega_2)t + \angle H_{2sym}(i\omega_1,i\omega_2)] \\ &+ 2A_1^2|H_{2sym}(i\omega_1,i\omega_1)|\cos[2\omega_1 t + \angle H_{2sym}(i\omega_1,i\omega_1)] \\ &+ 2A_2^2|H_{2sym}(i\omega_2,i\omega_2)|\cos[2\omega_2 t + \angle H_{2sym}(i\omega_2,i\omega_2)] \end{aligned} \tag{20}$$

Now suppose that ω_1 and ω_2 are such that the frequency $\omega_1 + \omega_2$ is distinct from the other frequencies appearing in (20). Then amplitude and phase measurement of this component of the steady-state frequency response will give the (complex) value $H_{2sym}(i\omega_1,i\omega_2)$.

Postponing the discussion of what to do with this value, it should be clear how to proceed for higher-degree homogeneous systems. To outline the degree-3 case, consider the three-tone input

$$u(t) = 2A_1\cos(\omega_1 t) + 2A_2\cos(\omega_2 t) + 2A_3\cos(\omega_3 t) \tag{21}$$

Specializing (46) of Section 5.3 to $n = 3$, $L = M = N - 1$, the coefficient of $e^{i(\omega_1+\omega_2+\omega_3)t}$ is

$$3!A_1A_2A_3H_{3sym}(i\omega_1,i\omega_2,i\omega_3)$$

and the coefficient of $e^{-i(\omega_1+\omega_2+\omega_3)t}$ is

$$3!A_1A_2A_3H_{3sym}(-i\omega_1,-i\omega_2,-i\omega_3)$$

These give the real frequency term

$$3!2A_1A_2A_3|H_{3sym}(i\omega_1,i\omega_2,i\omega_3)|\cos[(\omega_1+\omega_2+\omega_3)t + \angle H_3(i\omega_1,i\omega_2,i\omega_3)] \tag{22}$$

If the frequencies ω_1, ω_2, and ω_3 are incommensurable, this frequency term will be distinct, and thus the amplitude and phase can be measured to obtain the value $H_{3sym}(i\omega_1,i\omega_2,i\omega_3)$. This result extends directly to the degree-n case, where the response to an n-tone input can be used to determine the value of $H(i\omega_1,\ldots,i\omega_n)$.

Finding these transfer function evaluations in the polynomial system case is greatly complicated by the fact that higher-degree homogeneous subsystems contribute steady-state response terms at the same frequencies as the lower-degree subsystems. As a simple example, suppose a polynomial system is composed of just degree-1 and degree-3 homogeneous subsystems. If the input $2A\cos(\omega t)$ is applied, then the steady-state response is

$$y_{ss}(t) = 2A|H_1(i\omega)|\cos[\omega t + \angle H_1(i\omega)]$$
$$+ 2A^3|H_{3sym}(i\omega,i\omega,-i\omega)|\cos[\omega t + \angle H_{3sym}(i\omega,i\omega,-i\omega)]$$
$$+ 2A^3|H_{3sym}(i\omega,i\omega,i\omega)|\cos[3\omega t + \angle H_{3sym}(i\omega,i\omega,i\omega)] \tag{23}$$

Of course, the two terms at frequency ω can be combined into one term using standard identities. But the point is that the degree-3 homogeneous subsystem contributes to the frequency components needed to determine $H_1(i\omega)$.

It is instructive to pursue this example a little further. The response of the system to the input

$$u(t) = 2A_1\cos(\omega_1 t) + 2A_2\cos(\omega_2 t) + 2A_3\cos(\omega_3 t) \tag{24}$$

will contain a term at the frequency $\omega_1+\omega_2+\omega_3$. Furthermore, this component will be distinct if the three input frequencies are incommensurable. This indicates that values of the degree-3 subsystem transfer function $H_{3sym}(i\omega_1,i\omega_2,i\omega_3)$ can be determined just as before. However, the reader can easily verify that the difficulty in determining values of $H_1(i\omega)$ remains. For instance, $H_{3sym}(i\omega_1,i\omega_2,-i\omega_2)$, $H_{3sym}(i\omega_1,i\omega_3,-i\omega_3)$, and $H_1(i\omega_1)$ all contribute to the frequency-ω_1 term in the steady-state response.

This situation brings up the problem of determining a symmetric transfer function from its evaluations. It is to be expected that certain assumptions will be needed on the structure of the transfer function, although just what these should be is unclear. In the linear case, it usually is assumed that the transfer function $H(s)$ is a strictly proper rational function, and sometimes $H(s)$ is assumed to be of known degree n. Then there are many methods for determining the transfer function from a set of evaluations of the form $H(i\omega)$. This approach is unrealistic when it is assumed that n is known, although it provides a simple starting point for further study. Unfortunately, such a general starting point is unavailable at present in the nonlinear case. Thus, I will abandon the general situation and illustrate one approach with a simple class of polynomial systems. Suitably severe restrictions will be imposed on the form of the homogeneous-subsystem transfer functions so that they can be easily determined from evaluations of the type that arise from frequency-response measurements.

Suppose an unknown nonlinear system is known to have the interconnection structure shown in Figure 7.2, where it is assumed that the linear subsystems are

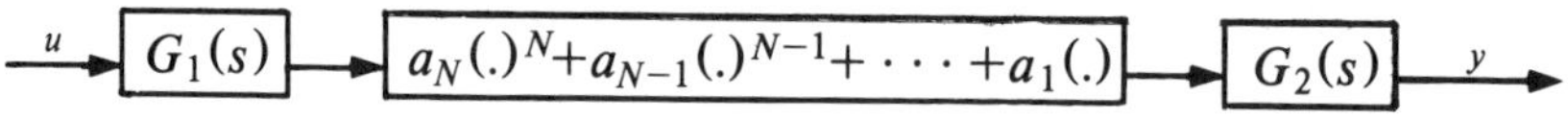

$$\xrightarrow{u} \boxed{G_1(s)} \longrightarrow \boxed{a_N(.)^N + a_{N-1}(.)^{N-1} + \cdots + a_1(.)} \longrightarrow \boxed{G_2(s)} \xrightarrow{y}$$

Figure 7.2. A cascade interconnection structure.

stable. Furthermore, since constant multipliers can be distributed throughout the cascade in any number of ways, it is assumed that $G_1(0) = G_2(0) = 1$. The interconnection structure is equivalent to assuming that the symmetric transfer functions for the system have the form

$$H_{nsym}(s_1,\ldots,s_n) = a_n G_1(s_1)\cdots G_1(s_n) G_2(s_1+\cdots+s_n), \quad n = 1,2,\ldots,N \tag{25}$$

I hardly need repeat that this structural assumption is quite severe. However, it will permit the determination of the subsystem transfer functions, at least in principle, from simple measurements of the steady-state frequency response. In fact, only single-tone inputs will be required, regardless of the value of N.

The results of Section 5.2 can be applied to easily calculate the steady-state frequency response of a system of the form shown in Figure 7.2. For an input

$$u(t) = 2A\cos(\omega t)$$

the steady-state response can be written in the form

$$y_{ss}(t) = f_0(A,i\omega) + 2\sum_{n=1}^{N} |f_n(A,i\omega)| cos[n\omega t + \angle f_n(A,i\omega)] \tag{26}$$

where

$$f_0(A,i\omega) = \sum_{k=1}^{[N/2]} \binom{2k}{k} A^{2k} a_{2k} G_1^k(i\omega) G_1^k(-i\omega)$$

$$f_n(A,i\omega) = \sum_{k=0}^{[(N-n)/2]} \binom{n+2k}{n+k} A^{n+2k} a_{n+2k} G_1^{n+k}(i\omega) G_1^k(-i\omega) G_2(in\omega), \tag{27}$$

$$n = 1,2,\ldots,N$$

where $[x]$ indicates the greatest integer $\leqslant x$.

There are several approaches that can be used to determine the linear-subsystem transfer functions and the coefficients in the polynomial nonlinearity. I will discuss a very simple method that requires only single-tone inputs (including constant inputs), and that does not require the measurement of relative phase. However, for reasons that will become clear shortly, it must be assumed that $G_1(s)$ and $G_2(s)$ are minimum-phase transfer functions.

The first step is to determine the coefficients $a_1, a_2, \ldots, a_N$ by measuring the steady-state response to step-function inputs at various amplitudes. The steady-state response of the system to $u(t) = A\delta_{-1}(t)$ is

$$y_{ss}(t) = a_1 A + a_2 A^2 + \cdots + a_N A^N$$

Therefore, measuring the (constant) value of $y_{ss}(t)$ for N different input amplitudes gives the coefficient values by polynomial interpolation.

The determination of the linear-subsystem transfer functions $G_1(s)$ and $G_2(s)$ will be accomplished from amplitude measurements on the fundamental frequency component of the steady-state response to inputs of the form $u(t) = 2Acos(\omega t)$. In other words, measurements of $|f_1(A,i\omega)|$ for various values of A and ω will be used.

For definiteness it is assumed that N is odd so that $f_1(A,i\omega)$ can be written out in the form

$$f_1(A,i\omega) = G_1(i\omega)G_2(i\omega)[Aa_1 + \binom{3}{2}A^3a_3|G_1(i\omega)|^2 + \cdots + \binom{N}{\frac{N+1}{2}}A^Na_N|G_1(i\omega)|^{N-1}] \tag{28}$$

Since $f_1(A,i\omega)$ is given in the form of a product of a simple complex function of ω and a complicated real function of ω, it is a simple matter to compute the corresponding squared-magnitude function:

$$\begin{aligned} |f_1(A,i\omega)|^2 &= |G_1(i\omega)|^2|G_2(i\omega)|^2[Aa_1 + \binom{3}{1}A^3a_3|G_1(i\omega)|^2 \\ &\quad + \cdots + \binom{N}{1}A^Na_N|G_1(i\omega)|^{N-1}]^2 \\ &= A^2a_1^2|G_1(i\omega)|^2|G_2(i\omega)|^2 + 2\binom{3}{2}A^4a_1a_3|G_1(i\omega)|^4|G_2(i\omega)|^2 \\ &\quad + \cdots + \binom{N}{\frac{N+1}{2}}^2 A^{2N}a_N^2|G_1(i\omega)|^{2N}|G_2(i\omega)|^2 \end{aligned} \tag{29}$$

Now, an identification strategy can be outlined as follows, assuming $a_1,a_3 \neq 0$ for convenience. For fixed frequency ω_1, $|f_1(A,i\omega_1)|^2$ is a polynomial in A^2. Thus, measuring the amplitude of the fundamental of the responses to a suitable number of different amplitude inputs with frequency ω_1 permits calculation of the coefficients

$$P_1(\omega_1) = a_1^2|G_1(i\omega_1)|^2|G_2(i\omega_1)|^2$$

$$P_2(\omega_1) = a_1a_3|G_1(i\omega_1)|^4|G_2(i\omega_1)|^2$$

by polynomial interpolation. Therefore,

$$|G_1(i\omega_1)|^2 = \frac{a_1^2}{a_1a_3}\,\frac{P_2(\omega_1)}{P_1(\omega_1)}$$

$$|G_2(i\omega_1)|^2 = \frac{a_1a_3}{a_1^4}\,\frac{P_1^2(\omega_1)}{P_2(\omega_1)}$$

This process can be repeated for various values of ω_1 so that the squared-magnitude functions for the linear subsystems can be determined as functions of ω. Then using the minimum-phase assumption, and the normalization $G_1(0) = G_2(0) = 1$, the transfer functions $G_1(s)$ and $G_2(s)$ can be computed using well known methods in linear system theory.

7.4 Identification Using Gaussian White Noise Excitation

This technique is an extension of a well known cross-correlation technique for the identification of a stationary linear system. To review briefly, suppose that the input to a linear system described by

$$y(t) = \int_{-\infty}^{\infty} h(\sigma)u(t-\sigma)\, d\sigma \tag{30}$$

is real, stationary Gaussian white noise with mean zero and intensity A. Then forming the product

$$y(t)u(t-T_1) = \int_{-\infty}^{\infty} h(\sigma)u(t-\sigma)u(t-T_1)\, d\sigma, \quad T_1 \geq 0 \tag{31}$$

and taking the expected value of both sides gives

$$\begin{aligned} E[y(t)u(t-T_1)] &= \int_{-\infty}^{\infty} h(\sigma)E[u(t-\sigma)u(t-T_1)]\, d\sigma \\ &= \int_{-\infty}^{\infty} h(\sigma)A\delta_0(\sigma-T_1)\, d\sigma \\ &= Ah(T_1) \end{aligned} \tag{32}$$

Thus, values of the kernel can be obtained from the obvious kind of input/output experiment based on (32). Of course, it is crucial from an implementation viewpoint that the ergodicity assumption be satisfied. For then the expected value is given by a time average, and (32) can be rewritten in the form

$$h(T_1) = \frac{1}{A} \lim_{T\to\infty} \frac{1}{2T} \int_{-T}^{T} y(t)u(t-T_1)\, dt \tag{33}$$

An implementation of this identification approach is diagramed in Figure 7.3.

A very similar analysis leads to a very similar procedure for determining values of the symmetric kernel of a degree-n homogeneous system. The salient features are

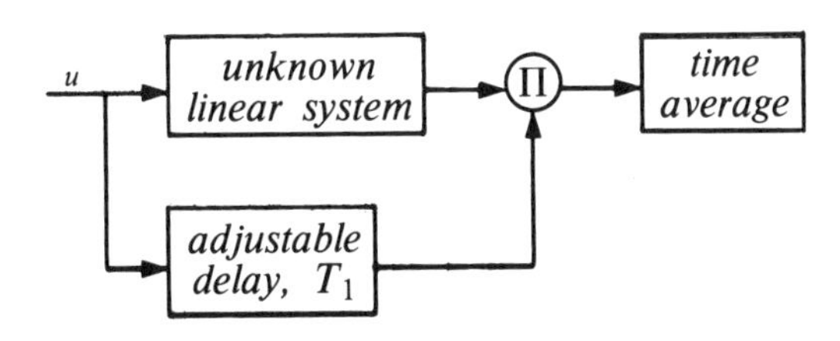

Figure 7.3. Cross-correlation identification of a linear system.

made apparent by the degree-2 case, so suppose the input to the system

$$y(t) = \int_{-\infty}^{\infty} h_{2sym}(\sigma_1,\sigma_2)u(t-\sigma_1)u(t-\sigma_2)\, d\sigma_1 d\sigma_2 \tag{34}$$

is Gaussian white noise just as before. I assume that the kernel is symmetric for reasons that will become apparent when terms are added up (below). Now, for $T_1,T_2 \geqslant 0$, $T_1 \neq T_2$,

$$E[y(t)u(t-T_1)u(t-T_2)]$$
$$= \int_{-\infty}^{\infty} h_{2sym}(\sigma_1,\sigma_2)E[u(t-\sigma_1)u(t-\sigma_2)u(t-T_1)u(t-T_2)]\, d\sigma_1 d\sigma_2 \tag{35}$$

The expectation on the right side can be expanded to give

$$\begin{aligned} E[y(t)u(t-T_1)u(t-T_2)] &= A^2\int_{-\infty}^{\infty} h_{2sym}(\sigma_1,\sigma_2)\delta_0(\sigma_2-\sigma_1)\delta_0(T_2-T_1)\, d\sigma_1 d\sigma_2 \\ &+ A^2\int_{-\infty}^{\infty} h_{2sym}(\sigma_1,\sigma_2)\delta_0(T_1-\sigma_1)\delta_0(T_2-\sigma_2)\, d\sigma_1 d\sigma_2 \\ &+ A^2\int_{-\infty}^{\infty} h_{2sym}(\sigma_1,\sigma_2)\delta_0(T_2-\sigma_1)\delta_0(T_1-\sigma_2)\, d\sigma_1 d\sigma_2 \\ &= A^2\delta_0(T_2-T_1)\int_{-\infty}^{\infty} h_{2sym}(\sigma,\sigma)\, d\sigma \\ &+ 2A^2 h_{2sym}(T_1,T_2) \end{aligned} \tag{36}$$

Since $T_1 \neq T_2$, (36) yields

$$h_{2sym}(T_1,T_2) = \frac{1}{2A^2}\, E[y(t)u(t-T_1)u(t-T_2)] \tag{37}$$

Imposing the ergodicity assumption permits (37) to be written in the time-average form

$$h_{2sym}(T_1,T_2) = \frac{1}{2A^2}\lim_{T\to\infty}\frac{1}{2T}\int_{-T}^{T} y(t)u(t-T_1)u(t-T_2)\, dt,\quad T_1 \neq T_2 \tag{38}$$

An implementation of (38) is diagramed in Figure 7.4.

For $T_1 = T_2 \geqslant 0$, this approach breaks down because for white noise $E[u^2(t)]$ does not exist. Traditionally, this is sidestepped by either of the claims: 1) in any implementation of the method, $u(t)$ is actually not white, 2) the values $h_{2sym}(T,T)$ can be obtained by continuous extension of values $h_{2sym}(T_1,T_2)$ for $T_1 \neq T_2$. Either claim can be valid under appropriate circumstances, but it will be seen in due course that this "diagonal value" issue can cause important difficulties.

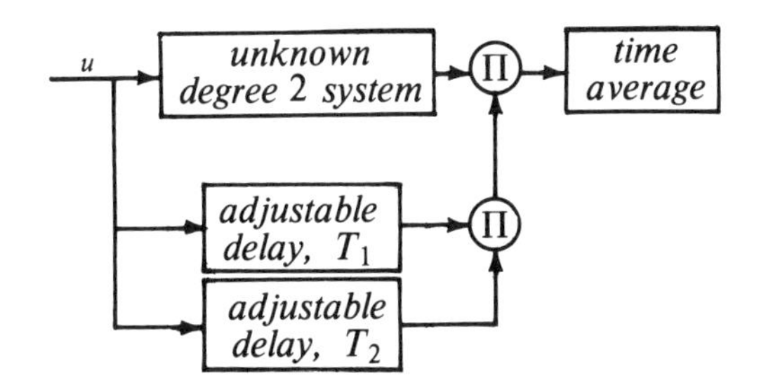

Figure 7.4. Cross-correlation identification method for a degree-2 system.

For general degree-n homogeneous systems, the cross-correlation identification method is based on the relationship

$$h_{nsym}(T_1, \ldots, T_n) = \frac{1}{n!A^n} E[y(t)u(t-T_1) \cdots u(t-T_n)] \tag{39}$$

where $T_1, \ldots, T_n$ are distinct, nonnegative numbers. The derivation of this formula is left as an exercise, the solution to which is essentially contained in a calculation later in this section.

To consider the application of the cross-correlation approach to polynomial systems, a degree-3 polynomial system will be used:

$$y(t) = \int_{-\infty}^{\infty} h_1(\sigma_1)u(t-\sigma_1)\, d\sigma_1$$
$$+ \int_{-\infty}^{\infty} h_{3sym}(\sigma_1, \sigma_2, \sigma_3)u(t-\sigma_1)u(t-\sigma_2)u(t-\sigma_3)\, d\sigma_1 d\sigma_2 d\sigma_3$$

Computing the input/output cross-correlation $E[y(t)u(t-T_1)u(t-T_2)u(t-T_3)]$ gives

$$E[y(t)u(t-T_1)u(t-T_2)u(t-T_3)] = Ah_1(T_1)\delta_0(T_3-T_2) + Ah_1(T_2)\delta_0(T_3-T_1)$$
$$+ Ah_1(T_3)\delta_0(T_2-T_1) + 3!A^3 h_{3sym}(T_1, T_2, T_3)$$

Thus, for T_1, T_2, T_3 distinct, the degree-3 polynomial case is just like the degree-3 homogeneous case in giving

$$h_{3sym}(T_1, T_2, T_3) = \frac{1}{3!A^3} E[y(t)u(t-T_1)u(t-T_2)u(t-T_3)]$$

Computing the cross-correlation $E[y(t)u(t-T_1)]$ gives

$$E[y(t)u(t-T_1)] = Ah_1(T_1) + 3A^2 \int_{-\infty}^{\infty} h_{3sym}(\sigma, \sigma, T_1)\, d\sigma \tag{40}$$

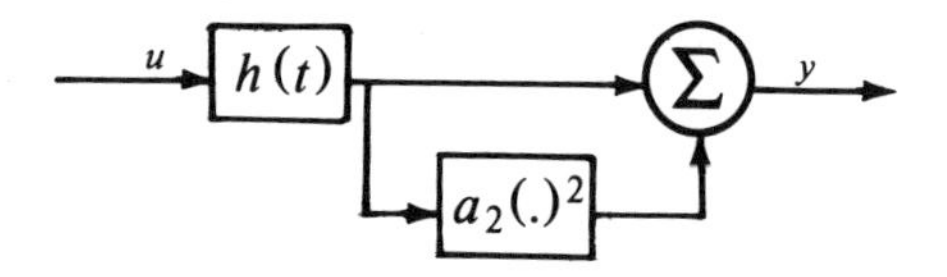

Figure 7.5. A degree-2 polynomial system.

Therefore, determining values of the degree-1 kernel involves the degree-3 kernel with not all arguments distinct. Unless the integral term in (40) can be approximated accurately using appropriate approximate values of $h_{3sym}(\sigma,\sigma,T_1)$, the degree-1 kernel values cannot be isolated. Of course, there are hypotheses, usually quite restrictive, that can ameliorate the situation. Often these hypotheses can be conveniently phrased in terms of an assumed interconnection structure for the unknown system.

Example 7.2 Suppose a system is known to have the interconnection structure shown in Figure 7.5. Then the input/output representation can be written in the form

$$y(t) = \int_{-\infty}^{\infty} h(\sigma)u(t-\sigma)\,d\sigma + \int_{-\infty}^{\infty}\int_{-\infty}^{\infty} a_2h(\sigma_1)h(\sigma_2)u(t-\sigma_1)u(t-\sigma_2)\,d\sigma_1 d\sigma_2$$

With an input that is a sample function from a zero-mean, white Gaussian random process with intensity A, the mean of the response is

$$E[y(t)] = \int_{-\infty}^{\infty} h(\sigma)E[u(t-\sigma)]\,d\sigma$$

$$+ \int_{-\infty}^{\infty}\int_{-\infty}^{\infty} a_2h(\sigma_1)h(\sigma_2)E[u(t-\sigma_1)u(t-\sigma_2)]\,d\sigma_1 d\sigma_2$$

$$= a_2A \int_{-\infty}^{\infty} h^2(\sigma)\,d\sigma$$

The input/output cross-correlation is given by

$$R_{yu}(\tau) = E[y(t)u(t-\tau)]$$

$$= \int_{-\infty}^{\infty} h(\sigma)E[u(t-\sigma)u(t-\tau)]\,d\sigma$$

$$+ \int_{-\infty}^{\infty}\int_{-\infty}^{\infty} a_2h(\sigma_1)h(\sigma_2)E[u(t-\sigma_1)u(t-\sigma_2)u(t-\tau)]\,d\sigma_1 d\sigma_2$$

$$= Ah(\tau)$$

Thus, values of the kernel can be computed from input/output cross-correlations. And if a sufficient number of values are computed to approximate the integral, then the constant a_2 can be computed from the response mean.

The general difficulties encountered in the polynomial system case can be circumvented by adopting the Wiener orthogonal representation. (Another important reason for using the Wiener representation is suggested in Problem 7.5.) Suppose that a system can be described by

$$y(t) = \sum_{n=0}^{N} G_n[k_n,u(t)] \tag{41}$$

where, as given in Section 5.5,

$$G_n[k_n,u(t)] = \sum_{i=0}^{[n/2]} \frac{(-1)^i n! A^i}{2^i(n-2i)!i!} \int_{-\infty}^{\infty} k_n(\sigma_1,\ldots,\sigma_{n-2i},\tau_1,\tau_1,\ldots,\tau_i,\tau_i)$$
$$d\tau_1 \cdots d\tau_i u(t-\sigma_1) \cdots u(t-\sigma_{n-2i})\, d\sigma_1 \cdots d\sigma_{n-2i} \tag{42}$$

Following the notation in Section 5.5, the Wiener kernels are symmetric despite the absence of the subscript "sym". Now the identification problem can be viewed as the problem of determining the symmetric function $k_n(t_1,\ldots,t_n)$ which specifies $G_n[k_n,u(t)]$, $n = 0,1,\ldots,N$.

The procedure again involves products of delayed versions of the Gaussian white noise input. Such a product, $u(t-T_1) \cdots u(t-T_n)$, can be viewed as a degree-n homogeneous operator on the input, and this viewpoint allows use of the orthogonality property of the Wiener operators. (Recall that the homogeneous operator $u(t-T_1) \cdots u(t-T_n)$ can be written in integral form using impulsive kernels, but there seems to be little reason to do so for the following calculation.)

First note that the expected value of the output is

$$E[y(t)] = \sum_{n=0}^{N} E[G_n[k_n,u(t)]] \tag{43}$$

and, using the result established in the proof of Theorem 5.1, the degree-0 Wiener kernel is given by

$$k_0 = E[y(t)] \tag{44}$$

The value of the degree-1 Wiener kernel $k_1(t)$, at $t = T_1 \geqslant 0$ is found as follows. First,

$$E[y(t)u(t-T_1)] = E[\sum_{n=0}^{N} G_n[k_n,u(t)]\, u(t-T_1)]$$

$$= E[G_0[k_0,u(t)]u(t-T_1)] + E[G_1[k_1,u(t)]\, u(t-T_1)] \tag{45}$$

where the fact that Wiener operators of degree > 1 are orthogonal to any degree-1 operator has been used. In a more explicit notation,

$$E[y(t)u(t-T_1)] = k_0E[u(t-T_1)] + \int_{-\infty}^{\infty} k_1(\sigma)E[u(t-\sigma)u(t-T_1)]\,d\sigma$$

$$= Ak_1(T_1)$$

so that

$$k_1(T_1) = \frac{1}{A}\,E[y(t)u(t-T_1)] \tag{46}$$

Of course, under an ergodicity hypothesis this calculation can be implemented as a time average.

Now I press on to the determination of the degree-2 Wiener kernel. For distinct nonnegative numbers T_1 and T_2, the evaluation $k_2(T_1,T_2)$ can be found by noting that

$$E[y(t)u(t-T_1)u(t-T_2)] = E[\sum_{n=0}^{N} G_n[k_n,u(t)]u(t-T_1)u(t-T_2)]$$

$$= E[k_0u(t-T_1)u(t-T_2)$$

$$+ \int_{-\infty}^{\infty} k_1(\sigma)u(t-\sigma)u(t-T_1)u(t-T_2)\,d\sigma$$

$$+ \int_{-\infty}^{\infty} k_2(\sigma_1,\sigma_2)u(t-\sigma_1)u(t-\sigma_2)u(t-T_1)u(t-T_2)\,d\sigma_1d\sigma_2$$

$$- A\int_{-\infty}^{\infty} k_2(\sigma,\sigma)\,d\sigma\, u(t-T_1)u(t-T_2)]$$

$$= Ak_0\delta_0(T_1-T_2) + 2A^2k_2(T_1,T_2) \tag{47}$$

Thus, since $T_1 \neq T_2$,

$$k_2(T_1,T_2) = \frac{1}{2A^2}\,E[y(t)u(t-T_1)u(t-T_2)] \tag{48}$$

The degree-m ($\leqslant N$) Wiener kernel is evaluated in a similar fashion. For $T_1, \ldots, T_m$ distinct nonnegative numbers, the calculation can be outlined as follows:

$$E[y(t)u(t-T_1)\cdots u(t-T_m)] = E[\sum_{n=0}^{N} G_n[k_n,u(t)]u(t-T_1)\cdots u(t-T_m)] \tag{49}$$

By the orthogonality property,

$$E[y(t)u(t-T_1)\cdots u(t-T_m)] = \sum_{n=0}^{m} E[G_n[k_n,u(t)]u(t-T_1)\cdots u(t-T_m)] \tag{50}$$

Changing to a more explicit notation and using (42) gives

$$E[y(t)u(t-T_1)\dots u(t-T_m)] = \sum_{n=0}^{m} \int_{-\infty}^{\infty} k_n(\sigma_1,\dots,\sigma_n)E[u(t-\sigma_1)\cdots u(t-\sigma_n)$$

$$u(t-T_1)\cdots u(t-T_m)]\,d\sigma_1\cdots d\sigma_n$$

$$+\sum_{n=0}^{m}\sum_{i=1}^{[n/2]} \frac{(-1)^i n!A^i}{2^i(n-2i)!i!} \int_{-\infty}^{\infty} k_n(\sigma_1,\dots,\sigma_{n-2i},$$

$$\tau_1,\tau_1,\dots,\tau_i,\tau_i)E[u(t-\sigma_1)\cdots u(t-\sigma_{n-2i})u(t-T_1)$$

$$\cdots u(t-T_m)]d\tau_1\cdots d\tau_i d\sigma_1\cdots d\sigma_{n-2i} \tag{51}$$

In the first summation in (51), the expected value can be rewritten as a sum of products of impulses. When $n = m$ further analysis of the integrations indicates that two types of terms will arise: those that contain a factor $\delta_0(T_i-T_j)$, and those that contain no impulse, but rather an evaluation of the kernel for some permutation of arguments $T_1,\dots,T_m$. Since the T_j's are distinct, all those terms with impulse factors will be zero, and it can be shown that the remaining terms give, by symmetry of the kernel, $m!A^m k_m(T_1,\dots,T_m)$. When $n < m$ in the first summation in (51), there are two cases. If $n + m$ is odd, then the expected value is zero. If $n + m$ is even, then each term in the expected value will contain a factor of the form $\delta_0(T_i-T_j)$, and so again zero is obtained. For similar reasons, all the terms in the second summation in (51) yield zero. Thus,

$$k_m(T_1,\dots,T_m) = \frac{1}{m!A^m}\, E[y(t)u(t-T_1)\cdots u(t-T_m)] \tag{52}$$

under the hypothesis that the T_j's are distinct.

The reader undoubtedly is convinced by now of the crucial nature of the distinct T_j assumption. Unfortunately, this causes an important difficulty when it is the symmetric Volterra kernel that is of interest. To convert the Wiener representation (41) into a Volterra series representation, the various terms of like degree in (41) must be gathered together. Recalling Theorem 5.3, the degree-n symmetric kernel in a Volterra series representation of the system in (41) is given by

$$h_{nsym}(t_1,\dots,t_n) = \sum_{m=0}^{(N-n)/2} \frac{(-1)^m(n+2m)!A^m}{n!m!2^m}$$

$$\int_0^{\infty} k_{n+2m}(t_1,\dots,t_n,\sigma_1,\sigma_1,\dots,\sigma_m,\sigma_m)\,d\sigma_1\cdots d\sigma_m \tag{53}$$

It is clear that values of the symmetric Volterra kernel even for distinct arguments depend on values of the Wiener kernels for indistinct arguments.

A way to avoid the diagonal difficulty is to use the *residual*

$$y(t) - \sum_{n=0}^{m-1} G_n[k_n,u(t)] \tag{54}$$

rather than just the response $y(t)$ in the computation of kernel values. It can be shown (see Problem 7.4) that for any nonnegative values $T_1, \cdots, T_m$,

$$k_m(T_1,\ldots,T_m) = \frac{1}{m!A^m} E[(y(t) - \sum_{n=0}^{m-1} G_n[k_n,u(t)])u(t-T_1) \cdots u(t-T_m)] \tag{55}$$

Example 7.3 The difficulty in determining kernel values for nondistinct arguments does not arise in the discrete-time case. When the input is a stationary, zero-mean, white Gaussian random process with intensity A, Theorem 6.9 can be simplified using the results of Example 6.8 to give the following relationships.

$$k_0 = E[y(k_0)]$$

$$k_1(i_1) = \frac{1}{A} E[y(k_0)u(k_0-i_1)]$$

$$k_2(i_1,i_2) = \begin{cases} \dfrac{1}{2A^2} E[y(k_0)u(k_0-i_1)u(k_0-i_2)], & i_1 \neq i_2 \\ \dfrac{1}{2A^2} E[y(k_0)(u^2(k_0-i_1) - A)], & i_1 = i_2 \end{cases}$$

The higher-degree kernels are given by similar formulas.

Just as in the case of Volterra kernels, the question of how to use the values of the Wiener kernels is difficult. Again I will indicate one approach by investigating further the case where a particular interconnection structure is assumed. A side benefit is that in the course of the development there will be occasion to exercise a number of the tools that have been developed for manipulating Volterra series and Wiener series representations.

Suppose an unknown system is known to have the interconnection structure shown in Figure 7.6, where the two linear systems are assumed to be stable, minimum phase, and such that $G_1(0) = G_2(0) = 1$. Notice that here the Fourier transform notation is being used so that the linear subsystems are specified in terms of system functions.

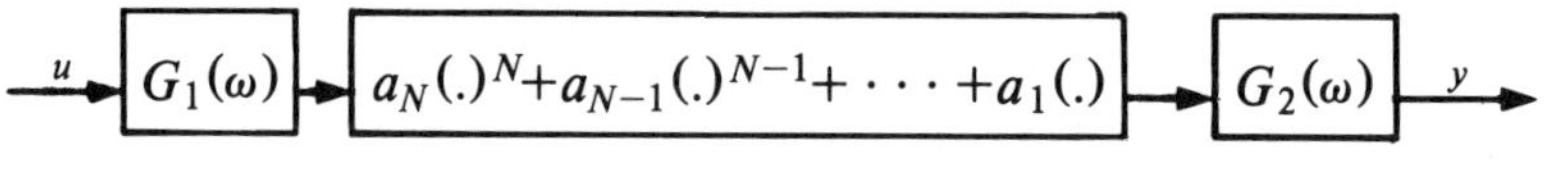

$u \rightarrow$ $G_1(\omega)$ $\rightarrow$ $a_N(.)^N + a_{N-1}(.)^{N-1} + \cdots + a_1(.)$ $\rightarrow$ $G_2(\omega)$ $\rightarrow y$

Figure 7.6. A familiar interconnection structure.

Proceeding as in the case of transfer functions, it is easy to show that in terms of the subsystem system functions, symmetric Volterra system functions are given by

$$H_{nsym}(\omega_1,\ldots,\omega_n) = a_n G_1(\omega_1)\cdots G_1(\omega_n) G_2(\omega_1+\cdots+\omega_n), \quad n = 1,2,\ldots,N \tag{56}$$

Then from Problem 5.14, the Wiener system functions, the Fourier transforms of the Wiener kernels, are given by

$$K_n(\omega_1,\ldots,\omega_n) = \sum_{j=0}^{[(N-n)/2]} \frac{(n+2j)!A^j a_{n+2j}}{n!j!2^j} G_1(\omega_1)\cdots G_1(\omega_n)$$

$$G_2(\omega_1+\cdots+\omega_n)[\frac{1}{2\pi}\int_{-\infty}^{\infty} G_1(\gamma)G_1(-\gamma)\,d\gamma]^j \tag{57}$$

Using Parseval's relation for single-variable Fourier transforms gives

$$K_n(\omega_1,\ldots,\omega_n) = \sum_{j=0}^{[(N-n)/2]} \frac{(n+2j)!A^j a_{n+2j}}{n!j!2^j} [\int_{-\infty}^{\infty} g_1^2(\tau)\,d\tau]^j$$

$$G_1(\omega_1)\cdots G_1(\omega_n)G_2(\omega_1+\cdots+\omega_n), \quad n = 1,\ldots,N \tag{58}$$

Now, from the results of the cross-correlation method it will be assumed that a sufficient number of values of the degree-1 Wiener kernel have been obtained to permit the computation of $K_1(\omega)$. Then (58) gives

$$K_1(\omega) = \sum_{j=0}^{[(N-1)/2]} \frac{(1+2j)!A^j a_{1+2j}}{j!2^j} [\int_{-\infty}^{\infty} g_1^2(\tau)\,d\tau]^j G_1(\omega)G_2(\omega) \tag{59}$$

That is, the product $G_1(\omega)G_2(\omega)$ is determined up to an unknown constant.

Suppose also that a sufficient number of values of the degree-2 Wiener kernel have been computed to permit the calculation of $K_2(\omega_1,\omega_2)$. Then (58) gives

$$K_2(\omega_1,\omega_2) = \sum_{j=0}^{[(N-2)/2]} \frac{(2+2j)!A^j a_{2+2j}}{2!j!2^j}$$

$$[\int_{-\infty}^{\infty} g_1^2(\tau)\,d\tau]^j G_1(\omega_1)G_1(\omega_2)G_2(\omega_1+\omega_2) \tag{60}$$

That is, the product $G_1(\omega_1)G_1(\omega_2)G_2(\omega_1+\omega_2)$ is determined up to a constant.

In order to show how to obtain $G_1(\omega)$ and $G_2(\omega)$ from the first two Wiener system functions, it is convenient to write

$$K_1(\omega) = \alpha_1 G_1(\omega)G_2(\omega)$$

$$K_2(\omega_1,\omega_2) = \alpha_2 G_1(\omega_1)G_1(\omega_2)G_2(\omega_1+\omega_2)$$

where α_1 and α_2 are unknown constants. Then it is easy to verify that for any ω,

$$\frac{K_2(-\omega/2,\omega)}{K_1(\omega/2)} = \frac{\alpha_2}{\alpha_1}\,\frac{G_1(-\omega/2)G_1(\omega)}{G_1(\omega/2)}$$

so that the magnitude spectrum of $G_1(\omega)$ is determined up to an unknown constant according to

$$|G_1(\omega)| = \frac{|\alpha_1|}{|\alpha_2|} \frac{|K_2(-\omega/2,\omega)|}{|K_1(\omega/2)|}$$

Of course, this implies that the magnitude spectrum of $G_2(\omega)$ is determined up to an unknown constant according to

$$|G_2(\omega)| = \frac{1}{|\alpha_1|} \frac{|K_1(\omega)|}{|G_1(\omega)|}$$

Using the minimum-phase and normalization assumptions, the calculation of $G_1(\omega)$ and $G_2(\omega)$ is a well known problem in linear system theory. Further consideration of the identification problem, in particular, the determination of the coefficients in the nonlinearity, is left to Problem 7.8. But it is important to notice how the linear subsystems in the degree-N polynomial system can be determined from just two kinds of input/output cross-correlations.

7.5 Orthogonal Expansion of the Wiener Kernels

Because of difficulties in the use of Wiener-kernel values, an orthogonal expansion approach can be an important alternative. The basic idea, similar to that briefly discussed in Section 7.1, is to represent the Wiener kernels of the unknown system in terms of an orthonormal basis for the Hilbert space $L_2(0,\infty)$, and then determine the coefficients in this orthonormal expansion. Again, the input signal to be used is a real, stationary, zero-mean, white Gaussian random process with intensity A.

Suppose the unknown system can be described in terms of the Wiener orthogonal representation. Furthermore, assume that each Wiener kernel $k_n(t_1,\ldots,t_n)$ can be represented in the following way. Let $\phi_1(t),\phi_2(t),\ldots$ be an orthonormal basis in $L_2(0,\infty)$. That is,

$$\int_0^\infty \phi_i(t)\phi_j(t)\,dt = \begin{cases} 0, & i \neq j \\ 1, & i = j \end{cases} \tag{61}$$

Then in terms of this basis write each Wiener kernel in the form

$$k_n(t_1,\ldots,t_n) = \sum_{i_1=1}^{\infty} \cdots \sum_{i_n=1}^{\infty} k_{i_1 \cdots i_n} \phi_{i_1}(t_1) \cdots \phi_{i_n}(t_n) \tag{62}$$

where

$$k_{i_1 \cdots i_n} = \int_0^\infty k_n(t_1,\ldots,t_n)\phi_{i_1}(t_1) \cdots \phi_{i_n}(t_n)\,dt_1 \cdots dt_n \tag{63}$$

Notice that for any permutation π of $i_1,\ldots,i_n$,

$$k_{i_1 \cdots i_n} = k_{\pi(i_1) \cdots \pi(i_n)} \tag{64}$$

by the symmetry hypothesis implicit in the use of the Wiener operators. Of course, the expansion (62) will be truncated to some finite number of terms in practice, thereby yielding an approximate representation. The identification problem now is posed in terms of determining the expansion coefficients $k_{i_1 \cdots i_n}$.

For the degree-0 Wiener kernel, there is nothing to discuss since $k_0 = E[y(t)]$. For the degree-1 Wiener kernel,

$$k_1(t) = \sum_{i=1}^{\infty} k_i \phi_i(t) \tag{65}$$

the ith coefficient can be identified according to the following cross-correlation calculation:

$$E[y(t) \int_0^{\infty} \phi_i(\sigma) u(t-\sigma) d\sigma] = E[\sum_{n=0}^{\infty} G_n[k_n, u(t)] \int_0^{\infty} \phi_i(\sigma) u(t-\sigma)\, d\sigma]$$

$$= k_0 \int_0^{\infty} \phi_i(\sigma) E[u(t-\sigma)]\, d\sigma + \int_0^{\infty} \int_0^{\infty} k_1(\tau) \phi_i(\sigma) E[u(t-\tau) u(t-\sigma)]\, d\tau d\sigma$$

$$= A \int_0^{\infty} k_1(\tau) \phi_i(\tau)\, d\tau = A k_i \tag{66}$$

In terms of the notation to be used for the higher-degree cases, (66) can be written as

$$k_i = \frac{1}{A} E[y(t) G_1[\phi_i, u(t)]] \tag{67}$$

If ergodicity is assumed, the cross-correlation can be computed by time-averaging. Then the identification method can be diagramed in terms of a multiplicative connection of the unknown system with the known system $G_1[\phi_i, u(t)]$ as shown in Figure 7.7.

The determination of the coefficient $k_{i_1 i_2}$ for the Wiener kernel $k_2(t_1, t_2)$ can be diagramed as shown in Figure 7.8, where the unknown system is connected in multiplicative parallel with the known system described by the Wiener operator

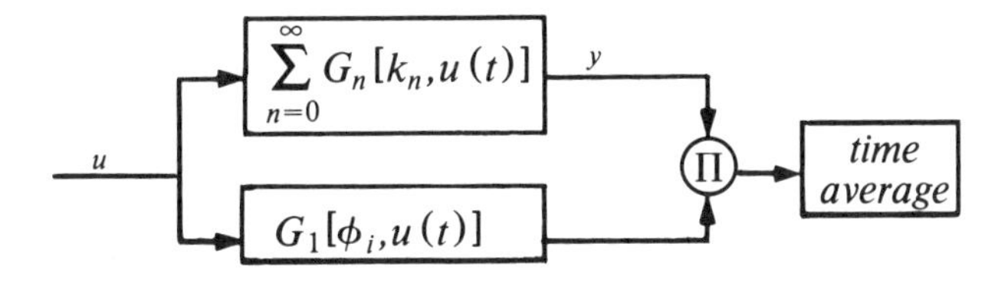

Figure 7.7. Coefficient identification method for $k_1(t)$.

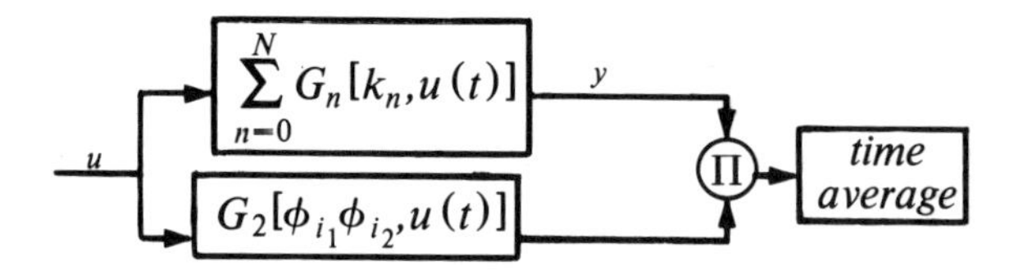

Figure 7.8. Coefficient identification method for $k_2(t_1,t_2)$.

$$G_2[\phi_{i_1}\phi_{i_2},u(t)] = \int_0^\infty \frac{1}{2}[\phi_{i_1}(\tau_1)\phi_{i_2}(\tau_2)+\phi_{i_1}(\tau_2)\phi_{i_2}(\tau_1)]u(t-\tau_1)u(t-\tau_2)\,d\tau_1 d\tau_2$$

$$- A\int_0^\infty \phi_{i_1}(\tau)\phi_{i_2}(\tau)\,d\tau \tag{68}$$

(Notice that the Wiener-operator notation is being abused slightly to avoid writing out the symmetric version of $\phi_{i_1}\phi_{i_2}$.)

Using the orthogonality properties of the Wiener operator,

$$E[y(t)\ G_2[\phi_{i_1}\phi_{i_2},u(t)]] = E[\sum_{n=0}^{\infty} G_n[k_n,u(t)]G_2[\phi_{i_1}\phi_{i_2},u(t)]]$$

$$= E[G_2[k_2,u(t)]G_2[\phi_{i_1}\phi_{i_2},u(t)]]$$

$$= E[G_2[k_2,u(t)]\int_0^\infty \frac{1}{2}[\phi_{i_1}(\tau_1)\phi_{i_2}(\tau_2)+\phi_{i_1}(\tau_2)\phi_{i_2}(\tau_1)]u(t-\tau_1)u(t-\tau_2)\,d\tau_1 d\tau_2$$

$$= \int_0^\infty \frac{1}{2}\, k_2(\sigma_1,\sigma_2)[\phi_{i_1}(\tau_1)\phi_{i_2}(\tau_2) + \phi_{i_1}(\tau_2)\phi_{i_2}(\tau_1)]$$

$$E[u(t-\sigma_1)u(t-\sigma_2)u(t-\tau_1)u(t-\tau_2)]\,d\sigma_1 d\sigma_2 d\tau_1 d\tau_2$$

$$- A\int_0^\infty k_2(\sigma,\sigma)\phi_{i_1}(\tau_1)\phi_{i_2}(\tau_2)E[u(t-\tau_1)u(t-\tau_2)]\,d\sigma\, d\tau_1 d\tau_2 \tag{69}$$

Computation of the expected values goes in the usual manner to yield

$$E[y(t)\ G_2[\phi_{i_1}\phi_{i_2},u(t)]] = 2A^2\,\frac{1}{2}[k_{i_1i_2} + k_{i_2i_1}] = 2A^2 k_{i_1i_2} \tag{70}$$

That is,

$$k_{i_1i_2} = \frac{1}{2A^2}\, E[y(t)\ G_2[\phi_{i_1}\phi_{i_2},u(t)]] \tag{71}$$

The identification procedure for the expansion coefficients for the degree-n Wiener kernel $k_n(t_1,\ldots,t_n)$ proceeds in just the same way. The calculations corresponding to (69) are much more complicated, but these can be avoided by invoking earlier results. The starting point is shown in Figure.7.8.

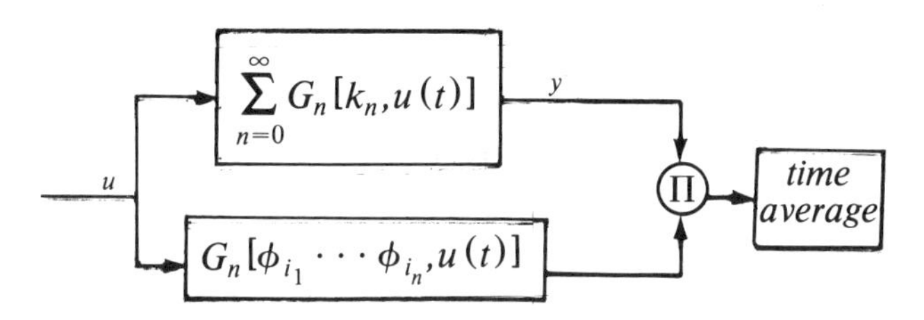

Figure 7.8. Coefficient identification method for $k_n(t_1,\ldots,t_n)$.

Application of the orthogonality property immediately gives

$$E[y(t)\ G_n[\phi_{i_1}\cdots\phi_{i_n},u(t)]] = E[G_n[k_n,u(t)]G_n[\phi_{i_1}\cdots\phi_{i_n},u(t)]] \tag{72}$$

Now using a slight variant of the proof of Theorem 5.2, it is easy to show that

$$E[G_n[k_n,u(t)]G_n[\phi_{i_1}\ldots\phi_{i_n},u(t)]]$$

$$= n!A^n\int_0^\infty k_n(t_1,\ldots,t_n)\frac{1}{n!}\sum_{\pi(.)}\phi_{i_1}(t_{\pi(1)})\cdots\phi_{i_n}(t_{\pi(n)})\,dt_1\cdots dt_n$$

$$= n!A^n\int_0^\infty k_n(t_1,\ldots,t_n)\phi_{i_1}(t_1)\cdots\phi_{i_n}(t_n)\,dt_1\cdots dt_n$$

$$= n!A^n k_{i_1\cdots i_n} \tag{73}$$

This gives the general formula

$$k_{i_1\cdots i_n} = \frac{1}{n!A^n}E[y(t)\ G_n[\phi_{i_1}\cdots\phi_{i_n},u(t)]] \tag{74}$$

7.6 Remarks and References

Remark 7.1 The fact that the determination of kernel values is a linear problem has been discussed by many authors working from several different viewpoints. For a treatment in a general continuous-time setting, see

W. Root, "On the Modeling of Systems for Identification Part I: ϵ-Representations of Classes of Systems," *SIAM Journal on Control,* Vol. 13, pp. 927-975, 1975.

The polynomial-system identification problem can be viewed as fitting a polynomial system to a given set of input/output pairs. An operator-theoretic study of this formulation is given in

W. Porter, "Synthesis of Polynomic Systems," *SIAM Journal on Mathematical Analysis,* Vol. 11, pp. 308-315, 1980.

An elementary discussion of the material of Section 7.1 along with an interesting application can be found in

J. Amarocho, A. Bandstetter, "Determination of Nonlinear Rainfall-Runoff Processes," *Water Resources Research,* Vol. 7, pp.1087-1101, 1971.

Remark 7.2 A more complete discussion of the use of impulse inputs for identification can be found in the paper

M. Schetzen, "Measurement of the Kernels of a Nonlinear System of Finite Order," *International Journal of Control,* Vol. 1, pp. 251-263, 1965.

For the discrete-time case, see

S. Clancy, W. Rugh, "A Note on the Identification of Discrete-Time Polynomial Systems," *IEEE Transactions on Automatic Control,* Vol. AC-24, pp.975-978, 1979.

Remark 7.3 An elementary review of the structural aspects of linear-system identification using rational interpolation theory is given in

W. Rugh, *Mathematical Description of Linear Systems,* Marcel Dekker, New York, 1975.

This treatment includes the topic of identification from steady-state frequency response. The steady-state response to single-tone inputs also can be used for identification in a class of interconnection structured systems somewhat more general than the linear-polynomial-linear sandwich. See

S. Baumgartner, W. Rugh, "Complete Identification of a Class of Nonlinear Systems from Steady-State Frequency Response," *IEEE Transactions on Circuits and Systems,* Vol. CAS-22, pp. 753-759, 1975.

E. Wysocki, W. Rugh, "Further Results on the Identification Problem for the Class of Nonlinear Systems S_M," *IEEE Transactions on Circuits and Systems,* Vol. CAS-23, pp. 664-670, 1976.

J. Sandor, D. Williamson, "Identification and Analysis of Nonlinear Systems by Tensor Techniques," *International Journal of Control,* Vol. 27, pp. 853-878, 1978.

Identification based on the steady-state response to multi-tone inputs is discussed in

K. Shanmugam, M. Jong, "Identification of Nonlinear Systems in Frequency Domain," *IEEE Transactions on Aerospace and Electronics,* Vol. AES-11, pp. 1218-1225, 1975.

The following paper on nonlinear system identification does not use the Volterra or Wiener representations, but it should be consulted by the serious reader.

L. Zadeh, "On the Identification Problem," *IRE Transactions on Circuit Theory,* Vol. 3, pp. 277-281, 1956.

Remark 7.4 The method in Section 7.5 for obtaining Wiener-kernel orthogonal expansion coefficients using a Gaussian white noise input is the original identification procedure suggested by Wiener. Wiener's results are presented in terms of a Laguerre function expansion basically because the Laguerre functions can be realized using electrical circuits. However, any orthogonal expansion can be used. For a detailed analysis of the Wiener model, see

M. Schetzen, *The Volterra and Wiener Theories of Nonlinear Systems,* John Wiley, New York, 1980.

Remark 7.5 The cross-correlation technique for determining Wiener kernel values was proposed in

Y. Lee, M. Schetzen, "Measurement of the Wiener Kernels of a Nonlinear System by Cross-correlation," *International Journal on Control,* Vol. 2, pp. 237-254, 1965.

Further discussion of the cross-correlation method from a more mathematical point of view can be found in

S. Klein, S. Yasui, "Nonlinear Systems Analysis with Non-Gaussian White Stimuli: General Basis Functionals and Kernels," *IEEE Transactions on Information Theory,* Vol. IT-25, pp. 495-500, 1979.

G. Palm, T. Poggio, "The Volterra Representation and the Wiener Expansion: Validity and Pitfalls," *SIAM Journal on Applied Mathematics,* Vol. 33, pp. 195-216, 1977.

S. Yasui, "Stochastic Functional Fourier Series, Volterra Series, and Nonlinear Systems Analysis," *IEEE Transactions on Automatic Control,* Vol. AC-24, pp. 230-242, 1979.

The first of these papers discusses the difficulties involved in finding equal-argument kernel values by the cross-correlation method. This method has been used much more widely than the original Wiener method, in large part because of the often great number of expansion coefficients that must be found in the Wiener method. Applications of the cross-correlation method have been particularly numerous in the biological modeling field. See for example

P. Marmarelis, V. Marmarelis, *Analysis of Physiological Systems,* Plenum, New York, 1978.

This book contains discussions of a number of important issues that arise in applications. These issues include the problem of approximating white noise, and computational methods for the cross-correlation method. The identification of cascade structured systems using the cross-correlation method has been treated in

M. Korenberg, "Identification of Biological Cascades of Linear and Static Nonlinear Systems," *Proceedings of the Sixteenth Midwest Symposium on Circuit Theory,* pp. 1-9, 1973.

M. Korenberg, "Cross-correlation Analysis of Neural Cascades," *Proceedings of the Tenth Annual Rocky Mountain Bioengineering Symposium,* pp. 47-52, 1973.

This work is also discussed in the book by Marmarelis and Marmarelis.

Remark 7.6 The identification problem for polynomial systems using Gaussian inputs can be formulated in terms of Fourier transforms. This leads to an expression for the system function in terms of higher-order cumulant spectra of the response. This formulation and methods for estimating cumulant spectra are discussed in

D. Brillinger, "Fourier Analysis of Stationary Processes," *Proceedings of the IEEE,* Vol. 62, pp. 1628-1643, 1974.

D. Brillinger, "The Identification of Polynomial Systems by means of Higher Order Spectra," *Journal of Sound and Vibration,* Vol. 12, pp. 301-313, 1970.

I should note that these papers require a deeper background in statistics than that presumed in Section 7.4.

7.7 Problems

7.1. Suppose a discrete-time, degree-n, homogeneous, system is such that

$$h_{reg}(i_1, \ldots, i_n) = 0 \text{ , if } any \ i_j = 0$$

Show that for the set of positive integers $I_1, \ldots, I_n$, $h_{reg}(I_1, \ldots, I_n)$ can be determined from the system response to

$$u(k) = \delta_0(k) + \delta_0(k-I_1) + \delta_0(k-I_1-I_2) + \cdots + \delta_0(k-I_1-\cdots-I_n)$$

7.2. For a degree-n homogeneous system with the cascade structure shown below, analyze the possibility of identification using the steady-state response to single-tone inputs.

$u \rightarrow G_1(s) \rightarrow (.)^{m_1} \rightarrow G_2(s) \rightarrow (.)^{m_2} \rightarrow y$

7.3. For the system shown in Figure 7.2, devise a single-tone identification strategy that does not require step function inputs.

7.4. Derive (55) for $m = 0, 1, 2, 3$, and discuss the limitations of the corresponding modified cross-correlation approach to identification.

7.5. For identification in the infinite series case using Gaussian white noise, discuss the advantages of the Wiener representation over the Volterra representation.

7.6. Develop a simple cross-correlation technique for the identification of cascade structured systems of the form shown below. Do not assume that $G(\omega)$ is minimum phase.

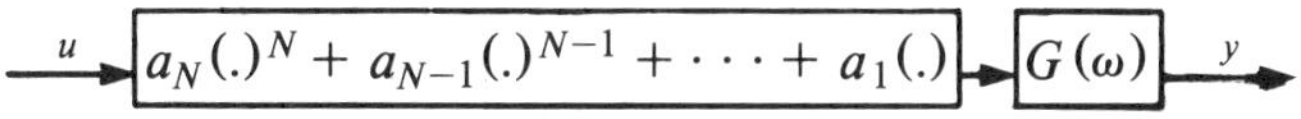

$u \rightarrow a_N(.)^N + a_{N-1}(.)^{N-1} + \cdots + a_1(.) \rightarrow G(\omega) \rightarrow y$

7.7. For the class of systems considered in Problem 7.6, develop an identification approach based on steady-state responses to single-tone inputs.

7.8. For the system shown in Figure 7.6, show how to determine the coefficients in the nonlinearity via cross-correlation.

Author Index

Subject Index